# 2nd International Conference on Life and Applied Sciences for Sustainable Rural Development (LASSRD 2019)

IOP Conference Series: Earth and Environmental Science
Volume 406

Purwokerto, Indonesia
20 - 22 November 2019

**Editors:**

**Mukhtar Effendi**
**Norman Arie Prayogo**
**R. Taufan Harisam**

ISBN: 978-1-7138-1304-0
ISSN: 1755-1307

**Printed from e-media with permission by:**

Curran Associates, Inc.
57 Morehouse Lane
Red Hook, NY 12571

**Some format issues inherent in the e-media version may also appear in this print version.**



Institute of Physics
Dirac House, Temple Back
Bristol BS1 6BE UK

Phone: 44 1 17 929 7481
Fax: 44 1 17 920 0979

techtracking@iop.org

**Additional copies of this publication are available from:**

Curran Associates, Inc.
57 Morehouse Lane
Red Hook, NY 12571 USA
Phone: 845-758-0400
Fax: 845-758-2633
Email: curran@proceedings.com
Web: www.proceedings.com

# TABLE OF CONTENTS

**LAND RESOURCE**



**ENVIRONMENTAL SCIENCE**

## FISHERIES

# Preface

On behalf of Head of The Research and Society Service Institute of Jenderal Soedirman University, I would like to inform that we organize the second *International Conference Life and Applied Science for sustainable rural development 2019.* We have succesfully held the 1st *International Conference Life and Applied Science in 2018* with 100 participaant and publish in IOP EES.

This year, to improve the publication quality of the conferences, we arrange the 2st *International Conference Life and Applied Science.* The registered participants in this conference were about 59 titles for the international conference.

The given oral and poster presentation would show outputs for future need as indicated in the conference theme of "*Life and Applied Science*".

The purposes of the conference are:

- The first to provide a forum for scientific discussion, professional networking, research collaboration, education, and dissemination of scientific research, innovation and industrial products.
- The second to increase the quality of research and development in the *Life and Applied Science* approach for sustainable rural development.
- And the third to encourage the local and regional young scientists to attend and present their works at the international level.

The success of the Conference would not have been attained without strong supports from contributing scientists as well as both national and international conference Committee.

I would like to thank all of them for helping to make a very successful conference.

Editor

Mukhtar Effendi

Norman Arie Prayogo

R. Taufan Harisam

IOP Conf. Series: Earth and Environmental Science **406** (2019) 011002 doi:10.1088/1755-1315/406/1/011002

# Peer review statement

All papers published in this volume of *IOP Conference Series: Earth and Environmental Science* have been peer reviewed through processes administered by the proceedings Editors. Reviews were conducted by expert referees to the professional and scientific standards expected of a proceedings journal published by IOP Publishing.

# Growth Response and Yield of Saline Tolerant Rice Varieties to Bio-fertilizer Application at Central Java North Coastal Saline Paddy Field

**Suprayogi, Purwanto dan Jinan Salsabila**

Faculty of Agriculture, Jenderal Soedirman University

suprayogi2004@gmail.com, purwanto.unsoed@gmail.com, jinansalsabila35@gmail.com

**Abstract.** Extension of rice production to coastal area of some islands in Indonesia is subjectd to problems that include: high salinity, poor soil fertility and limited number of available saline tolerant varieties. A field experiment was carried out to study the growth response and yield of several saline tolerant rice varieties to bio-fertilizer application at northern coastal area paddy field of Pemalang, Central Java, Indonesia. Four rice varieties that consisted of saline tolerant varieties Inpari Unsoed 79 Agritan, Inpari 34 Salin, and Banyuasin, and one saline intolerant variety Ciherang, were grown and treated with foliar application of commercial bio-fertilizers Bio-P60, Herbafarm and Biofarm which are known to contain active microorganisms. These rice varieties and bio-fertilizers were arranged in Randomized Complete Blocked Design with three replicates. Rice crop management were carried out following recommendation of locale rice management practice. The results demonstrated that BioP60 and Biofarm have greater effect on leaf chlorophyll content as compared to Herbafarm. The lowest percentage of empty grains per panicle was found on Biofarm application (19.6%). No significant diference was observed on the effect of bio-fertilizers on other growth variables and yield. Growth performance and yield variation were observed among the four rice varieties when grown on saline paddy field (EC of 0.65 dS/m, 6.7 pH, 5.37 cmol/kg exchangeable-Na). Ciherang and Inpari Unsoed 79 Agritan demonstrated greater grain weight per effective plot and greater grain weight per hectare. No interaction effect between bio-fertilizer application and varieties was observed on all growth variables and yield of the rice.

**Keywords:** bio-fertilizer, rice, varieties, salinity, saline paddy field

## 1. Introduction

As the stapple food of the majority of Indonesian people, rice and its production has got a serious attention from the government. Despite intensification and extensification that have been carried out to meet rice self-sufficiency, rice production is challenged by the increasing land convertion from agriculture to non agricultural uses, such as housing and industrial area, as well as highway. It is predicted that not less than 40,000 ha per year of wetland in Java are converted to non agricultural uses [2]. This phenomenon could threaten food security because Java is well known as the main rice producing area in Indonesia. Therefore, rice production is now extended to sub-optimal area that includes saline wetland area.

Salinity is among a stress condition which is found in about 20 percent of total agricultural land and 50 percent of irrigated wetland in the world. With a total of wetland area of about 8.09 million Hectare, about 965,262 Hectare is located in Central Java and some of which is spread along the north coastal area. During dry season, the irrigation water in this area is affected by seawater intrusion such that the water and soil salinity rises [9]. Soil salinity affects the nutrients availability and transport in plant. Excessive accumulation of $Na^+$ and $Cl^-$ in the soil influences ionic homeostatic results in nutrients deficiency [7] due to compeyiyion between $Na^+$ and $Cl^-$ with $K^+$, $Ca^{2+}$, $Fe^{2+}$ and $Mg^{2+}$ [3].

Addition of bio-fertilizer is expected to be a solution for nutrient deficiency as it contains a more readily available nutrients. One ml bio-fertilizer is equal to one Kg five month-fermented solid organic fertilizer [10]. Foliar application of bio-fertilizer could prevent loss due to leaching so that it is effective solution for plant nutrient deficiency.

The objectives of this research are: 1) to know the effect of different bio-fertilizers applicatiom on the growth and yield of some rice varieties, 2) to know the growth and yield performance of several rice varieties on salin paddy field of Petarukan, Pemalang, Central Java, and 3) To know if there is an interaction effect between bio-fertilizer and rice varieties on growth and yield of rice.

## 2. Research Method

This research was carried out at paddy field of Loning Village of Petarukan District, Pemalang, Central Java during the month December-April 2019. The average salinity during the experiment was 0.65 dS/m ($EC_{1:5}$) and categorized as moderate saline [4][5], 5.37 exchangeable Na, ph pf 6.76 and average daily temperature of $32^0$C day-time and $28^0$C night-time. Saline tolerant rice varieties of Inpari Unsoed 79 Agritan, Inpari 34 Salin Agritan, Banyuasin, and a control non-tolerant variety of Ciherang wer evaluated under field condition upon application with three different commercial bio-fertilizers, namely: Bio-P60, Herbafarm and Biofarm. Bio-fertilizer was applied foliarly on 20, 40, 60, and 80 days after planting, with the application dosage was following application recommendation of the respective bio-fertilizer. A regular basic fertilizers of 250 Kg Urea (46% N), 50 Kg KCl (60% $K_2O$), and 75 Kg SP-36 (36% $P_2O_5$) per Ha were also applied.

The experimental treatments (rice varieties and bio-fertilizer) were arranged using Randomized Complete Block design (RCBD) with three replications. Each experimental plot covered an area of 4x3 $m^2$. The observed variables included: plant height, leaves area, leaves area index, leaves chlorophyll content, total tillers number, productive tiller number, panicle length, seeds number per panicle, filled seeds number per panicle, percent of empty seeds per panicle, thousand seeds weight, seed dryweight per effective plot and seed dryweight per Hectare. The observed data were analyzed using ANOVA, followed pairwise comparison using Duncan Multiple Range Test (DMRT) with confidence level of $P \leq 0.05$.

## 3. Result and Discussion

### 3. 1. Effect of Bio-fertilizer Application

Bio-fertilizer application affected the cholorophyll content (mg $g^{-1}$) (CV 11.15%) and percent of empty seeds number per panicle (CV 17.13%) (Figure 1). BioP60 and Herbafarm had 3.32% greater effect on chlorophyll content than Biofarm. Effect of Bio-fertilizer was not significant on plant height, leaves area and leaves area

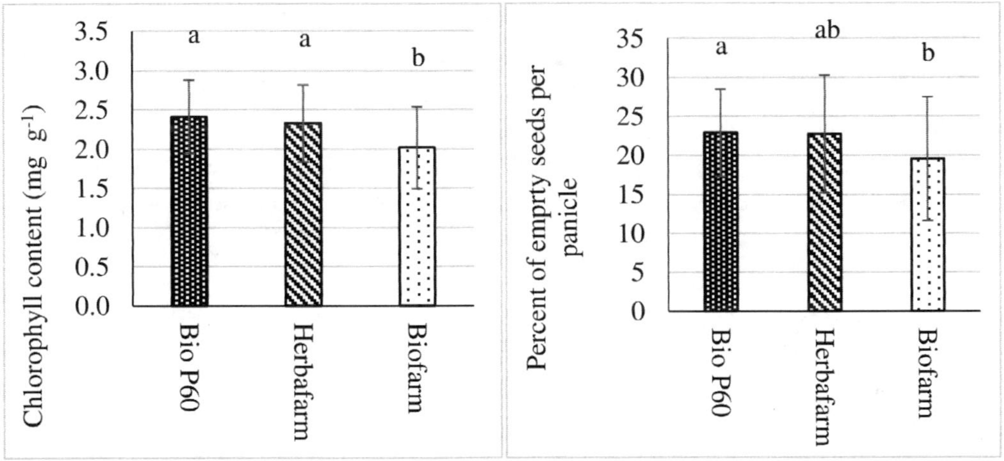


Figure 1. Effect of Bio-fertilizer application on chlorophyll content (mg $g^{-1}$) and percent of empty seeds per panicle

Bio-P60 is known to content *Psedomonas fluorescent* bacteria. *P. fluorescent* produces *Indole Acetic Acid* (IAA) which has a positive effect on plant growth. *Psedomonas* has also been reported to result in significant increase of chlorophyll content of basil leaves [6].

Plant treated with Biofarm demonstrated lower percentage of empty seeds per panicle, as compared to Herbafarm and Bio-P60. It was stated [8] that percentage of empty seed per panicle could be reduced by Nitrogen application or organic matter amendment.

Table 1. Effect of Bio-fertilizer application on plant height (cm), leaves area ($cm^2$) and leaves area index (LAI)

| Bio-fertilizer | Plant height | leaves area | LAI |
|---|---|---|---|
| Bio-P60 | 131,49 | 44,04 | 1,76 |
| Herbafarm | 131,45 | 48,43 | 1,94 |
| Biofarm | 132,77 | 45,39 | 1,82 |

### 3. 2. Effect of varieties

Among the evaluated varieties, significant different on plant growth and yield were observed. Depending on each genetic background, each variety carries different inherent growth and yield, as well response to environment [1]. Table 2 shows different growth performances among varieties.

Table 2. Different growth performances among varieties

| Variety | Plant height | Panicle length | Leaves area | LAI |
|---|---|---|---|---|
| Inpari Unsoed 79 Agritan | 130,63 b | 26,96 a | 51,19 b | 2,05 b |
| Inpari 34 | 146,46 a | 27,03 a | 65,96 a | 2,64 a |
| Banyuasin | 131,12 b | 23,74 b | 55,89 b | 2,24 b |
| Ciherang | 119,42 c | 23,67 b | 50,16 b | 2,01 b |
| CV (%) | 2,94 | 2,70 | 12,02 | 12,02 |

Note: numbers in the same column followed by similar alphabet indicate no significant different ($P \leq 0.05$)

Among the evaluated varieties, yield performances were significantly different. This correlated well with the different in yield components i.e.: total tiller numbers, productive tiller numbers, percentage of empty seed per panicle, thousand seed weight, seed dryweigth per effective plot and seed dryweigth per hectare (Table 3). Greater yield in Inpari Unsoed 79 Agritan and Ciherang obviously correlated with greater seed dryweight per effective plot (1.0**), and this was because greater thousand seed weight (0.6**), productive tiller numbers (0.63**), but not much by lower percentage of empty seed numbers (0.17) (Table 4).

Table 3. Different yield components and yield performances of the evaluated varieties

| Variety | Total tiller numbers | Productive tiller numbers | Percentage of empty seeds per panicle | Thousand seed weight | Seed dryweigth per effective plot | Seed dryweigth per hectare |
|---|---|---|---|---|---|---|
| Inpari Unsoed 79 Agritan | 21,38 a | 18,18 a | 20,63 b | 21,61 b | 7,51 a | 6,27 a |
| Inpari 34 | 19,98 ab | 16,47 b | 25,09 a | 21,81 b | 5,53 b | 4,61 b |
| Banyuasin | 19,82 b | 16,73 b | 28,56 a | 21,12 b | 5,89 b | 4,90 b |
| Ciherang | 20,58 ab | 17,56 ab | 12,62 c | 24,55 a | 8,22 a | 6,85 a |
| KK (%) | 6,71 | 7,11 | 17,13 | 4,47 | 15,64 | 15,64 |

Note: numbers in the same column followed by similar alphabet indicate no significant different ($P \leq 0.05$)

Table 4. Correlation coefficient among growth and yield components as responses to bio-fertilizer application

| Variable | PH | CC | TTN | PTN | PL | LA | LAI | TSW | SWEP | SWPH | FSN | PESP | SNPH |
|---|---|---|---|---|---|---|---|---|---|---|---|---|---|
| PH | 1 | -0,32 | -0,01 | -0,15 | 0,57** | 0.66** | 0,66** | -0,49 | -0,59 | -0,59 | -0,49 | -0,59 | -0,07 |
| CC | | 1 | 0,23 | 0,16 | 0,07 | -0,34* | -0,34* | 0,27 | 0,37* | 0,37* | 0,29 | -0,46** | -0,21 |
| TTN | | | 1 | 0,90** | 0,12 | -0,33* | -0,33* | -0,07 | 0,52** | 0,52** | 0,33* | -0,15 | 0,22 |
| PTN | | | | 1 | 0,08 | -0,31 | -0,31 | 0,10 | 0,63** | 0,63** | 0,36* | -0,14 | 0,22 |
| PL | | | | | 1 | 0,31 | 0,31 | -0,26 | -0,09 | -0,09 | 0,08 | 0,16 | 0,18 |
| LA | | | | | | 1 | 1** | -0,12 | -0,51** | -0,51** | -0,52** | 0,42** | -0,31 |
| LAI | | | | | | | 1 | -0,12 | -0,51** | -0,51** | -0,52** | 0,43** | -0,31 |
| TSW | | | | | | | | 1 | 0,6** | 0,6** | 0,28 | -0,67 | -0,21 |
| SWEP | | | | | | | | | 1 | 1** | 0,84** | -0,63** | 0,38* |
| SWPH | | | | | | | | | | 1 | 0,84** | -0,63** | 0,38* |
| FSN | | | | | | | | | | | 1 | -0,51** | 0,70** |
| PESP | | | | | | | | | | | | 1 | 0,17 |
| SNPH | | | | | | | | | | | | | 1 |

Note: **= highly significant; *= significant; PH= Plant height; CC= Chlorophyll content; TTN= Total tiller numbers; PTN= Productive tiller number; PL= Panicle length; LA= leaves area; LAI= Leaves Area Index; TSW= Thousand Seed Weight; SWEP= Seed dryweight per effective plot; SWPH= Deed dryweigth per hectare; FSN= Filled seed numbers; PESP= Percentage of empty seed per plot; SNPH=Seed numbers per hill.

## 4. Conclusion

Based on the discussion on research data analysis, the following conclusion may be drawn:

4. 1 . Bio- P60 could effectively be used to increase chlorophyll content of rice. If in this experiment the bio-fertilizer effect on the growth and yield performances of the evaluated variety was not significant, it could be because dosages of application need to be increased.
4. 2 . The evaluated varieties have different growth and yield performance when grown on moderate field saline condition of Loning Village, Petarukan, Pemalang, Central Java. Inpari Unsoed 79 Agritan and Ciherang demonstrate the greatest yield performances as compared to Inpari 34 and Banyuasin.
4. 3 . No interaction effect between Bio-fertilizer application and varieties on the growth and yield of the evaluated rice could be observed in the experiment.

**Acknowledgement**
This research was part of the Incentive Grant of Research Innovation Commercialization 2018 of Directorate Industry Innovation, Directorate General Innovation Strengthening, the Ministry of Research, Technology and Higher Education, Indonesia. The Author thanks to Ms. Endang and Dr. Wayan of National Research Council for the support for Research Development of Saline Tolerant Rice Variety, Inpari Unsoed 79 Agritan through the above grant.

**References:**
[1] Alavan, A., R. Hayati., and E. Hayati. 2015. Pengaruh pemupukan terhadap pertumbuhan beberapa varietas padi gogo (*Oryza sativa* L.). *J. Floratek* 10: 61-68.
[2] Ashari, 2003. Tinjauan Tentang Alih Fungsi Lahan Sawah ke Non Sawah dan Dampaknya di Pulau Jawa. Forum Penelitian Agro Ekonomi. Vol 21. No.2: 83-98.
[3] Cakmak I. 2005. The Role of Potassium in Alleviating Detrimental Effects of Abiotic Stresses in Plants. *J. Plant Nutr. Soil Sci.* 168: 521–530.
[4] Eviati and Sulaeman. 2009. *Analisis Kimia Tanah, Tanaman, Air, dan Pupuk*. Petunjuk Teknis Edisi 2. Bogor: Balai Penelitian Tanah.
[5] Hardie, M. and R. Doyle. 2012. Measuring soil salinity. *Methods in Molecular Biology*. 913: 415-425.
[6] Heidari, M., S.M. Mousavinik and A. Golpayegani. 2011. Plant Growth Promoting Rhizobacteria (PGPR) Effect on Physiological Parameters and Mineral Uptake in Basil (Ociumum basilicm L.) under Water Stress. *Journal of Agricultural and Biological Science* 6(5): 6-11.
[7] Kristiono, A., R.D. Purwaningrahayu and A. Taufiq. 2013. Respon Tanaman Kedelai, Kacang Tanah dan Kacang Hijau terhadap Cekaman Salinitas. *Buletin Palawija* 26: 45-60.
[8] Makarim, A. K. and E. Suhartatik. 2009. *Morfologi dan Fisiologi Tanaman Padi*. Subang: Balai Besar Tanaman Padi, Sukamandi.
[9] Marwanto, S., A. Rachman, D. Erfandi and I.G.M. Subiksa. 2009. Tingkat Salinitas Tanah pada Lahan Sawah Intensif di Kabupaten Indramayu, Jawa Barat. p. 175-190. In: Kurnia U., F. Agus, D. Setyorini and A. Setiyanto (eds). Pros. Seminar Nasional. Multifungsi dan Konversi Lahan Pertanian. Bogor: Balai Penelitian Tanah.
[10] Nezarat, S. and A. Gholami. 2009. Screenig Plant Growth Promoting Rhizobacteia for Improving Seed Germination, Seedling Growth and Yield of Maize. *Pak. J. Biol. Sci.* 12:26–32.

IOP Conf. Series: Earth and Environmental Science **406** (2019) 012002 doi:10.1088/1755-1315/406/1/012002

# Impact Of The Implementation Of Strategic Cultivars On Traditional Cultivars In The Technical Efficiency Of Farmers And Welfare (The Case Of Strawberry Crop Farming In Serang Village)

Adi Poernomo[1]*, Lilis Sitibadriah[2], Diah Setyorini Gunawan[2], Suharno[2]

[1] The Faculty of Economics and Business, Wijayakusuma University, Purwokerto, Indonesia.
[2] The Faculty of Economics and Business, Jenderal Soedirman University, Purwokerto, Indonesia.
Corresponding author: ebotpurnama@gmail.com

**Abstract**: The implementation of strategic strawberry cultivars in Serang village agriculture has caused some farmers to switch from vegetable farming to strawberry farming, this phenomena is interesting to find out whether the transfer to strawberry cultivar plants is followed by increased production efficiency and farmer welfare. The aim of this study is to analyze the level of technical efficiency of strawberry farming and vegetable farming, and to measure the level of welfare of farmers through the "farmer exchange rate index". The data used in this study are primary data, with the method of "in-depth interviews". Quantitative analysis was carried out to measure the level of technical efficiency of strawberry and vegetable farming using the Data Envelopment Analysis approach. The measurement of technical efficiency of strawberry and vegetable production uses input variables which consist of "land rent, labor, fertilizer, seeds, medicines, irrigation, and agricultural equipment", and the output variable is production. Measurements of the efficiency of strawberry distribution and vegetable farming use variable input number of distributors and shipping costs, and the output variable is the distributor's selling price. To measure the level of welfare of farmers, the farmers' exchange rate index is used, which is a measure of the ability of the exchange rate of products produced and sold by farmers with products consumed by farmers, both those which are household needs and agricultural production costs. The results of the Data Envelopment Analysis show that from the level of efficiency of production and distribution, strawberry farming and vegetable farming on average are already efficient. The results of measuring the level of welfare of farmers indicate that strawberry and vegetable farmers are in prosperous conditions.

Keywords: New Cultivar, Economic Efficiency, Data Envelopment Analysis, Farmer Exchange Rate, Farmer Welfare

## 1. Introduction

Serang Village Karangreja Subdistrict Purbalingga Regency is a highland area with an altitude of about 650 to 1300 meters above sea level with an average air temperature of 20 degrees Celsius. Very strategic location for vegetable farming. Climate conditions that meet the requirements for farming types of horticultural agriculture, so that the villagers use it to farm vegetables, as a basic farmer for local farming.

In the late 1990s, farmers in Serang Village began to recognize and develop the "Cultivar Oso grande" strawberry plant which has a larger size and sweeter taste [16], this cultivar is also an introduced cultivar from California United States, with the characteristics of the fruit size is very large, dense, the middle part is textured like foam, and the yield is high[2]. With the inclusion of strawberry crop farming in Serang Village, it is hoped that it can improve the welfare of the community, especially farmers who previously planted vegetables and turned into strawberry farmers (strategic cultivation with high economic value).

The problem of strawberry crop farming in Indonesia is the availability of high-quality and disease-resistant seeds. Provision of seeds has been carried out conventionally by using stolon (modification of stems that grow sideways and in the growth of new plants). The weakness of this method is to have limited multiplication capacity and disease resistance, due to the infection of endogenous pathogens that are transmitted from the parent plant. Seedlings that are exposed to this pathogen infection cause the quality of production to decline after about three periods of planting [42].

In the dry season strawberries planted in open land are capable of maximum production, but in the rainy season, strawberry plants are not able to produce well, because the production of many strawberries is rotten, so special cultivation techniques are needed so that strawberries will continue to produce optimally during the rainy season. Strawberry plants are generally planted in open land such as in beds, or in plastic sacks, but at this time, many strawberries are cultivated in greenhouses by planting in pots or polybags with wooden shelves or multilevel gutters and some are planted in polybag pads [6].

The government's involvement in supporting agricultural activities in Serang Village Karangreja Sub-District, Purbalingga District, is realized by developing road infrastructure to Serang Village, such as widening the road to the village, and "rest area" around strawberry farming in Serang Village, to facilitate the emergence of agrotourism rural area. Consumers who initially came to buy strawberries directly, now they not only buy strawberries, but also can enjoy the beauty of mountain scenery and strawberry farming directly. Responding to these conditions, the farmers, in addition to selling strawberry crops, farmers also develop strawberry-based agro-tourism such as direct strawberry picking, making roadside stands around agro-tourism to sell agricultural products as souvenirs [20,31,37].

This study aims to analyze and compare the technical efficiency of production and distribution, "strategic cultivar" (strawberry) farming with "traditional cultivars" (vegetables). To analyze the technical efficiency of farming, the Data Envelopment Analysis model was used, while to find out the level of farmer welfare was measured by the Farmer Exchange Rate index. Measuring the level of welfare of farmers is intended to determine the impact of the implementation of strategic strawberry cultivar farming on traditional vegetable cultivar farming at the level of farmers' welfare.

The difference of this research with other research that has been done is on the analysis and comparison of the level of technical efficiency of strawberry and vegetable farming with a mathematical method using the Data Envelopment Analysis model, which can measure and compare productivity well between comparable units [10]. While other studies in the same case in the Serang village, only analyzed at the level of agricultural productivity strawberries and vegetables partially and not using the same analysis model. In addition to technical efficiency, this study also conducted an analysis of both cultivars distribution efficiency, the above limitations that researchers know, have not been analyzed by other researchers on the case and the same location, so it is a novelty of this study.

## 2. Literature Review

Implementing strategic (commercial) cultivars to farmers who are used to traditionally carrying out traditional crop farming activities, often carried out by the government in an effort to accelerate rural (traditional) agricultural development, because it is considered to have high economic value, can immediately change traditional cropping patterns to modern, which is accompanied by an increase in the welfare of rural farmers [25], but these efforts require hard work and a long process of adaptation, besides being faced with a high risk of failure, because strategic cultivars need agricultural knowledge with specific modern technologies while rural farmers are accustomed with cropping patterns and traditional types of plants, so the level of success is often not as expected[3,30].

Transformation of traditional farming patterns into strategic cultivar agriculture if not followed by increased knowledge of farmers [19], also often not followed by increased farm efficiency and farmer welfare [5], another fact explains that strategic cultivars economically it is estimated to be profitable, in fact it faces the reality of instability in output markets 24]. The instability of the strategic cultivar output market often occurs because the product is not a consumer item that is favored by the local market (although it has good benefits and quality) or an uncompetitive export-oriented product in the global market, so the government and private sector need to overcome the problem [23].

The traditional cropping pattern of rural farmers who are considered uneconomical and the cause of prolonged rural farmers' poverty by the government, has proven to be capable of surviving despite having to deal with imported agricultural products, global climate change [11] and the deteriorating quality of rural agricultural environments [38], in other words rural traditional agriculture is actually able to show the form of sustainable agriculture based on local wisdom [41].

Other research results show that rural peasant poverty is not absolutely due to agricultural patterns and traditional types of crops or inefficiencies in production, but it is largely caused by economic instability [27] which is a dynamic of economic activity and economic policy government macro [1,22]. Regional barriers in the form of rural infrastructure are very limited, and the cultural attitudes of rural communities that tend to live in a simple and modest way play a role in creating rural ethical poverty. Rural farmers are also often used as a medium for government political interests in running pro-poor development programs [14, 33] to get public sympathy as a potential constituent[9,29].

## 3. Method

This research uses primary data, collected by observation, interview, and discussion methods. The observation method is carried out by making direct observations during the research activities in Serang Village to obtain supporting information and data from farmers, extension workers, or government officials. The interview method is done by asking several questions that have been compiled in the concept of the questionnaire.

Purposive sampling method by selecting farmers who deliberately sampled the appropriate criteria in this study. The sample criteria in this study were strawberry farmers and vegetable farmers who had the scale of the area they worked on as relatively wide. The total filtered respondents were 60 farmers consisting of 30 vegetable farmers and 30 strawberry farmers.

Quantitative analysis was carried out to analyze the technical efficiency of vegetable and strawberry farming using the Data Envelopment Analysis (DEA) approach. Data is processed using computers (Microsoft Excel 2013) for each type of agriculture. The results of processing primary data are presented in the form of tables which are then interpreted to explain the true event. DEA is a strong methodology for measuring the efficiency of multiple inputs in several output production units [7]. Therefore, DEA is relatively an efficient measure that accommodates several inputs, several outputs and other factors in one model [15]. Technically, DEA analysis is comparing input and output data from a DMU (Decision Making Units) data organization with other input and output data in a similar DMU. This comparison is done to get an efficiency value [12]. In addition to producing efficiency values for each DMU, DEA also shows units that are references for inefficient units [17,26].

The Banker Charnes Cooper [8] model or Variable Returns to Scale (VRS) used in this study [26], this model assumes that farming does not or does not yet operate at an optimal scale. This model assumes that the comparison that the ratio between the addition of input and output is not the same, namely VRS, which means the addition of inputs x times, can be smaller or even larger. The VRS formula can be mathematically written as follows:

$$\max_{\mu_k, v_i} \quad \Sigma_K^P \mu_k y_{k0} - U_0$$
$$s.t. \quad \Sigma_{i=1}^m v_i x_{i0} = 1$$
$$\Sigma_K^P \mu_k y_{kj} - \Sigma_{i=1}^m v_i x_{ij} - U_0 \leq 0 \quad j = 1, \ldots, n$$
$$\mu_k \geq \in, v_i \geq \in \quad k = 1, \ldots p$$
$$i = 1, \ldots m$$

[28]

Where:
Ykj = Number of output r produced by farmer j,
Xij = Amount of input i used by farmer j,
μk = Weight given to output r, (r = 1 …. t and t is the number of outputs),
Vi = Weight given to input i, (i = 1 …. m and m is the number of inputs),
N = Number of farmers,
$i_0$ = Farmers who are given an assessment

Usually the result of calculation of output and input-oriented DEA, will identify an efficient DMU is exactly same. The efficiency value for the output-oriented model will be the same as the efficiency value of the input-oriented model. The average efficiency value for the VRS model input orientation will generally be greater than the input oriented CRS model [17], [13].

The welfare level of play is measured using the "farmer exchange rate index", which is a comparison between the price index received by farmers and the price index paid by farmers [4]. The price index received by farmers illustrates the price movements of the products produced by farmers, which also reflects the level of income of the agricultural sector. The price index paid by farmers reflects an illustration of the price movements of goods consumed by farmers, including the amount of agricultural production costs incurred, the movement of the index paid by farmers is also a picture of inflation in the rural agricultural sector. Farmer Exchange Rate is a measure of the ability of the exchange rate of products produced and sold by farmers with products consumed by farmers, both those which are household needs and costs of agricultural production. Farmer Exchange Rate also shows the competitiveness of rural farmers' production compared to other products [4,35,36].

The criteria of the Farmer Exchange Rate index are as follows:

- NTP > 100, means that farmers experience a surplus or farmers' income is greater than their consumption.
- NTP = 100, meaning farmers experience break even if farmer income is the same as consumption.
- NTP< 100, means farmers experience a deficit or farmer's income is smaller than their expenditure [4].

## 4. Result

### 4.1. Production Technical Efficiency

Technical efficiency of agricultural production of strawberries and vegetables are calculated by comparing the number of inputs and outputs that are used in agricultural production systems strawberry and vegetable farm. The results obtained from these calculations will be presented by the table below.

Tabel 1: Production Technical Efficiency.

| | **Production Technical Efficiency** | |
|---|---|---|
| | **Strawberry** | **Vegetables** |
| Mean | 1.36 | 1.48 |
| Median | 1.13 | 1.58 |
| Minimum | 0.47 | 0.08 |
| Maximum | 3.64 | 2.41 |

Source: Primary data processed

From table 1, it can be seen that from each of the 30 samples of strawberry and vegetable farmers, it is known that both (strawberry and vegetable farming) are efficient in "technical production". The average level of technical efficiency of strawberry farming is 1.36 and vegetables are 1.48. Where is the technical efficiency of agricultural production> 1; it means that the use of inputs on strawberry farming and vegetables is efficient, but when viewed from the level of efficiency of each respondent, then, of the 30 respondents in strawberry farming, farmers who produce efficiently are 18 farmers or around 60 percent, and 12 farmers or around 40 percent, it is not efficient in its production. Whereas in vegetable farming there were 22 farmers or around 73 percent were declared efficient and 8 farmers or around 27 percent were declared inefficient. the explanation above illustrates that Serang village farmers, are still more efficient in producing vegetables, both in terms of their level of efficiency, as well as from the number of farmers who are able to produce efficiently.

The information above also explains, the difference in the level of efficiency of strawberry and vegetable farming is very narrow, this illustrates that the effort to introduce strawberry plants to vegetable farmers requires more effort, because strategic cultivars should get a higher level of efficiency.

4.2. Distribution Efficiency

The results of calculating the technical efficiency level of distribution of strawberry farming and vegetable farming are as follows:

Tabel 2: Distribution Efficiency

| | **Distribution Efficiency** | |
|---|---|---|
| | **Strawberry** | **Vegetables** |
| Mean | 1.45 | 1.20 |
| Median | 1.44 | 1.22 |
| Minimum | 1.30 | 1.12 |
| Maximum | 1.59 | 1.27 |

Source: Primary data processed

From table 2, it is known that the distribution of strawberries and vegetable farming is technically efficient. The distribution of strawberry farming has an average efficiency of 1.45, while vegetable farming has an average value of 1.20, distribution efficiency> 1: shows that the distribution channels of both agricultural products are efficient, the highest average value is obtained by strawberry farmers, who means that the distribution channel of strawberry agriculture is better, this is very likely to occur because of the strategic nature of the product (having high economic value), and local government support in creating strawberry agro-tourism that can boost the price level and the sale of strawberry agricultural products.

The distribution efficiency of vegetable farming is lower than strawberry farming because, distribution channels, vegetable products, still use traditional distribution channels, which are only channeled through local collectors, to meet the needs of the local market and sell directly to consumers.

4.3. Farmer Welfare

The results of the calculation of the level of farmers' welfare are measured using the farmer exchange rate index

Tabel 3: Welfare Level of Strawberry Farmers

| | **Welfare Level of Strawberry Farmers** | |
|---|---|---|
| | < 100 | > 100 |
| Mean | 72.82 | 158.45 |
| Median | 80.15 | 129.30 |
| Minimum | 46.79 | 102.42 |
| Maximum | 99.66 | 323.60 |
| **Number of Farmers** | **12** | **18** |

Source: Primary data processed

From table 3, it is known that strawberry farmers who have a farmers' exchange rate index <100 are 12 people or around 40 percent, with an average index value of 72.82, which means that 40 percent of farmers are not yet prosperous. While 18 other farmers or around 60 percent obtained a farmer exchange rate index> 100, with an average value of 158.45, which means 60 percent of farmers prospered, in general these differences were quite moderate, but what was very worrying was that there were farmers who got the lowest index, with a value of 46.79, which can be declared a poor farmer, on the other hand there are farmers who get the highest index, with a value of 323.60, which can be declared by rich farmers. This very high gap in the level of welfare of farmers is a serious problem related to the social and economic problems of farmers, this problem requires a separate study to solve it.

Tabel 4: Welfare Level of Vegetables Farmers

| | **Welfare Level of Vegetables Farmers** | |
|---|---|---|
| | < 100 | > 100 |
| Mean | 52.69 | 144.32 |
| Median | 55.74 | 140.53 |
| Minimum | 6.47 | 100.16 |
| Maximum | 98.81 | 238.29 |
| **Number of Farmers** | **8** | **22** |

Source: Primary data processed

From table 4, it is known that vegetable farmers who have a farmers' exchange rate index <100 are 8 farmers or around 20 percent, with an average index value of 52.69, which means that 20 percent of farmers are not yet prosperous. While 22 other farmers or around 80 percent obtained a farmer exchange rate index of> 100, with an average value of 144.32, which means that 80 percent of farmers prospered, this indicates that vegetable farmers generally are in a prosperous state. The same problem is also faced by vegetable farmers, where, it is found that farmers get the lowest index, with a value of 6.47, which can be stated as very poor farmers, on the other hand there are farmers who get the highest index, with a value of 238.29, who can say as farmers rich. So that when viewed from the level of farmers' welfare, the two agricultural models are still faced with the classic problem of income inequality among farmers.

## 5. Conclution

The empirical facts explain that, the implementation of strategic cultivars (strawberries) on traditional cultivar agriculture (vegetables) actually cannot be fully adopted by rural farmers. if the measure of success comes from technical efficiency, it turns out there are still 40 percent of farmers who adopt strawberry cultivars, are not efficient in production, while for vegetable farmers only around 27 percent, this explains that the failure rate of strawberry cultivar production is still higher than vegetables , It should be a strategic cultivar with the adoption of more modern agricultural technology, has a high competitive value and more economical, able to create a high level of technical efficiency. In this problem the effort to implement strategic cultivars in rural traditional farming needs to be carefully planned by all parties concerned to introduce strategic cultivars to rural farmers, so that rural farmers are no longer faced with the uncertainty of farming certain crops, but instead have confidence strong to keep on producing cultivars that really have good production techniques.

From the results of analysis efficiency and distribution of agricultural products show the facts that all parties in the distribution channel have a high level of efficiency. For strawberry products the average has an efficiency level of 1.45, for vegetables an average of 1.20, the highest average value of distribution efficiency is in strawberry products, which means that the distribution channels for strawberry farming are better, this is very likely because the nature of the product is strategic (has high economic value), and local government support in creating strawberry agro-tourism that can boost price levels, accelerate distribution, which ultimately increases the value of selling strawberry agricultural products. For vegetable products, they still survive using traditional distribution channels that run from generation to generation, without definite conditions. If the vegetable distribution channel gets the same touch, maybe you can talk differently.

From the results of analysis of farmer exchange rates, the average welfare level of farmers for both cultivars can be said to be prosperous. But in strawberry farming there are still 12 (40 percent) respondents of farmers who are not prosperous, and in vegetable farming there are only 8 (27 percent) farmers who are not prosperous. This fact also proves the implementation of strategic cultivars, does not directly improve the welfare of farmers evenly or still faces the classic problem of sustainable income inequality of rural farmers. The very high problem of income inequality in the level of welfare of farmers is a serious problem related to the social and economic problems of farmers, this problem requires a separate study to solve it.

## References

[1]. Andrew Dabalen, A. d. (2017, July ). *Pathways to Prosperity in Rural Malawi.* International Bank for Reconstruction and Development / The World Bank. doi:https://doi.org/10.1596/978-1-4648-0997-2

[2]. Ani Widiastuti, M. Y. (2013). Heat shock-induced resistance in strawberry against crown rot fungus Colletotrichum gloeosporioides. *Physiological and molecular plant pathology*(84), 86-91.

[3]. Anoush Ficiciyan, J. L.-G. (2018, August Thuesday, 09). More than Yield: Ecosystem Services of Traditional versus Modern Crop Varieties Revisited . *Sustainability , 10*(8), 01-15. doi:https://doi.org/10.3390/su10082834

[4]. Badan Pusat Statistik. (2017, November 23). *https://www.bps.go.id/subject/22/nilai-tukar-petani.html*. (Badan Pusat Statistik (BPS - Statistics Indonesia)) Retrieved Descember 22, 2018, from www.bps.go.id: https://www.bps.go.id/subject/22/nilai-tukar-petani.html#subjekViewTab2

[5]. Bernard Vanlauwe, R. C. (2016, May 31). Beyond Averages: New Approaches to Understand Heterogeneity and Risk of Technology Success or Failure in Smallholder Farming. *Experimental Agriculture, 55*(s1), 84-106. doi:https://doi.org/10.1017/S0014479716000193

[6]. Budiman, S. d. (2008). *Berkebun Stroberi Secara Komersial.* Jakarta: Penebar Swadaya.

[7]. Cook WD, S. L. (2009). Data envelopment anaysis. *Thirty years on. European Journal of Operating Research*(192), 1-17.

[8]. Cooper, W. S. (2006). *Introduction to data Envelopment Analysis and its uses: with DEA-solver software and references.* springer Science & Business Media.

[9]. David L. Weimer, A. R. (2017). *Policy Analysis* (6th ed.). New York, USA. doi:https://doi.org/10.4324/9781315442129

[10]. Dula, J. H. (2002). *Data Enveelopmen Analisis (DEA) In Massive Data Sets.* Kluwer Academic.

[11]. Dung, N. M. (2018). Climate Change and Crop Production in Vietnam. In P. E. Maria Lourdes T. Munárriz (Ed.), *Developing Sustainable and Resilient Rural Communities in the Midst of Climate Change : A Challenge to Disaster Preparedness and Mitigation Strategies. Third International Conference*, pp. 168-182. Quezon City, Metro Manila, Philippines: University of the Philippines - School of Urban and Regional Planning.

[12]. Effendi, A. T. (2017). DEA Optimization with Neural Network in Benchmarking. *Annual Applied Science and Engineering Conference.*

[13]. Fajar Firmana, R. N. (2016). Efisiensi Teknis Usahatani Padi di Kabupaten Karawang Dengan Pendekatan Data Envelopment Analysis (DEA). *Forum Agribisnis (Agribusiness Forum), 6*(2), 213-226. doi:https://doi.org/10.29244/fagb,6,2,%25p

[14]. Gascón, J. (2015, October ). Pro Poor Tourism as a Strategy to Fight Rural Poverty: A Critique. *Journal of Agrarian Change, 15*(4), 499-518. doi:https://doi.org/10.1111/joac.12087

[15]. Halkos George, a. S. (2014). Efficiency measurement of the Greek commercial banks with the use of financial ratios. *A Data Envelopment Analysis approach, Management Accounting Research, II*(15), 201-224.

[16]. Hanif, Z. (2015). Penerapan Inovasi Teknologi Holtikultura dalam Mendukung Pembangunan Holtikultura Yang Berdaya Saing dan Berbasis Sumber Daya Genetik Lokal.

[17]. Indrawati, Y. (2009). Aplikasi Metode Data Envelopment Analysis. *Analisis Efisiensi Bank Umum Di Indonesia Periode 2004-2007*, 25-41.

[18]. Khomsan, E. S. (2006). Kesejahteraan Keluarga Petani Mengapa Sulit Di Wujudkan? *Institut Pertanian Bogor.*

[19]. Kirtti Ranjan Paltasingh, P. G. (2018, May Wendsday, 02). Impact of farmer education on farm productivity under varying technologies: case of paddy growers in India. *Agricultural and Food Economics, 6*(7), 01-19. doi:https://doi.org/10.1186/s40100-018-0101-9

[20]. Lucyana Trimo, I. N. (2018, September 25). Study of development of agro-tourism potential on tea small-holder. *Jurnal Penelitian Teh dan Kina, 20*(1). doi:https://doi.org/10.22302/pptk.jur.jptk.v20i1.122

[21]. M. H. Clark, W. R. (2008). *Encyclopedia of Survey Research Methods.* (P. J. Lavrakas, Ed.) Thousand Oaks, US: Sage Publications, Inc. doi:http://dx.doi.org/10.4135/9781412963947.n540

[22]. M.Dzanku, F. (2015, August). Transient rural livelihoods and poverty in Ghana. *Journal of Rural Studies, 40*, 102-110. doi:https://doi.org/10.1016/j.jrurstud.2015.06.009

[23]. Manuela Coromaldi, G. P. (2015). Adoption of modern varieties, farmers' welfare and crop biodiversity: Evidence from Uganda. *Ecological Economics, 119*, 346-358. doi:https://doi.org/10.1016/j.ecolecon.2015.09.004

[24]. Mica Jenkins, C. B. (2018, Descember). Factors affecting farmers' willingness and ability to adopt and retain vitamin A-rich varieties of orange-fleshed sweet potato in Mozambique. *Food Security, 10*(6), 1501–1519. doi:https://doi.org/10.1007/s12571-018-0845-9

[25]. Raju Ghimire, H. W.-c. (2015, Januari Thuesday, 15). Factors Affecting Adoption of Improved Rice Varieties among Rural Farm Households in Central Nepal. *Rice Science, 22*(1), 35-43. doi:https://doi.org/10.1016/j.rsci.2015.05.006

[26]. Ramesh Bhat, B. B. (2001). Data Envelopment Analysis (DEA). *Journal of Health Management, 3*(2), 309-328. doi:https://doi.org/10.1177%2F097206340100300207

[27]. Rana Ejaz Ali Khan, H. u.-u.-H. (2015). Determinants of Rural Household Poverty:The Role of Household Socioeconomic Empowerment. *American-Eurasian Journal Agriculture & Environmental Science, 15*(1), 93-98. doi:DOI: 10.5829/idosi.aejaes.2015.15.1.1050

[28]. Rusydiana, A. S. (2018). Efisiensi Dan Stabilitas Bank Umum Syariah Di Indonesia. *Jurnal Ilmu Akuntansi*(11), 203 - 222. doi:10.15408/7033

[29]. S.Blakeslee, D. (2018, January 9). Politics and public goods in developing countries: Evidence from the assassination of Rajiv Gandhi. *Journal of Public Economics, 163*, 1-19. doi:https://doi.org/10.1016/j.jpubeco.2017.12.011

[30]. Shiferaw Felekea, T. Z. (2006, October). Adoption of improved maize varieties in Southern Ethiopia: Factors and strategy options. *Food Policy, 3*(5), 442-457. doi:https://doi.org/10.1016/j.foodpol.2005.12.003

[31]. Siwi Harning Pambudi, n. S. (2018). Strategi Pengembangan Agrowisata dalam Mendukung Pembangunan Pertanian - Studi Kasus di Desa Wisata Kaligono (Dewi Kano) Kecamatan Kaligesing Kabupaten Purworejo. *Analisis Kebijakan Pertanian, 16*(2). doi:http://dx.doi.org/10.21082/akp.v16n2.2018.165-184

[32]. Soekartawi. (2003). *Teori Ekonomi Produksi.* Jakarta: PT Raja Grafindo Persada.

[33]. Sok, S. (2017, September). Pro-poor growth development and income inequality: Poverty-related Millennium Development Goal (MDG 1) on banks of the Lower Mekong Basin in Cambodia. *World Development Perspectives*, 1-8. doi:https://doi.org/10.1016/j.wdp.2017.10.001

[34]. Statistik, B. P. (2008). *Nilai Tukar Petani.* Jakarta: Badan Pusat Statistik.

[35]. Tjetjep Nurasa, M. R. (2013, October). NILAI TUKAR PETANI PADI DI BEBERAPA SENTRA PRODUKSI PADI. *Jurnal Agro Ekonomi, 31*(2), 161-179. Retrieved Descember 2018, from http://repository.pertanian.go.id/handle/123456789/252

[36]. Tri Wardati Khusniyah, S. (2016, May). Prediksi Nilai Tukar Petani Menggunakan Jaringan Syaraf Tiruan Backpropagation. *Scientific Journal of Informatics, 3*(2), 11-18. doi:https://doi.org/10.15294/sji.v3i1.4970

[37]. Umi Lestariningsih, A. S. (2018). Analisis Pengaruh Agrowisata Terhadap Peningkatan Pendapatan Petani Bunga Krisan Di Kecamatan Bandungan Kabupaten Semarang. *AGRISAINTIFIKA Jurnal Ilmu-ilmu Pertanian, 2*(1), 51-59. Retrieved Descember 20, 2018 , from http://jurnal.agrisaintifika-fpunivet.ac.id/index.php/AGRISAINTIFIKA/article/view/39

[38]. Utpal Kumar De, K. B. (2017, November Saturday, 25). *Adaptability of Farmers in Assam Towards Extreme Climate Effects: An Empirical Investigation.* Singapore: Springer. doi:https://doi.org/10.1007/978-981-10-6274-2_24

[39]. Velayati, R. (2013). Analisis Efisiensi Dengan Pendekatan Data Envelopment Analysis (DEA) Dan Pendapatan Usahatani Talas Di Kecamatan Cijeruk Kabupaten Bogor. *Departemen Ekonomi Sumberdaya Dan Lingkungan Bogor.*

[40]. Wulansari, R. (2010). Efisiensi Relatif Operasional Puskesmas dengan menggunakan Data Envelopment Analysis (DEA). *Ekonomi Universitas Indonesia.*

[41]. Ye Jingzhong, P. L. (2016). *Concepts and realities of family farming in Asia and the Pacific.* United Nations Development Programme, International Policy Centre for Inclusive Growth (IPC-IG). Brasília, DF - Brazil: http://www.fao.org/publications. Retrieved Descember Thuesday, 13, 2018, from http://hdl.handle.net/10419/173804

[42]. Zebrowska. (2004). Micropropagaon In The Strawberry. *Inbredlines, Food, Ag & Environ, 2,* 253-255.

IOP Conf. Series: Earth and Environmental Science **406** (2019) 012010 doi:10.1088/1755-1315/406/1/012010

# Analysis of Stomata, Chloroplast, and Chromosome of Local Mandarin Citrus (*Citrus reticulata*) Plants Grown from Endosperm Culture

**Mia Kosmiatin[1], Chaireni Martasari[2], Frizky Amelia Kurniawati[3], Rinanda Gandhi Ningrum Prasetia[3], and Prita Sari Dewi[3]***

Mia Kosmiatin[1], Chaireni Martasari[2], Rinanda Gandhi Ningrum Prasetia[3], Frizky Amelia Kurniawati[3], dan Prita Sari Dewi[3]*

[1] Indonesian Center for Agricultural Biotechnology and Genetic Resources Research and Development (ICABIOGRAD) Bogor, Indonesia

[2] Indonesian Citrus and Subtropical Fruits Research Institute (ICSFRI) Malang, Indonesia

[3] Faculty of Agriculture, Jenderal Soedirman Karangwangkal, Purwokerto Utara 53123, Indonesia

*Corresponding author: p_saridewi@yahoo.com

**Abstract.** This research aimed to identify accessions of jeruk keprok Garut plants (*Citrus reticulata*) grown from endosperm culture with stomata analysis, chloroplasts analysis, and chromosome analysis. This research was conducted at the Plant Breeding Laboratory of Indonesian Citrus and Subtropical Fruits Research Institute (ICSFRI), Malang. This research used seven accessions of jeruk keprok Garut (local mandarin citrus) from endosperm culture and mother plant of Garut mandarin. Data analysis was carried out with the Statistical Package for the Social Sciences (SPSS) application Application 22.0. The alignment analysis showed that all one accession (GT 224) out of seven accessions of Garut mandarin grown from the endosperm cultures is consistent with the mother plant according to chromosome number observation, stomata number observation, and chloroplast number observation.

**Keywords:** local mandarin, endosperm culture, stomata, chloroplast, and chromosomes

## 1. Introduction

Citrus plants are one of the most important horticultural crops. Citrus plants are generally cultivated to produce fruit for fresh consumption (Purwito *et al.*, 2015). Local commercial citrus commodities in Indonesia (citrus, mandarin, and pamelo) in production and quality have not been able to meet the need for domestic consumption (Martasari, 2008). The need for fresh consumption of citrus fruits in Indonesia is 3.26 kg/capita/year (assuming that the consumption of oranges is 10 percent of the fruit consumption, Food and Agricultural Organization/FAO standard). The consumption of citrus in Indonesia in the coming years is estimated to reach 11.85 million tonnes per year or equivalent to 580,000 ha with a production of 20 ton/ha/year. The prospect and potential of the citrus market are very large, so it is necessary to increase the quantity, quality, and continuity of the citrus administration in Indonesia (Martasari, 2008).

Central Bureau for Statistics and Directorate General of Horticultural (2015) stated that the growth of Indonesian citrus production in 2014 to 2015 decreased by 3.6 percent. The decline in domestic citrus production made Indonesia the second-largest citrus importing country in ASEAN after Malaysia. The growth of orange imports by 11% annually in five years has made Indonesia a promising market share for other countries in marketing its products. The production of Indonesian citrus that has not enough consumption needs of domestic oranges to provide challenges and opportunities for both farmers, citrus entrepreneurs, and governments in the effort to increase the production of citrus (Soelarso, 1996).

Central Bureau for Statistics (2015) reported that North Sumatera province, East Java, and West Kalimantan are three of the largest mandarin citrus provinces to national production. The production of mandarin citrus in North Sumatra province amounted to 483,006 tonnes (27.68 percent), East Java amounted to 480,396 tons (27.54 percent), and last West Kalimantan by 147,371 tonnes (8.44 percent).

Citrus fruits that are in the interest of society generally have a sweet, easily peeled, attractive and smooth fruit skin color, and no seeds (seedless) (Purwito *et al.*, 2015). One of plant breeding systems that is applied to create citrus plants without seeds is an endosperm culture (Purnama *et al.*, 2017). The endosperm culture is an alternative technique to produce a triploid plant directly by taking one stage. The endosperm tissue is triploid because it is a fusion result between two polar cores and one sperm (Sukamto, 2010). The condition can be utilized to form a new crop of triploid by conforming and generalizing endosperm cells. Plants that have ploidi triploid will usually be a crop of no-seed or seed but sterile (Purnama *et al.*, 2017).The method used to identify mandarin citrus as a result of endosperm culture by performing cytological analysis. The cytological analyses carried out include stomata analysis, chloroplast analysis, and chromosome analysis. The results of each cytological analysis will be obtained by the number of stomata, the number of chloroplasts, and the number of chromosomes.

This research aimed to 1) identify mandarin citrus (*Citrus reticulata*) plants grown from endosperm culture with stomata analysis, chloroplasts analysis, and h chromosome analysis.

## 2. Materials and methods

This research was conducted at the Plant Breeding Laboratory of Indonesian Citrus and Subtropical Fruits Research Institute (ICSFRI), Malang.

### *2.1. Plant Materials and tools*

Plant materials used in this study included seven accessions of Garut mandarin (*C. reticulata*) from endosperm culture and one mother plant of Garut mandarin (Table 1). Materials were glacial acetic acid ($CH_3COOH$) 45 percent solution, 1 N hydrochloric acid solution (HCl), Orsein, emersion oil, distilled water, clear nail polish, 70 percent alcohol, and 1 percent silver nitrate ($AgNO_3$) solution. The tools used in this study were dropper pipettes, cover glass, glass preparations, razor blades, scalpels, tweezers, 2 μl tubes, Olympus BX 51 microscopes, memert ovens, fume hoods, refrigerators, stationery, markers, tissue, scissors, sticky tape, masks, gloves, and cameras.

Table 1. Accessions of Garut mandarin used

| Accession Code | Accession Code |
|---|---|
| Mother plant of Garut mandarin | GT 113 |
| GT 12 | GT 224 |
| GT 110 | GT 109 |
| GT 2 | GT 6 |

### 2.2. *Data analysis*

Data analysis was performed based on the results of photographs obtained from the samples. Then the data is tabulated into an observation table and averaged. After obtaining the mean value, then it is calculated using standard deviations and an analysis of the kinship by using the Statistical Package for the Social Sciences (SPSS) application Application 22.0.

## 3. Results and discussion

### 3.1. *Stomata analysis*

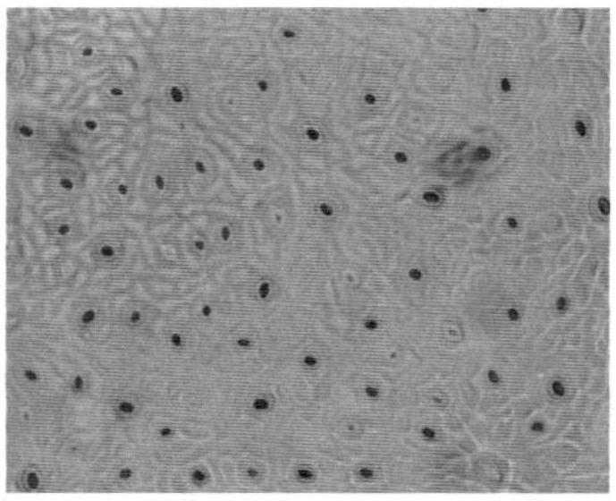

Figure 1. Triploid accession GT 224 stomata with a magnification of 400x.

Table 2. Number of stomata of seven accessions of Garut mandarin grown from the endosperm culture

| Accession Code | Number stomata/fields of view | Accession Code | Number stomata/fields of view |
|---|---|---|---|
| Mother plant of Garut mandarin | 26,92 ± 3,37 | GT 113 | 30,52 ± 5,46 |
| GT 12 | 39,20 ± 7,43 | GT 224 | 37,60 ± 6,19 |
| GT 110 | 39,88 ± 6,30 | GT 109 | 39,88 ± 6,35 |
| GT 2 | 20,36 ± 3,43 | GT 6 | 19,56 ± 3,47 |

Note: Average ± Standard Deviation

The average number of stomata results from the seven accessions of mandarin citrus from endosperm culture and mother plant of Garut mandarin can be obtained in terms of the similarity of the values of each accession as presented in Figure 2.

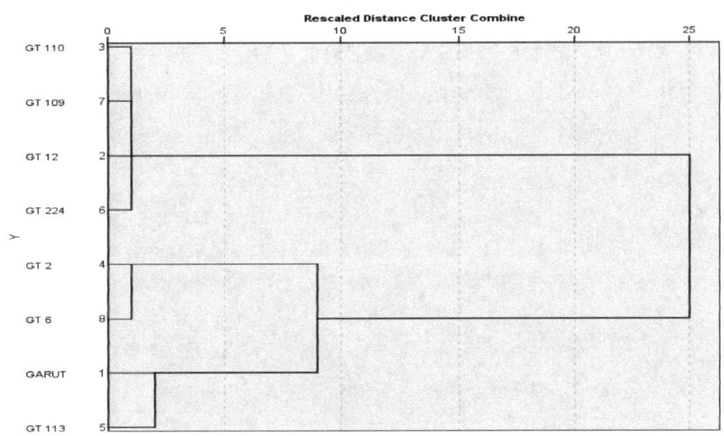


Figure 2. Dendogram of stomata accessions of Garut mandarin

Figure 2 shows that the accessions of Garut mandarin were consisted of three groups that have similar values, namely group 1 consisting of accessions GT 110, GT 109, GT 12, and GT 224 with an average number of stomata 37-39; group 2 consisted of GT 2 and GT 6 accessions with an average number of stomata of 20-26, and group 3 consisted of accessions of Garut mandarin and GT 113 with an average number of stomata of 26-30. Accessions in group 2 and group 3 have similar values because group 2 has an average stomata number of 20-26 and group 3 has an average number of stomata of 26-30. Group 1 is the accession that has the highest number of stomata compared to accessions in other groups.

The number of stomata in the arrowroot mother plant of Garut mandarin has the least amount when compared to the accession of mandarin citrus from endosperm culture, this is thought to be influenced by the number of chromosomes because mother plant of Garut mandarin has a number of chromosomes 2n and the accession of mandarin citrus resulting from endosperm culture has a chromosome number of 3n. Increasing the number of chromosomes will be followed by an increase in the size of the stomata which results in an effect on the density of the stomata. Also, the amount of stomata will increase in line with the increase in the number of chromosomes.

Haryanti (2010) stated that the number of stomata per unit area varies not only between types but also within one type, due to their relationship with environmental factors during growth. The number and size of stomata are influenced by genotype and environment. Closing cells that surround the stomata control the opening and closing of the stomata. Closure of stomata is important to prevent water loss when the water supply is limited while limiting $CO_2$ uptake for photosynthesis. Stomata open during the day and close at night. The process of opening and closing the stomata is affected by turgor pressure on the closing cells. Increasing and decreasing the size of guard cell openings is a result of changes in turgor pressure on guard cells (Meriko and Abizar, 2017).

According to Yulianti *et al.*, (2015), the length of the stomata is related to the amount of chloroplast in the guard cell. Increasing the amount of chloroplasts in guard cells causes the size of the stomata to become larger. Increasing stomata size results in lower stomata density. Diploid plants have smaller stomata size compared to tetraploid derivatives.

### *3.2. Chloroplast analysis*

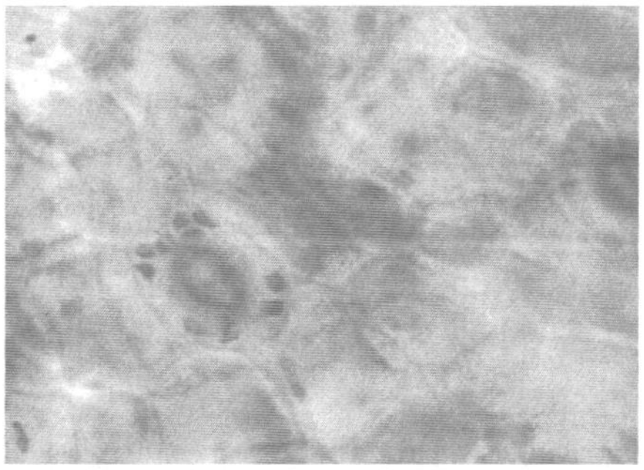

Figure 3. Chloroplast and stomata of mandarin citrus from endosperm accession GT 224 with 1000x magnification.

Table 3. Number of chloroplast 7 accessions of Garut mandarin from endosperm culture

| Accession Code | The number of chloroplasts/guard cells | Accession Code | The number of chloroplasts/guard cells |
|---|---|---|---|
| Mother plant of Garut mandarin | 16 ± 0 | GT 113 | 18 ± 0 |
| GT 12 | 13,66 ± 2,08 | GT 224 | 16,66 ± 1,15 |
| GT 110 | 14,66 ± 1,53 | GT 109 | 16,66 ± 1,15 |
| GT 2 | 14,66 ± 2,31 | GT 6 | 15 ± 0 |

Note: Average ± Standard Deviation

The average result of the number of chloroplasts from the seven accessions of mandarin citrus from endosperm culture with the parent plant of Garut mandarin can be obtained from the similarity dendogram in Figure 4.

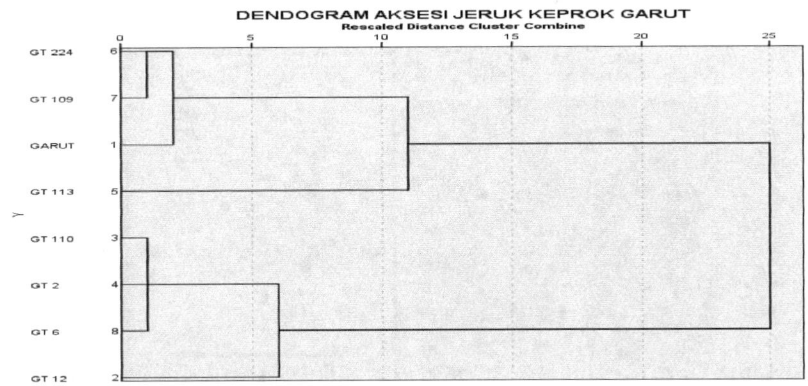


Figure 4. Dendogram amount of chloroplast accession of Garut mandarin.

Figure 4 shows that the accessions of Garut mandarin have 2 groups of accessions that have similar values, namely group 1 consisting of accessions GT 224 and GT 109 with an average number of chloroplasts 16 and group 2 consisting of accessions GT 110, GT 2, and GT 6 with an average number chloroplast 13-15. Accessions in group 1 and accessions of mother plant of Garut mandarin have similar values with the average number of chloroplasts 16. Accessions in group 2 with accessions GT 12 have similar values because group 2 has an average number of chloroplast 13-15 and GT 12 accessions have the average number of chloroplasts 13. Accessions in group 1 are accessions that have the highest mean number of chloroplasts when compared to accessions in other groups. The Accession of the mother plant of Garut mandarin was in group 1 with a mean value of 16 chloroplasts.

The amount of chloroplasts in mother plant of Garut mandarin has the least amount compared to accessions of mandarin citrus from endosperm culture. This shows that the number of chloroplast mandarin citrus resulting from endosperm culture will increase a lot followed by increasing the size of the stomata. This is due to the increasing size of the stomata resulting in the number of chloroplasts becoming more. The size of the stomata increases in harmony with the increase in the number of chromosomes. According to Kartiman *et al.*, (2018), stomatal size is related to the number of chloroplasts in stomatal guard cells. Increasing the amount of chloroplasts causes the size of the stomata (length and width) to be larger. The greater the size of the stomata, the better it absorbs $CO_2$ for photosynthesis. Changes that occur in the number of chloroplasts in stoma guard cells and stomata density can be accurate, fast, and easy indicators for the identification of various levels of plant ploidy. The number of chloroplasts guarding the stoma of the tetraploid plant is approximately two times more than that of the original diploid plant. Changes in stomata density negatively correlated with the level of ploidy, the higher the level of ploidy, the lower the stomata density (Yulianti *et al.*, 2015).

### *3.3. Chromosome analysis*

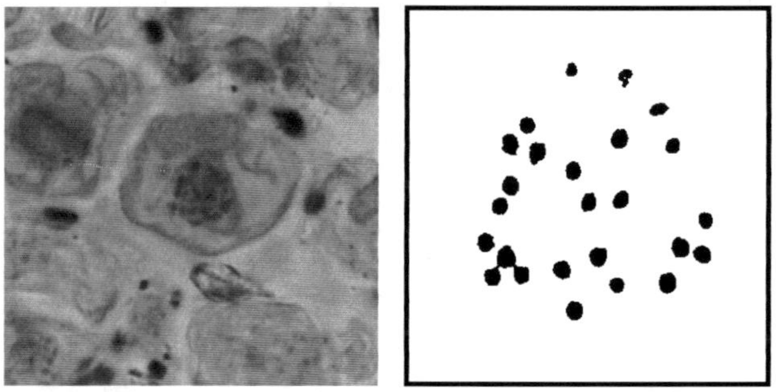

Figure 5. A mandarin citrus chromosome from the endosperm of GT 224 accession with 1000x magnification.

Table 4. Number of chromosomes the seven accessions of Garut mandarins from endosperm culture

| Accession Code | Number of Chromosomes | Accession Code | Number of Chromosomes |
|---|---|---|---|
| Mother plant of Garut mandarin | 17,80 ± 0,63 | GT 113 | 26,60 ± 1,35 |
| GT 12 | 24 ± 1,76 | GT 224 | 25,10 ± 2,64 |
| GT 110 | 23,70 ± 2 | GT 109 | 22,50 ± 3,89 |
| GT 2 | 17,70 ± 2,67 | GT 6 | 22,80 ± 2,10 |

Note: Average ± Standard Deviation

The results of the average number of chromosomes from 7 accessions of mandarin citrus from endosperm culture with mother plant of Garut mandarin can be obtained by the similarity of dendogram values in Figure 6.

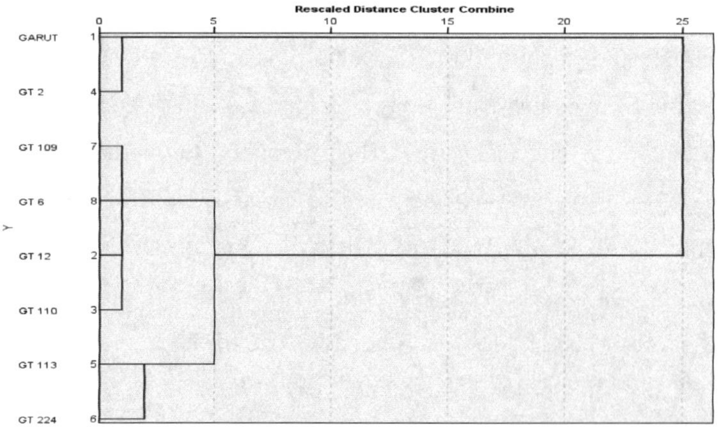


Figure 6. Dendogram of chromosome accessions of Garut mandarin.

Figure 6 shows that the accessions of Garut mandarin have 3 groups of accessions that have similar values, namely group 1 consisting of accessions of Garut and GT 2 with an average number of chromosomes 18; group 2 consisted of accessions GT 109, GT 6, GT 12, and GT 110 with a mean number of chromosomes 22-24, and group 3 consisted of accessions GT 113 and GT 224 with a mean number of chromosomes 25-26. Accessions in group 2 and group 3 have similar values because group 2 has an average number of chromosomes 22-24 and group 3 has an average number of chromosomes 25-26. Accessions in group 3 are accessions that have the highest number of chromosomes compared to accessions in other groups. Mother plant of Garut mandarin are in one group with accession of GT 2 with an average number of chromosomes 18, this is because the accession of GT 2 should have an average number of chromosomes $3n = 2x = 27$ but only have an average number of chromosomes $2n = 2x = 18$ so that accessions of GT 2 has the same rate as the mother plant of Garut mandarin.

The results of the calculation of the number of chromosomes can describe the genetics of the plants used where endosperm culture plants have several chromosomes 3n and the parent plant has several chromosomes 2n. The number of these chromosomes will also be related to the number of stomata and the number of chloroplasts, because with the addition of the number of chromosomes will be in harmony with the increasing number of chloroplasts and increasing the size of the stomata. This is by following the opinion of Nurwanti (2010) which states that cells that have increased the number of chromosomes will experience an increase in the number of chloroplasts in the *Anthurium plowmanii* plant. Calculation of the number of chloroplasts in stomatal guard cells is reported to be more stable in describing genetic background. The chromosome ploidy level of a plant is in harmony with the size of its leaf stomata. The higher the level of ploidy chromosomes, the greater the leaf stomata (Sukamto *et al*., 2010).

### *3.4. Aligment between Number of Chromosomes, Number of Stomata, and Number of Chloroplasts*

Alignment between the number of stomata, the number of chloroplasts, and the number of chromosomes is done to find out the consistency of the accessions used in this study. This alignment refers to the number of chromosomes obtained because observations of the number of chromosomes can determine the level of ploidy from the accession.

Table 5. Alignment between the number of chromosomes, the number of stomata, and the number of chloroplast accessions of Garut mandarin.

| Number of chromosomes | | Number of stomata | | Number of chloroplasts | |
|---|---|---|---|---|---|
| Group (Number of chromosomes) | Accession Code | Group (Number of stomata) | Accession Code | Group (Number of chloroplasts) | Accession Code |
| 1 (18) | Mother plant of Garut mandarin | **1 (37-39)** | GT 110 | **1 (16)** | **GT 224** |
| | GT 2 | | GT 109 | | GT 109 |
| 2 (22-24) | GT 109 | | GT 12 | 2 (13-15) | GT 110 |
| | GT 6 | | **GT 224** | | GT 2 |
| | GT 12 | 2 (20-26) | GT 2 | | GT 6 |
| | GT 110 | | GT 6 | | |
| **3 (25-26)** | GT 113 | 3 (26-30) | Garut | | |
| | **GT 224** | | GT 113 | | |

Note: **Group** = has the most number
**GT 224** = accession that has consistency

The results of alignment between the number of chromosomes, the number of stomata, and the number of chloroplast accessions of Garut mandarin are presented in (Table 15) shows that the accessions that have the most number of chromosomes are GT 113 accessions with chromosome number 26, 60 ± 1.35 and GT 224 with the number of chromosome 25.10 ± 2.64. Accessions that have the most number of stomata are GT 110 accessions with 39.88 ± 6.30 stomata; GT 109 with stomata of 39.88 ± 6.35;

GT 12 with the stomata number 39.20 ± 7.43; GT 224 with the number of stomata 37.60 ± 6.19. The accession which has the most number of chloroplasts is GT 224 with chloroplast 16.66 ± 1.15 and GT 109 with chloroplast 16.66 ± 1.15.

## 4. Conclusions

Alignment results from 7 accessions of Garut mandarin from endosperm culture and one mother plant of Garut mandarin indicate that the accession GT 224 is an accession that has consistency on the number of chromosomes, the number of stomata, and the number of chloroplasts.

## Acknowledgement

We are grateful and also thank to Director of Indonesian Citrus and Subtropical Fruits Research Institute (ICSFRI) Malang, East Java. This study was supported by Ministry of Agriculture of Indonesia.

**References**

[1] Badan Pusat Statistik dan Direktorat Jenderal Hortikultura. (2015). Produksi buah-buahan di Indonesia, 2011-2015. (*On-Line*). Http://www.Pertanian.Go.Id/Data5tahun/Pdf-Horti2016/2-Produksi%20nasional%20buah.Pdf diakses 11 November 2017

[2] Badan Pusat Statistik. (2015). Statistik tanaman buah-buahan dan sayuran tahunan Tahun 2015. (*On-Line*) Https://media.neliti.com/media/publications/48342-ID-statistik-tanaman buahbuahan-dan-sayuran-tahunan-indonesia-2015.pdf diakses 11 Januari 2019

[3] Kartiman., R, D. Sukma, S. I. Aisyah, dan A. Purwito. (2018). Multiplikasi *in vitro* anggrek hitam (*Coelogyne pandurata* Lindl.) pada perlakuan kombinasi NAA dan BAP. *Jurnal Bioteknologi dan Biosains Indonesia* 5 (1): 75-87; http://jurnal.untan.ac.id/index.php/jprb/article/viewFile/29078/75676578798

[4] Martasari, C. (2008). Variasi Jumlah Kloroplas dan Kromosom Tanaman Jeruk Siam Pontianak Hasil Perlakuan Colchicin. Balai Penelitian Tanaman Jeruk dan Buah Subtropika, Malang; https://jurnal.unikal.ac.id/index.php/biofarm/article/view/264

[5] Meriko., L dan Abidzar. (2017). Struktur stomata daun beberapa tumbuhan kantong semar (*Nepenthes* spp.). *Berita Biologi* 16 (3): 325-330; http://e-journal.biologi.lipi.go.id/index.php/berita_biologi/article/view/2398/0

[6] Nurwanti, L. (2010). Induksi Mutasi Kromosom dengan Kolkisin pada *Anthurium Wave of Love* (*Anthurium plowmanii* Croat.) *secara In Vitro*. *Skripsi*. Departemen Agronomi dan Hortikultura, Fakultas Pertanian, Institut Pertanian Bogor

[7] Purnama., I. C. G, C. Martasari, N. Kendarini, dan D. Saptadi. (2017). Analisis sitologis jeruk siam madu (*Citrus nobilis* L.) hasil kultur endosperma. *Jurnal Produksi Tanaman* 5 (5): 847-850; http://protan.studentjournal.ub.ac.id/index.php/protan/article/view/451/453

[8] Purwito., A, M. Prayogi, M. Kosmiatin, dan A. Husni. (2015). Embriogenesis somatik jeruk keprok (*Citrus reticulata* L. cv Batu 55) asal hasil perlakuan kolkisin. *Jurnal Hortikultura Indonesia* 6 (3): 161-171; http://journal.ipb.ac.id/index.php/jhi/article/download/12684/9673

[9] Soelarso. (1996). *Budidaya Jeruk Bebas Penyakit*. Kanisius, Yogyakarta.

[10] Sukamto., A, F. Ahmad, dan A. H. Wawo. (2010). Pengaruh oryzalin terhadap tingkat ploidi tanaman garut (*Maranta arundinacea* L.). *Buletin Littro* 21 (2): 93-102; http://ejurnal.litbang.pertanian.go.id/index.php/bultro/article/view/1865

[11] Haryanti., S (2010). Jumlah dan distribusi stomata pada daun beberapa spesies tanaman dikotil dan monokotil. *Buletin Anatomi dan Fisiologi* 18 (2): 21-28; https://ejournal.undip.ac.id/index.php/janafis/article/view/2600/2308

[12] Yulianti., F, A. Purwito, A. Husni, dan D. Dinarti. (2014). Induksi tetraploid tunas pucuk jeruk siam simadu (*Citrus nobilis* Lour) menggunakan kolkisin secara *in vitro*. *Jurnal Agronomi Indonesia* 43 (1): 66-71; https://www.researchgate.net/publication/316211528_Induksi_Tetraploid_Tunas_Pucuk_Jeruk_Siam_Simadu_Citrus_nobilis_Lour_Menggunakan_Kolkisin_secara_In_Vitro

IOP Conf. Series: Earth and Environmental Science **406** (2019) 012011 doi:10.1088/1755-1315/406/1/012011

# *ANTIBACTERIAL ACTIVITY OF KECOMBRANG FRUIT SIMPLICIA EKSTRACT (Nicolaia speciosa) AGAINST GRAM POSITIVE BACTERIA Staphylococcus aureus FNCC 0047 IN VITRO*

*Muhammad Ashif* [1*], *IDSAP Peramiarti* [2], *Afifah* [3]

[1]Departement of Medical Program, Faculty of Medicine, Jenderal Sudirman University, e-mail: fk@unsoed.ac.id
[2] Department of Microbiology Microbiology, Faculty of Medicine, Jenderal Sudirman University
[3] Department of Pharmacology, Faculty of Medicine, Jenderal Sudirman University

## *ABSTRACT*

*The survival of microorganisms depends on the environment. Unhealthy environment causes food intoxication problems resulting in food poisoning. One alternative that can be done to prevent the occurrence of food poisioning by using traditional medicines made from natural ingredients, especially kecombrang fruit simplicia extract. This study aims to determine the activity of kecombrang fruit simplicia extract against Staphylococcus aureus FNCC 0047 bacteria. The design used was a Completely Randomized Design (CRD) which was arranged with 10 treatment factors and repeated 3 times, resulting in 30 experimental units. The factors studied were the concentration of the kecombrang fruit simplicia extract consisting of concentrations of 10%, 20%, 30%, 40%, 50%, 60%, 70%, 80%, 90%, 100%. The variable observed in this study was the antibacterial activity of S.aureus FNCC 0047 with the well-diffusion method. The data obtained were analyzed by parametric analysis of one way analysis of variance (ANOVA) at the level of 5% then significant results were continued with Duncan Multiple Range Test (DMRT) at the level of 5%. Kecombrang fruit simplicia extract has antibacterial activity against the growth of Gram positive S. aureus FNCC 0047 bacteria with the most effective concentration in inhibiting the growth of S.aureus FNCC 0047 bacteria is 40%. In this case, the extract of the kecombrang fruit simplicia has the ability to inhibit the growth of S. aureus FNCC 0047 bacteria.*

***Keywords***: *Antibacterial activity, Kecombrang fruit simplicia extract, S. aureus FNCC 0047*

## Introduction

Microorganisms are very small living things. It depends on the surroundings which are formed by nature and human intervention and can affect human life [1]. Humans always live interacting with the environment both in a healthy environment and in an unhealthy environment. Poor human lifestyles can cause an unhealthy environment, leading to more interactions with bacteria, viruses, fungi, and various forms of parasitic life. This is also supported by the condition of dusty air, warm and humid temperatures that cause microbial fertility [2].

An unhealthy environment can cause food intoxication. Intoxication occurs when hygiene aspects are not considered in the processing of food ingredients, resulting in food poisoning (poisoning through food). One of the bacteria that often causes food poisoning is *S. aureus*. *S. aureus* is capable of producing enterotoxins which are toxic substances that affect the small intestine, causing excessive secretion of fluid into the intestinal cavity. Although not all *S. aureus* strains produce enterotoxins but all strains have the potential to cause poisoning [3]. The Food and Nutrition Culture Collection (FNCC), UGM Center for Food and Nutrition

Studies has isolated and identified about 500 bacterial isolates from various sources namely fermented foods, vegetables and fruits, processed meat products and gastrointestinal materials, one of which is the *S. aureus* FNCC 0047 bacteria. *S. aureus* bacteria FNCC 0047 are pure bacteria that have been isolated and characterized by microbiologists and are free of contamination [4].

Research on antimicrobial activity with natural ingredients has been carried out mainly from various types of local spice plants. Plants that are often used as natural sources of antimicrobials generally have bioactive compounds such as saponins, flavonoids and 25 types of phenolic compounds. Kecombrang is a group of zingiberaceae which has long been known and used by the community as a food flavor, as well as traditional medicines [5,6,7,8]. Based on community experience, the use of kecombrang fruit as a traditional medicine is applied by the people of Central Java by using a stew of kecombrang as a medicine for stomach ache and further processing into a kecombrang fruit syrup product. Another is used as a cleansing wound on the skin and red eyes. The experience associated with traditional medicine is certainly still traditionally empirical. Therefore, research is needed to prove the truth.

Each part of the kecombrang plant contains different dominant bioactive compounds [8,9,10]. The active compounds dominant in the kecombrang fruit are alkaloids, flavonoids and phenols [7;11]. The part of the stem, rhizome and leaves of kecombrang is dominated by other different compounds [9]. The use of kecombrang fruit is still rarely studied, so an idea emerged to utilize natural ingredients found in kecombrang fruit which can be used as bioactive alkaloid compounds, flavonoids and phenols as antimicrobials [11]. Bioactive components can be obtained from plant extraction. The extraction process can be carried out by various methods including maceration, percolation, soxhlet, reflux, digestion, and simplicia [1].

## Research Methodology

### *Materials and Treatment Design*

The materials used in this study are, kecombrang fruit, *S. aureus* bacteria FNCC 0047, Aquadest, Nutrient Agar (NA), Nutrient Broth (NB), MSA, Lamb blood, $H_2O_2$, Crystal violet, Lugol, 95% alcoholic acid, and Fuschine water. The tools used in this study are knives, fruit chopper machines, cabinet dryers "Teguh Wijaya Teknik", Erlenmeyer "Pyrex", "ACIS BC-500" digital scales, glassware, spatulas and filter cloth (250 mesh), Laminers, "All American" autoclave, 37°C "Memmert" incubator, shaker, "Pyrex" measuring cup, petri dish, vortex, stirrer, "Gilson" micropipette, blue tip, eyebrow needle, pyramid lamp, "LG" refrigerator, spatula , mortar, sorong term and other laboratory equipment for analysis. This study uses an experimental method with Completely Randomized Design (CRD). The treatment that was tried consisted of 1 factor with 10 treatment levels, repeated 3 times against the test bacteria to obtain 30 experimental units.

### *Research Analysis*

The research includes several stages, namely simplicia powder making [11], making simplicia extract [12,13], preparation of research tools, preparation of agar media, preparation of test bacteria, identification of test bacteria, analysis of antibacterial activity of well diffusion

IOP Conf. Series: Earth and Environmental Science **406** (2019) 012011 doi:10.1088/1755-1315/406/1/012011

method [9,11] and determination of MIC (Minimum Inhibitory Concentration) of simplicia extract kecombrang fruit [11].

## Result and Discussion

*Colonial Morphology*

The bacterial morphology can be divided into two, namely colony morphology and cell morphology. Examination of the morphology of the *S. aureus* FNCC 0047 bacterial colonies can be done by observing bacterial growth in nutrient agar (NA) after bacteria are grown for 2 x 24 hours on a petri dish. The morphology of microorganism colonies was assessed from several things, including form, smell, size, edge of colonies, surface, color and elevation [11]. Observations of morphology of bacterial colonies of research trials are presented in Table 1. and Figure 1.

Table 1. The results of morphological observations of colonies bacterial test *S. aureus* FNCC 0047

| No. | Colony morphology | Research observation result | Observation result [11] |
|---|---|---|---|
| 1 | form | Round, clustered | round, 1-2mm |
| 2 | smell | Bad Smell | Bad smell |
| 3 | type | medium | medium |
| 4 | side | flat | flat |
| 5 | surface | glowing | glowing |
| 6 | color | White yellowish | Dark yellow to brown |
| 7 | Elevation | Convex | Convex |

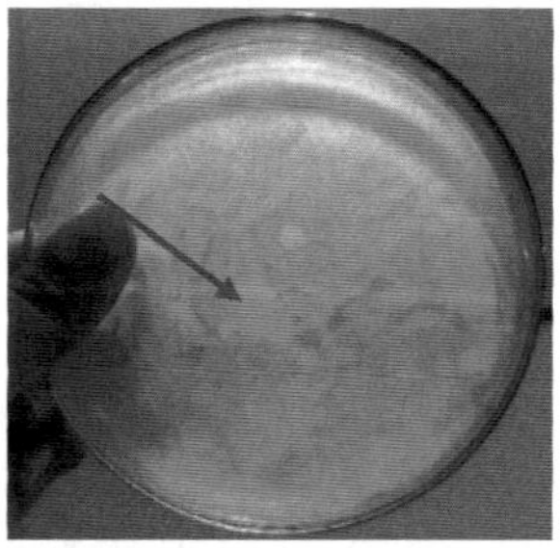

Figure 1. Morphology of *S. aureus* FNCC 0047 colonies on agar nutrient media (primary data processed, 2018).

Based on direct observations, the morphology of *S. aureus* FNCC 0047 colonies is in accordance with [14] statement, that colonies formed by *S. aureus* bacteria in nutrient growth media to be round, convex, opaque, shiny, soft consistency, and colored brown to yellowish brown but the young colonies of bacteria are colorless.

*Cell Morphology*

Cell morphology examination is done by testing Gram stain. Gram staining is to find out the bacteria tested including in Gram positive or Gram negative bacteria with the help of a microscope [15]. The results of the Gram staining method under a microscope with 1000x magnification carried out in this study are shown in Figure 2.

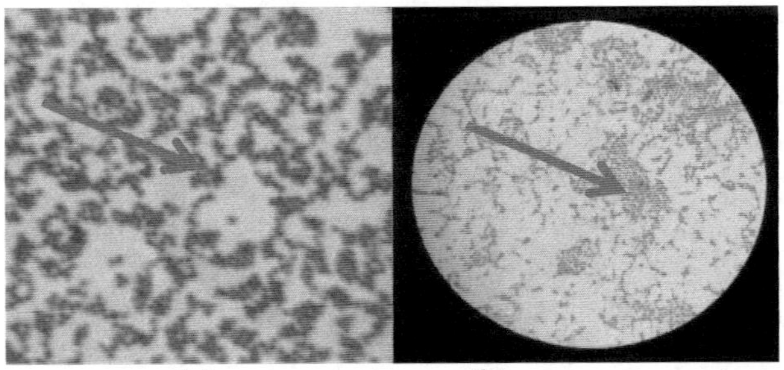

(a) (b)

Figure 2. (a) Gram staining of *S. aureus* in the literature [15]; (b) The direct observation of Gram stain of *S. aureus* FNCC 0047 researchers (primary data processed, 2018).

In Figure 2 (b) it is shown that the test bacteria used by researchers have purple characteristics, shaped like kokus, clustered and formed strands like grapes. These characteristics are in accordance with the statement [16], that *S. aureus* bacteria have cell morphological features in purple, kokus-shaped, and group like grapes on the Gram staining method. The direct observation of the test bacteria by the Gram staining method of the researcher was in accordance with the results of Gram staining in the literature (Figure 2a). Based on the results of Gram staining, the identification of the type of bacteria detected was *S. aureus* FNCC 0047 with a description of the purple cell morphology, kokus-shaped, as well as groups like grapes as infectious microorganisms in invasive skin and soft tissue [16]. Thus, bacteria test can be used to test the antibacterial activity of kecombrang fruit simplicia extract.

*Biochemical Test*

The biochemical testing of *S. aureus* FNCC 0047 was carried out by conducting catalase and oxidase tests. The catalase test results of *S. aureus* FNCC 0047 bacteria can be seen in Figure 3.

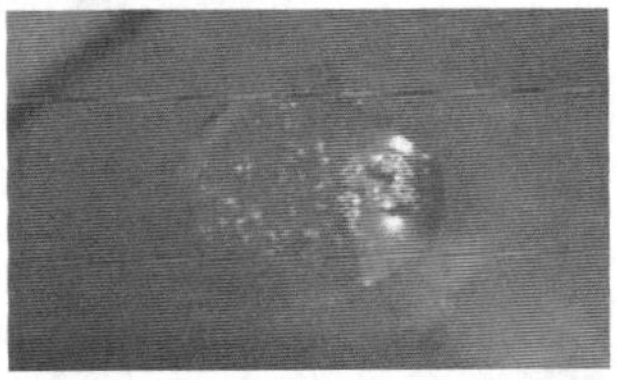

Figure 3. Air bubbles as a result of catalase test of *S. aureus* FNCC 0047 bacteria (primary data processed, 2018).

Based on Figure 3, the results of testing the catalase of *S. aureus* FNCC 0047 bacteria obtained positive results which were marked by the formation of air bubbles in the glass of the object being observed. The air bubbles formed prove that the *S. aureus* FNCC 0047 bacteria can produce the catalase enzyme which is able to hydrolyze hydrogen peroxide ($H_2O_2$) into water ($H_2O$) and gas bubbles ($O_2$). This is consistent with the research conducted by [15], regarding the identification and characterization of *Staphylococcus* sp. and *Streptococcus* sp.

The oxidase test is carried out to determine the presence or absence of oxidase enzymes in bacteria by using paper oxidase which can be seen from the color changes that occur on paper oxidase. The results of oxidase testing carried out on test bacteria using paper oxidase are presented in Figure 4.

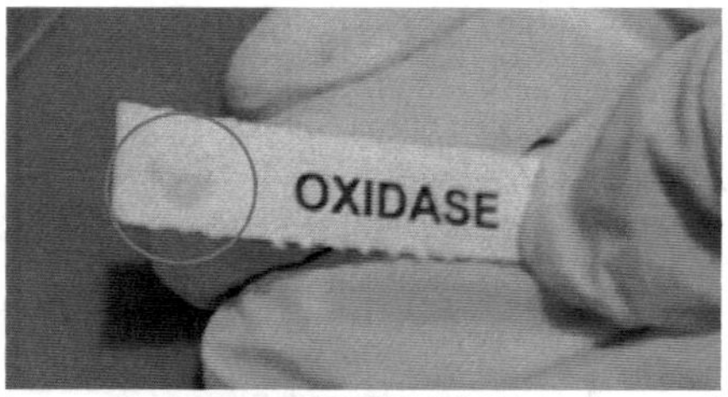


Figure 4. Oxidase test results on *S. aureus* FNCC 0047 (primary data processed, 2018)

Figure 4 shows that the *S. aureus* FNCC 0047 bacteria used in this study cannot change the color of the oxidase paper to purplish blue which shows negative oxidase. This is consistent with the research conducted by [17], that *S. aureus* has a negative oxidase result, because it does not have an oxidase enzyme.

*Staphylococcus aureus test FNCC 0047 with MSA Media and Lamb Blood*

The *S.aureus* FNCC 0477 test was carried out using the modified Mannitol Salt Agar (MSA) media with 5% added lamb blood from the total volume of MSA media. The addition of 5% lamb blood aims to determine the strength of the test bacteria in lysing blood. The test results of *S. aureus* FNCC 0047 bacteria can be seen in Figure 5.

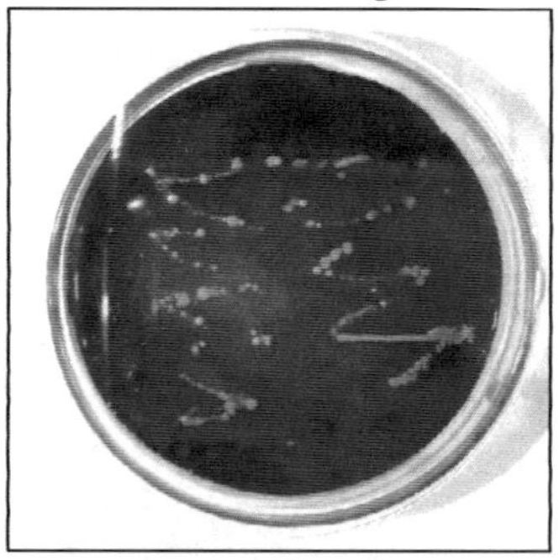

Figure 5. *S. aureus* FNCC 0047 test results on blood MSA media (primary data processed, 2018).

According to [18], *S. aureus* pathogens have the following characteristics: medium size, yellowish white, and have colonies with patterns of hemolysis (clear zone) on blood agar. Figure 5 shows only the discoloration of the MSA media by adding lamb blood to the area around the bacterial colonies from black red to pink but does not form a clear zone around the bacterial colonies. This shows that the *S. aureus* FNCC 0047 bacteria used in this study cannot lyse lamb blood and enter into the category of $\gamma$-hemolysis. $\gamma$-hemolysis is defined as the absence of lysis in red blood cells.

*Antibacterial Activity of Kecombrang Fruit Simplicia Extract with the Well Diffusion Method*

Research conducted to determine the antibacterial of kecombrang fruit simplicia extract to *S. aureus* FNCC 0047 test bacteria in vitro showed inhibitory zone diameter of kecombrang fruit simplicia extract against *S. aureus* FNCC 0047. Extraction of kecombrang fruit simplicia in this study was conducted using maceration method, so that the extract of the kecombrang simplicia was obtained with a concentration of 100%, the extraction results were then diluted using aquadest solvents to obtain a concentration of 10%, 20%, 30%, 40%, 50%, 60%, 70%, 80% and 90%.

The overall results showed that all concentrations of kecombrang simplicia could inhibit the growth of *S. aureus* FNCC 0047 bacteria. The lowest concentration of 10% kecombrang fruit simplicia had been able to inhibit the growth of *S. aureus* FNCC 0047 bacteria with a clear zone in the diameter of the inhibition zone. 7.26 mm. At a concentration of 20% and 30% of the kecombrang fruit simplicia, the inhibition zone diameter was 7.54 mm and 7.51 mm. The best inhibition occurred at a concentration of 40% by obtaining a inhibitory zone diameter of 7.83 mm, but for concentrations of 50% to 100% it showed a decrease in the gradual effectiveness of kecombrang fruit simplicia compared with a concentration of 40%. This condition can be caused by the speed factor of diffusion of antibacterial compounds on the media so that the higher the concentration of antibacterial compounds causes the antibacterial compound to become thicker and thus affects the speed of diffusion 4. The average value of the inhibition zone diameter of the kecombrang fruit simplicia concentrates in the S test bacteria aureus FNCC 0047 can be seen in Figure 6.

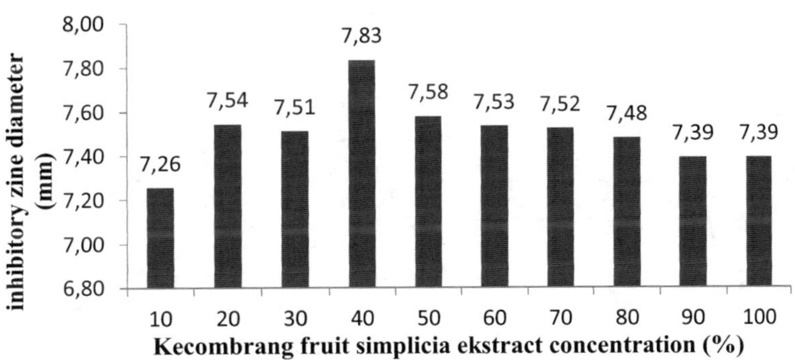


Figure 6. The average value of the inhibitory zone diameter of the kecombrang fruit simplicia with various concentrations in the test bacteria (primary data processed, 2018)

Based on Figure 6 shows that at a concentration of 40% is an effective and efficient concentration in inhibiting the growth of *S. aureus* FNCC 0047 bacteria compared to other concentration treatments with the strength of inhibition reaching 7.83 mm. Treatment can be regarded as effective concentration when the concentration reaches the most extensive inhibition zone compared to other concentration inhibition zones [19]. While efficient concentration is the concentration with the best antimicrobial power with fewer resources [20]. According to [21], the antibacterial provisions are with the formation of 20 mm or more resistance area means very strong, the resistance area 10 - 20 mm means strong, 5-10 mm means medium, and the resistance area 5 mm or less means weak. Based on the results obtained and the underlying literature, it can be concluded that the kecombrang simplicia extract has a moderate antibacterial effect on the *S. aureus* FNCC 0047 bacteria, because the average diameter is in the range of 5-10 mm.

In general, the diameter of the inhibitory zone tends to increase in proportion to the increase in extract concentration. However, in this study there was a decrease in the diameter of the inhibition zone at an extract concentration of 50% to 100%. Similar results also occur in the results of [22], which states that the diameter of the inhibition zone does not always increase in proportion to the increasing concentration of antibacterial compounds, this is possible because of differences in the speed of diffusion of antibacterial compounds in the agar media and the different types and concentrations of antibacterial compounds gives a different diameter of the inhibition zone at a certain length of time. Another literature mentions also when low extract concentrations are more effective than concentrations above them because they are influenced by the thickness of the extract solution, making it easier for the diffusion process in experimental media where bacteria grow around it [23].

*Microbial Inhibitory Concentration* (MIC) Analysis in this study based on Bloomfield method 1. The method uses linear regression techniques by plotting between Ln kecombrang fruit simplicia concentration on the X axis with the square of the inhibitory zone on the Y axis. Linear regression arising from the formula applied in this study is y = 22.34 x -9,783; $R^2$ = 0.1981. Determine the MIC value of the research that has been done by multiplying the value of X (Y = 0) with the method constant 0.25. MIC values obtained based on this study can be seen in Table 2.

Table 2. MIC values based on linear regression between the concentration of simplicia of kecombrang fruit and the diameter of the inhibitory zone

| Bacteria | Linear regression | $R^2$ | X(Y=0) | MIC (%) |
|---|---|---|---|---|
| *S. aureus* | **Y** = 22.34 **X** - 9.7833 | 0.1981 | 2.2834 | 0.5708 |

Based on the results of the MIC analysis of the [24] method, MIC results were obtained with a number of 0.5708% or 0.6%. Minimum Inhibition Concentration (MIC) is the lowest concentration (ingredient content in one solution) of an antibacterial which inhibits the growth of a microorganism after incubation [25]. These results can be explained that the kecombrang simplicia at a concentration level of 0.6% can inhibit bacterial growth *S. aureus* FNCC 0047.

The diameter of the inhibitory zone arising from the addition of kecombrang fruit simplicia extract to the test bacteria in the nutrient medium so that there are bioactive compounds or active substances contained in the kecombrang fruit. Active substances with antimicrobial effects contained in kecombrang fruit include phenols, flavonoids and alkaloids [11]. Alkaloids are an organic base containing elements of Nitrogen (N) generally derived from plants, which have strong physiological effects on humans. The usefulness of alkaloid compounds in the field of pharmacology is to stimulate the nervous system, increase blood pressure, and fight microbial infections [26]. The mechanism of action of alkaloids as antibacterial is by disrupting the constituent components of peptidoglycan in bacterial cells, so that the cell wall layer is not formed intact and causes cell death [27]. In this study, alkaloid became the most effective active substance in the inhibition mechanism of *S. aureus* FNCC

0047 because in its mechanism alkaloids can interfere with the components that make up the cell wall of bacteria which cause more bacterial cell death compared to living bacterial cells, resulting in inhibition bacterial growth.

Based on its mechanism of action, antimicrobials can be divided into five groups, namely antimicrobials that interfere with microbial cell metabolism, inhibit the synthesis of microbial cell walls, inhibit protein synthesis of microbial cells, inhibit the synthesis or damage of nucleic acids of microbial cells, and disrupt the permeability of microbial cell membranes. Underlying, the mechanism of inhibition of dominant bioactive compounds (phenols, flavonoids and alkaloids) contained in kecombrang fruit simplicia extract works on the second mechanism of action by inhibiting the synthesis of microbial cell walls, resulting in damage to microbial cell walls.

## Conclusion

Kecombrang fruit simplicia extract (*N. speciosa*) has antibacterial activity against the growth of Gram positive *S. aureus* FNCC 0047 bacteria with the most effective concentration of 40%. The MIC (Minimum inhibitory concentration) value of the kecombrang fruit simplicia extract against the growth of Gram positive *S. aureus* FNCC 0047 bacteria was 0.6%.

## References

[1] Mukhriani, M., Tahar, N., & Astha, A. S. W. 2014. Uji Aktivitas Antibakteri Hasil Fraksinasi dari Ekstrak Metanol Daun Katuk (*Sauropus androgynus*) terhadap beberapa Bakteri Patogen. *Jurnal Farmasi Universitas Islam Negeri Alauddin Makassar*. 2(1): 12-17.

[2] Yeni, Y. D., Djannah, S. N., dan Nurani, L. H. 2013. Uji Aktivitas Antibakteri Infusa Daun Sirsak (*Annona Muricata L.*) secara In Vitro terhadap *Staphylococcus aureus* ATCC 25923 dan *Escherichia coli* ATCC 35218 serta Profil Kromatografi Lapis Tipisnya. Jurnal Kesehatan Masyarakat. 4(3): 218-238.

[3] Nugroho, W. S. 2004. *Aspek Kesehatan Masyarakat Veteriner Staphylococcus, Bakteri Jahat yang Sering Disepelekan*. Universitas Gadjah Mada. Yogyakarta.

[4] Lestari, L. A. 2010. Probiotik dan Prebiotik dari Makanan Indonesia. Annual Scientific Meeting. Fakultas Ilmu Gizi dan Kesehatan. Universitas Gadjah Mada. Yogyakarta.

[5] Hidayat, S., dan J. R Hutapea. 1991. *Inventarisasi Tanaman Obat Indonesia*. Departemen Kesehatan RI. Jakarta.

[6] Naufalin, R., dan Herastuti, S. R. 2012. *Pengawet Alami Pada Produk Pangan*. UPT Percetakan dan Penerbitan Universitas Jenderal Soedirman. Purwokerto.

[7] Naufalin, R., 2019. Natural preservation opportunities and challenges in improving food safety. In *AIP Conference Proceedings*. Vol. 2094, No. 1, p. 020032. AIP Publishing.

[8] Naufalin, R., and S. R. Herastuti. 2018. Antibacterial activity of kecombrang flower extract (*Nicolaia speciosa*) microencapsulation with food additive materials formulation. IOP Conf. Series: Earth and Environmental Science. 102 012035. doi :10.1088/1755-1315/102/1/012035.

[9] Latifasari, N., R. Naufalin, R. Wicaksono. 2019. Edible coating application of kecombrang leaves to reduce gourami sausage damage. IOP Conf. Series: Earth and Environmental Science 250 (2019) 012055: 1-7.

[10] Hanifah, M.R., R. Naufalin, R. Wicaksono. 2019. The effect of edible coating contained Kecombrang leaves concentrate on gourami fish fillet quality. IOP Conf. Series: Earth and Environmental Science 250 (2019) 012057: 18.
[11] Naufalin, R., and Herastuti, S. R. 2017. Antibacterial activity of *Nicolaia speciosa* fruit extract. *International Food Research Journal* 24.1:379-385.
[12] Naufalin, R., Wicaksono, R., Arsil, P. And M.F. Salman. 2018. Antimicrobial coating on quality attributes of sausage during refrigerated storage. E3S Web of Conferences 47, 01002 (2018) https://doi.org/10.1051/e3sconf/20184701002
[13] Putri, F.A., R. Naufalin, R. Wicaksono. 2019. Antimicrobial edible coating application of Kecombrang flower concentrate to reduce microbial growth on gourami fish sausage. IOP Conf. Series: Earth and Environmental Science 250 (2019) 012056: 1-8.
[14] Jawetz, M. 2007. Microbiologia Medica. *Adelberg*. 18: 521-522.
[15] Feliatra., Efendi, I., dan Edwar Suryadi. 2004. Isolasi dan Identifikasi Bakteri Probiotik dari Ikan Kerapu Macan (*Ephinephelus Fuscogatus*) dalam Upaya Efisiensi Pakan Ikan. *Jurnal Natur Indonesia*. 6(2): 75-80.
[16] Rijayanti, R. P. 2014. Uji Aktivitas Antibakteri Ekstrak Etanol Daun Mangga Bacang (*Mangifera foetida L.*) terhadap *Staphylococcus aureus* secara In Vitro. *Jurnal Mahasiswa PSPD FK Universitas Tanjungpura*. 1(1).
[17] Toelle, N. N dan Lenda, V. 2014. Identifikasi dan Karakteristik *Staphylococcus* sp. dan *Streptococcus* sp. dari Infeksi Ovarium Pada Ayam Petelur Komersial. *Jurnal Ilmu Ternak*. 14(1): 32-37.
[18] Sugiri, Y. D., dan Anri, A. 2010. *Prevalensi patogen penyebab mastitis subklinis (Staphylococcus aureus dan Streptococcus agalactiae) dan patogen penyebab mastitis subklinis lainnya pada peternak skala kecil dan menengah di beberapa sentra peternakan sapi perah di Pulau Jawa*. BP3HK Lembang. Bandung.
[19] Oroh, S. B., Kandou, F. E., Pelealu, J., dan Pandiangan, D. 2015. Uji Daya Hambat Ekstrak Metanol *Selaginella Delicatula* dan *Diplazium Dilatatum* terhadap Bakteri *Staphylococcus aureus* dan *Escherichia coli*. *Jurnal Ilmiah Sains*. 15(1): 52-58.
[20] Bontjura, S. 2015. Uji Efek Antibakteri Ekstrak Daun Leilem (*Clerodendrum Minahassae L.*) terhadap Bakteri Streptococcus Mutans. *Pharmacon* 4(4): 96-101.
[21] Davis, W. W. dan Stout, T. R. 1971. Disc Plate Method of Microbiological Antibiotic Assay: II. Novel Procedure Offering Improved Accuracy. *Applied Microbiology*. 22(4): 666-670.
[22] Elifah, E. 2010. Uji Antibakteri Fraksi Aktif Ekstrak Metanol Daun Senggani (*Melastoma candidum*, D.Don) terhadap *Escherichia coli* dan *Bacillus subtilis* serta Profil Kromatografi Lapis Tipisnya. *Skripsi.* FMIPA UNS. Surakarta.
[23] Nurainy, F., Rizal, S., dan Yudiantoro, Y. 2012. Pengaruh Konsentrasi Kitosan terhadap Aktivitas Antibakteri dengan Metode Difusi Agar (Sumur). *Jurnal Teknologi dan Industri Hasil Pertanian*. 13(2): 117-125.
[24] Bloomfield, S. F. 1991. *Methods for Assessing Antimicrobial Activity. Di dalam* Denyer, S. P., dan Hugo, W. B., (eds). *Mechanism of Action of Chemical Biocides Their Study and Exploitation.* Blackwell Scientific Publication. London
[25] Purnamasari, D. A., Munadziroh, E., dan Yogiartono, R. M. 2010. Konsentrasi Ekstrak Biji Kakao sebagai Material Alam dalam Menghambat Pertumbuhan *Streptococcus Mutans*. *Jurnal Persatuan Dokter Gigi Indonesia*. 59(1): 14-8.
[26] Wullur, A. C., Schaduw, J., dan Wardhani, A. N. 2012. Identifikasi Alkaloid pada Daun Sirsak (*Annona Muricata L.*). *Jurnal Ilmiah Farmasi Poltekkes Manado*, *3*(2): 54-56.
[27] Pratiwi, D., Suswati, I., dan Abdullah, M. 2017. Efek Anti Bakteri Ekstrak Kulit Jeruk Nipis (*Citrus aurantifolia*) terhadap *Salmonella typhi* Secara In Vitro. *Saintika Medika*. 9(2): 110-115.

# Inulinase Activity of Thermophilic Bacteria isolated from Hot Springs of Penen Village, North Sumatera, Indonesia

[1]Edy Fachrial, [2]Harmileni, [3]Sari Anggraini, [4]Titania T. Nugroho, [4]*Saryono
[1]Laboratory of Molecular Biology, Faculty of Medicine, Universitas Prima Indonesia, Indonesia
[2]Department of Chemical Engineering, Politeknik Teknologi Kimia Industri, Indonesia
[3]Department of Agrotechnology, Faculty of Agro Technology, Universitas Prima Indonesia, Indonesia
[4]Laboratory of Biochemistry, Faculty of Mathematics and Natural Sciences, Universitas Riau, Indonesia
*corresponding author: saryono@lecturer.unri.ac.id

**Abstract**

Inulinase is an enzyme that catalyzes inulin to produce D-fructose as the end product which is considered as safe alternative sweetener to sucrose. This enzyme can be extracted from plants and vegetables but it is difficult to obtain it in high quantity and the production cost also high. The main problem in utilizing this enzyme in industrial is the instability against fluctuation of temperature that usually occurs in industrial process. Therefore thermophilic microbial inulinase has great potential for these industrial needs. The aim of this study was to isolate and characterize the thermophilic bacteria that possess the inulinase activity from hot springs in Penen Village, North Sumatera. To screening the inulinase producing bacteria, the modified Czapex Dox agar supplemented with 1,5% inulin powder was used. The temperature used for incubation is 48°C. The clear zone surrounds the bacteria after Lugol's iodine treatment on medium indicated the nucleolytic activity. As 4 isolates of thermophilic bacteria namely UTMP 11, UTMP 14, UTMP 15 and UTMP 17 showed the inulolytic activity. The biochemical characterization of these isolates showed negative in catalase test, negative on citrate test, non-motil, negative Gram staining, and rod-shaped morphology. Molecularly, this bacteria identified as *Bacillus subtilis* strain JCM 1465 and *Bacillus tequilensis* strain 10b. From the result, these isolates have potential as source of thermophilic inulinase production in food industry for the production of fructose.

Keywords: Thermophilic bacteria; inulinase; enzyme; hot springs

## Introduction

Inulin consisting of linear -2,1-linked polyfructose chains with terminal glucose units and also a good source for producing high fructose through hydrolysis ((1). Inulin is found in many plant families including dahlia tubers (Cichorium endivia) and Jerusalem artichoke (Helianthus tuberosus) and acts as reserve carbohydrates (2). To produce fructose, acid hydrolysis was carried out on inulin at high temperature, 80-100°C, however, the fructose easily degrades in this condition, producing difructose anhydrides difructose, a coloring product of fructose (3). The alternative approach to produce fructose-free from by-product can be done with enzymatic hydrolysis using inulinase. Inulinase is an enzyme that is able to produce fructose or fructooligosaccharides molecules through hydrolysis of inulin. Recently, inulinase has increased applications in industrial sector, especially in producing high fructose syrup. Inulinase has been reported from a variety of plants, and animals but they are present in very minute quantities. The main problem in the use of enzymes in industrial is instability of enzymes due to exposure of high temperature, changes in physical and chemical factors (4). Therefore, the best sources of inulinase are microorganism, because easy in cultivation, genetic manipulation, high production yield and usually inducible and extracellular in nature. Genetic changes using recombinant DNA technology can be easily done on microbial cells to increase enzyme production. Another advantage of microbial enzymes is the stability performance of the enzymes, so it can work in varied chemical and physical conditions that often occur during industrial processes (5). It is known that an organism's enzymes are conditioned to work optimally at its growth condition, including pH, salinity and temperature. One source of thermostable enzyme is thermophilic bacteria. Extremophilic microorganisms include members of all three domains of life, archaea, bacteria, and eukarya. Thermophilic microorganisms are growing optimally at temperature more than 45°C. The most enzyme produced by groups of thermophilic microorganisms is widely used because thermostability and resistance to change in physical and chemical factors (6). Based on this, the study on thermostable enzymes has attracted attention in recent years. Study regarding isolation and production of thermostable enzyme from microorganism have got wider attention. North Sumatera is one of the province in Indonesia which has many natural hot springs. Some of these hot springs have been commercialized as tourist attractions. The hot spring in Penen village is one of the spring which has not yet been explored. The studies on the exploration of thermophilic microbes and thermostable enzymes from hot springs in North Sumatera are still limited.The aim of this study is to isolate and identify inulinase producing thermophilic microorganisms, from hot springs in North Sumatera.

## Material and Methods

### Samples collection

The hot spring chosen in this study is the hot spring in Penen village, the regency of Deli Serdang, North Sumatera. This hot spring located at 3°18'5"N and 98°38'41"E. The location of sampling sites was shown in Figure 1.

IOP Conf. Series: Earth and Environmental Science **406** (2019) 012012 doi:10.1088/1755-1315/406/1/012012

Figure 1. The location of sampling sites

A sampling of water and sediment was conducted at three points, namely at the upstream, middle and downstream of location. During the sampling pH and temperature are measured. 10 mL of water from each sampling points was collected using well sampler, and stored in a sterile falcon tube. Sediment samples were collected using Ekman grab and stored in sterile sample bottles. All samples were stored in an icebox and analyzed further on the same day in the laboratory.

(a) (b)

(c)

Figure 2. sampling process. (a) sediment sampling at the downstream of hot springs, (b) sediment sampling at the middle stream of the hot spring and (c) sediment sampling at the upstream of hot spring. In each sampling process, Ekman grab is washed repeatedly with water in each sampling point to minimize the occurrence of microbial mixing from each sampling point.

Enrichment and Isolation of bacteria

1 mL of each water sample from each sampling point is mixed into a test tube and homogenized with the vortex. For sediment samples, each 1 g of sediment from each sampling point was mixed into a test tube containing 9 mL sterile distilled water and homogenized with the vortex. As 1 mL of mixed water sample and 1 g of mixed sediment sample were mixed in 9 mL of Tryptic Soy Broth, incubated for 24 hours at 49°C. The serial dilution was conducted up to $10^{-3}$dilution followed by inoculation on the surface of Tryptic Soy Agar. The Petri dish was incubated for 48 hours at 49°C. The growing colonies were characterized morphologically and biochemically.

Biochemical and morphological characterization

The growing bacteria were characterized biochemically. The biochemical characteristic of isolates was determined by gelatin assay, catalase test, citrate test, motility, and Gram staining.

Screening of inulinase activity

The inulinase activity of isolates was determined by inoculating on Czapex-Dox agar plates enriched with inulin (1% w/v) and incubated at 48°C for 48 hours. The plates were treated with Lugol's iodine for 3 to 5 minutes and washed with distilled H2O to remove excess Lugol's iodine. The clear zone around the colonies indicating inulin hydrolysis. Four bacterial with the largest clear zones if inulin hydrolysis were subcultured and kept for further studies.

Isolation, 16SrRNA gene amplification and sequencing

DNA isolation is carried out based on the protocol established by company (TianGen). The PCR solution consisted of 1µL forward Primer 16S, 1µL reverse primer 16S, 12,5µL GoTaq Green Master Mix 2X, 1 µL DNA template, 0,5 µL DMSO and 9 µL Nuclease free water. The program of PCR used in this study was : pre denaturation at 95° for 90 seconds, continued with 30 cycles consists of denaturation at 95°C for 30 seconds, annealing at 55 °C for 30 seconds, extension at 72 °C for 90 seconds, and final extension at 72 °C for 3 minutes. The primers used in this study were Primer 27 F: 5`- AGA GTT TGA TCC TGG CTC AG - 3` dan Primer 1492 R: 5`- GGT TAC CTT GTT ACG ACT T – 3. The sequence result then trimmed and assembled using BioEdit program (http://www.mbio.ncsu.edu/BioEdit/bioedit.html). The sequence result then submitted into BLAST (Basic Local Alignment Search Tool) with genome data that has been registered at NCBI (National Center for Biotechnology Information) at http://www.ncbi.nlm.nih.gov/BLAST, to determine species that have closest similarity molecularly.

Result and Discussion

Biochemical and morphological characterization

From 11 bacterial isolates that successfully isolated from sediment and water from hot springs in Penen Village, 4 isolates showed good inulinase activity based on the formation of the colourless zone around the colonies. These bacteria were isolated from the sediment, namely UTMP 11, UTMP 14, UTMP 15 and UTMP 17. These isolates showed good growth on Saboraud Dextrose Agar media. The growth of the isolates was showed and Figure 3.

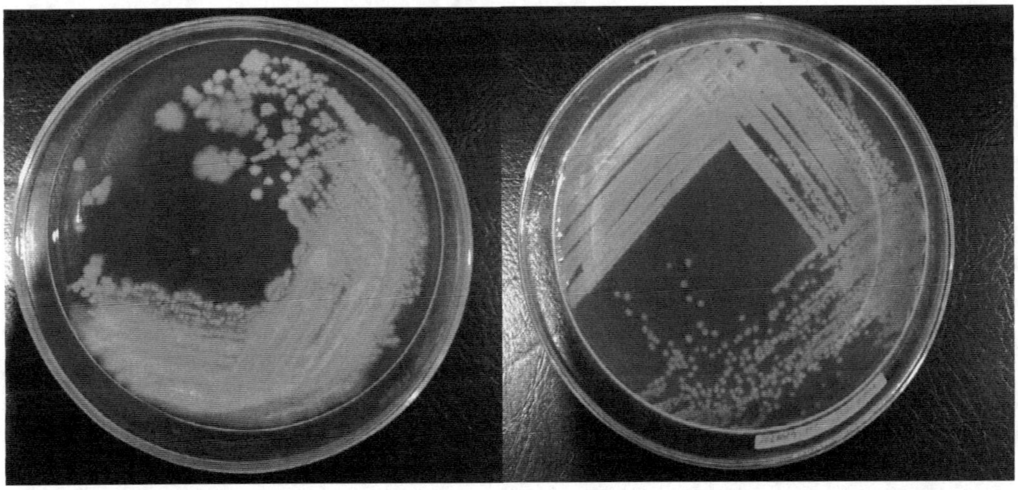

(a) UTMP 11 (b) UTMP 14

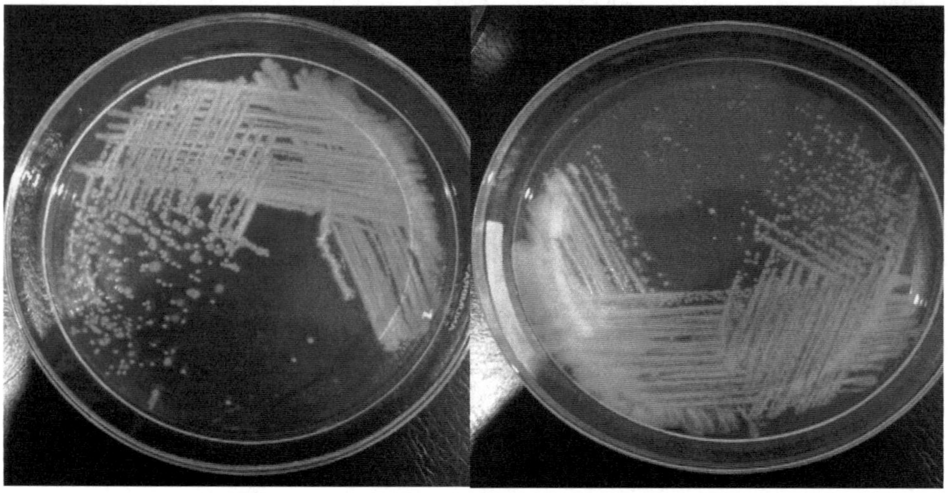

(c) UTMP 15 (d) UTMP 17

Figure 3. The growth of isolates UTMP 11, UTMP 14, UTMP 15 and UTMP 17 on Saboraud Dextrose Agar after 48 hours at 48° C

The biochemical characterization of the isolates was shown in table 1.

| Isolates code | Citrate test | Catalase test | Starch hydrolysis | Gelatin hydrolysis | motility |
|---|---|---|---|---|---|
| UTMP 11 | + | + | + | - | - |
| UTMP 14 | + | - | + | - | - |
| UTMP 15 | + | + | + | - | - |
| UTMP 17 | + | + | + | - | - |

From table 1, it is shown that all the isolates showed starch hydrolysis activity, indicating that the isolates also produced the amylase. Lugol's iodine solution can be used to detect the presence of starch. Iodine solution will form a blue to brown color. Hydrolyzed starch does not produce change of color (7). A similar result reported by Abejo et al (2018), which succeed to isolate amylase producing bacteria from hot springs in Southern Ethiopia (8). Out of 72 isolates, 10 amylases positive were screened on the basis of starch hydrolysis, indicated by the formation of clear zone around the colony on the growth medium supplemented by starch. The isolates also showed positive in citrate test. A citrate test was used to select the bacteria that have the ability to consume citrate as its sole source of carbon and ammonium as the sole nitrogen source (Ayitso). This is in accordance with Tariq et al (2018) which reported Bacillus species from sediment and swept clean flora, showed positive results in citrate utilization. 3 of 4 isolates resulted in positive in catalase test (9). Catalase is an oxidoreductase enzyme. This enzyme works as detoxification system in living cells against reactive oxygen species that are formed as a by-product of metabolic reactions (10). The Gram staining of these isolates showed that all the isolates are Gram-positive with rod-shaped bacteria. The result of Gram staining was showed in Figure 4.

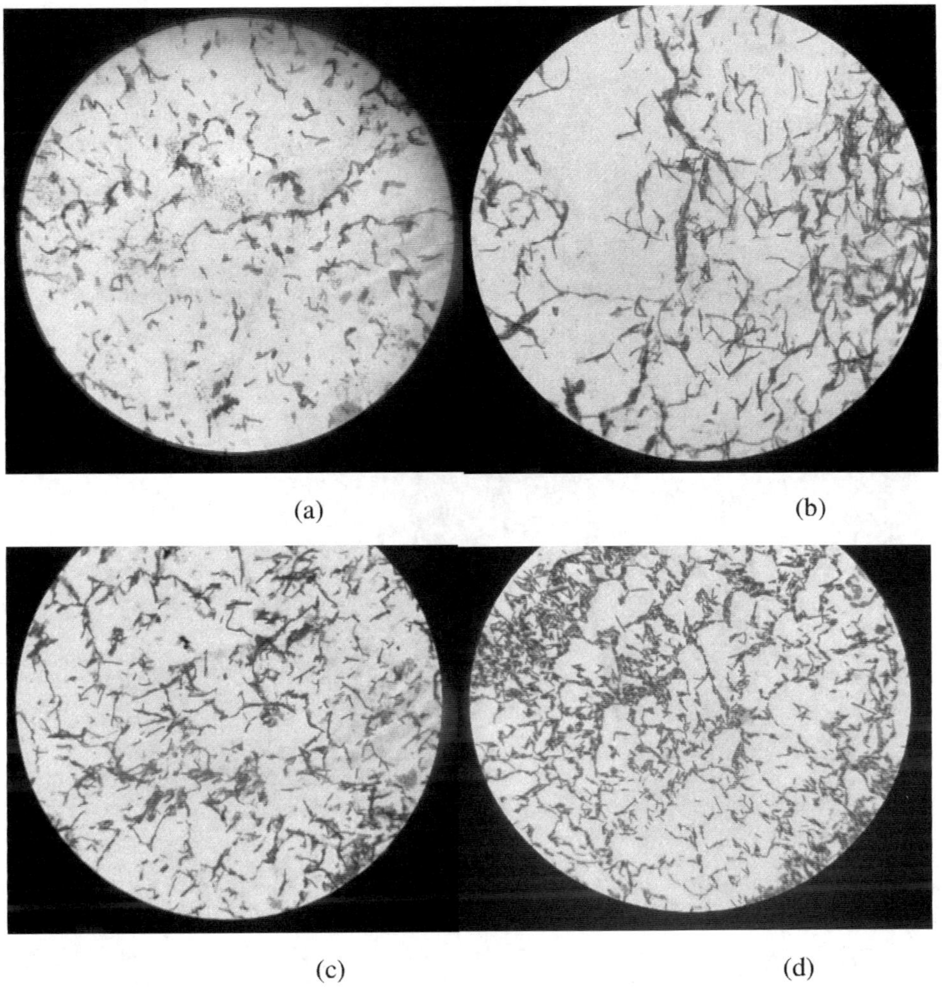


Figure 4. Result of Gram staining under light microscope with magnification 1000×. (a) UTMP 11; (b) UTMP; (c) UTMP 14; UTMP 15.

The aim of Gram staining is to divide the group of bacteria into positive and negative results based on their cell wall and cell wall permeability. Decolorization will cause significant damage to the surface of Gram-negative cells and only a small amount of damage to Gram-positive cells. Gram-negative bacteria are thought to have thin-walled cells and rich in lipid, causing cells to lose the first dye (crystal violet) and bind the second dye (safranin). Gram-positive cells are thick-walled cells and lipid-poor, appear blue from retaining the crystal violet

Inulinase activity

Inulinase activity of bacteria is indicated by the formation of a clear zone on Czapex-Dox media supplemented with 1% inulin. The inulinase activity of isolates UTMP 11, UTMP 14, UTMP 15 and UTMP 17 showed in Figure 5.

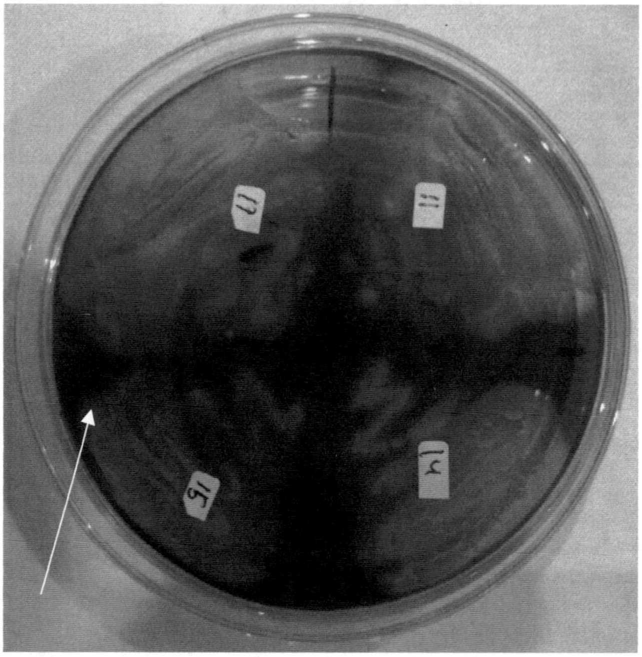

Figure 5. The inulinase activity of isolates UTMP 11, UTMP 14, UTMP 15 and UTMP 17. The formation of the clear zone due to inulin hydrolysis showed by the white arrow.

Studies on isolation of thermophilic bacteria producing inulinase are still very few reported. Most studies of inulinase-producing bacteria are mesophilic bacteria and fungal. Microbial inulinase has great potential in industrial to produce fructose from inulin. Inulinase produced by microorganism consist of exo-inulinase and endo-inulinase. Exo-inulinase splits off the terminal fructose units from non-reducing end of the inulin, and also hydrolyzed sucrose and raffinose. Endo-inulinase targets the β-2,1 linkage of inulin and hydrolizes it into fructose (11). Therefore, inulinase has a wide usage in industrial to produce fructose syrup by exo-enzymatic hydrolysis of inulin with D-fructose content over 95% or for the production of oligofructose syrups by endo-enzymatic hydrolysis (12). The plate screening method is generally used to detect extracellular hydrolytic enzymes produced by bacteria. The presence of inulinase on agar plates can be detected by formation of clear zone surrounding the bacterial colony after incubation. (5). The previous study reported bacterial inulinase isolated from the soil of sugarcane field. 4 microbes including *Bacillus subtilis, Lactobacillus casei, Pseudomonas aeruginosa* and *Achromobacter* sp were identified as inulinase producers. The optimum parameters of these bacteria were found at 40°C and pH 5 (2). Another study reported that inulinase producing bacteria obtained from rhizosphere of Taro, Dahlia, Garlic, and Onion. Among all the isolates, *Bacillus* sp considered as good inulinase produce and showed good enzyme production at 37°C and pH 6 using Dahlia as raw inulin source (13).

Conclusions

The study reported that thermophilic bacteria isolated from hot springs show inulinase activity. There are 4 isolates that show good inulinase activity at temperature 48°C, namely UTMP 11, UTMP 14, UTMP 15 and UTMP 17. All isolates are Gram-positive and rod-shaped bacteria. This revealed that thermophilic bacteria have potential as a source of enzyme inulinase. For further study, it is necessary to identify the inulinase producing thermophilic bacteria also production and characterization of crude inulinase produced by these bacteria.

Molecular Identification

The result of molecular identification of UTMP 11, UTMP 14, UTMP 15 and UTMP 17 was shown in table 2.

| Sequence code | The similarity result of homology-based on Basic Local Alignment Search Tool (BLAST) |
|---|---|
| UTMP 11 | ***Bacillus subtilis* strain *JCM 1465***<br>Accession no: **NR_113265.1**<br>Homology: **99.86%**<br>Query coverage: **100%** |
| UTMP 12 | ***Bacillus subtilis* strain *JCM 1465***<br>Accession no: **NR_113265.1**<br>Homology: **99.72%**<br>Query coverage: **100%** |
| UTMP 14 | ***Bacillus tequilensis* strain *10b***<br>Accession no: **NR_104919.1**<br>Homology: **99.79%**<br>Query coverage: **100%** |
| UTMP 15 | ***Bacillus subtilis* strain *JCM 1465***<br>Accession no: **NR_113265.1**<br>Homology: **99.64%**<br>Query coverage: **100%** |
| UTMP 17 | ***Bacillus tequilensis* strain *10b***<br>Accession no: **NR_104919.1**<br>Homology : **99.57%**<br>Query coverage: **100%** |

Based on the identification result, the isolates identified as *Bacillus subtilis* strain JCM 1465 and *Bacillus tequilensis* strain 10b. This is the first study to report that *Bacillus subtilis* strain JCM 1465 and *Bacillus tequilensis* strain 10b showed inulinase activity. Inulinase is produced by different types of bacteria including *Bacillus* sp, *Lactobacillus*, *Clostridium,* and *Pseudomonas*. From previous study, a total of ten isolates including *Bacillus subtilis*, isolated from sugarcane soil, showed inulinase activity. The molecular weight of this enzyme was estimated at 45 KDa and optimum conditions were found to be 40°C and pH 5 (2). The previous report for *Bacillus tequilensis* which isolated from coastal mud describes that this strain showed the amylolytic enzyme (14).

Conclusion

based on the result, this is the first study that reported that thermophilic Bacillus subtilus strain JCM 1465 and Bacillus tequilensis strain 10b showed the inulinase activity. these strains might be a source of industrially enzymes mainly inulinase

Acknowledgment

The author would say thanks to the Ministry of Research, Technology and Higher Education which funded this research through collaborative research between Universitas Prima Indonesia and Universitas Riau (contract number : T/61/L1.3.1/PT.01.03/2019)

References

1. Silvera D, Luthfin I, Aulia A, Wahyu PN, Saryono. Optimization of process parameters for inulinase production from endophytic fungi Fusarium solani LBKURCC67 , Humicola fuscoatra LBKURCC68 and Fusarium oxysporum LBKURCC69. Res J Chem Environ. 2018;22(II):71–8.

2. Angel SJ, Kavitha C, Vidyadharani G, Roy P, Dhandapani R. Isolation of Inulinase Producing Bacteria from Sugarcane Oil. Int J Appl Biol Pharm Technol. 2012;3(4):320–6.

3. Liu G, Chi Z, Chi Z. Molecular characterization and expression of microbial inulinase genes. Crit Rev Microbiol. 2013;39(5):152–65.

4. Mohammad BT, Daghistani HI Al, Jaouani A, Abdel-latif S, Kennes C. Isolation and Characterization of Thermophilic Bacteria from Jordanian Hot Springs : Bacillus licheniformis and Thermomonas hydrothermalis Isolates as Potential Producers of Thermostable Enzymes. Int J Microbiol. 2017;2017:1–12.

5. Singh R, Kumar M, Mittal A, Mehta PK. Microbial enzymes : industrial progress in 21st century. 3 Biotech. 2016;6(174):2–15.

6. Ifandi S, Alwi M. Isolation of Thermophilic Bacteria from Bora Hot Springs in Central Sulawesi. Biosaintifika. 2018;10(2):291–7.

7. Madhav K, Verma S, Tanta R. Isolation of amylase producing Bacillus species , from soil sample of different regions in Dehradun and to check the effect of pH and Temperature on their amylase activity Isolation of amylase producing Bacillus species , from soil sample of different reg. J Pharm Biomed Sci. 2018;12(03):1–8.

8. Albejo AL, Hamza TA. Isolation and Characterization of Thermostable Amylase Producing Bacteria from Hot Spring at Arba Minch Nech Sar National Park , Southern Ethiopia. Int J Nov Res Interdiscip Stud. 2017;4(5):9–16.

9. Tariq AL, Sudha S, Reyaz AL. Isolation and Screening of Bacillus Species from Sediments and Application in Bioremediation. Int J Curr Microbiol Appl Sci. 2016;5(6):916–24.

10. Babiker BM, Ahmed MAE, Ibrahim HM. Isolation & Identification of Catalase Producing Bacillus spp : A Comparative Study. Int J Adv Res. 2017;4(2):1206–11.

11. Saryono S, Chainulfiffah AM, Ardhi A, Pratiwi NW. Research Article Production , purification , and characterization of inulinase from dahlia rizhosphere-isolated Aspergillus clavatus. J Chem Pharm Res. 2015;7(9S):165–76.

12. Naidoo K, Kumar A, Sharma V, Permaul K, Singh S. Puri fi cation and Characterization of an Endoinulinase from Xanthomonas campestris pv . phaseoli KM 24 Mutant. Food Technol Biotechnol. 2015;53(2):146–53.

13. Tharun S, Guttikonda K, Vanitha S, Anuradha K. Screening of inulinase producing bacteria from diverse sources. Biosci Discov. 2017;8(4):695–701.

14. Manohar P, T S, Gothandam KM, Kannan VR, Ramesh N. ENHANCED AMYLOLYTIC ACTIVITY OF INTRACELLULAR alpha-AMYLASE PRODUCED BY BACILLUS TEQUILENSIS. J Micrpbiology, Biotechnol Food Sci. 2017;6(6):6–11.

# Novelties of laccase enzyme from bacteria of sago waste

**BS. Reksohadiwinoto**[1]
[1]Agency for the Assessment and Application of Technology, Indonesia
*Email: budhi.santoso@bppt.go.id

**Abstract.** Indonesia's sago industry yielded around 400 thousand tons per year, which was the second largest producer in the world. The Weaknesses of Indonesian sago, in term of its quality, included a brownish color, unpleasant odor and soft texture. All of those characteristics can be corrected by addition of Laccase enzyme. Wet sago still contains natural microbes that can be used as a source of beneficial germplasm of bacteria and fungi. In this study, conventional method for isolating bacteria was performed. Conventional isolation technique obtained 26 isolates from sago. It, then, followed by biochemical analysis to obtain 5 isolates of putative bacteria candidate for bioleaching. The sequencing result indicated 3 potential bacteria, which generate laccase enzymes, namely *Lactobacillus plantarum* (2) and *Lactobacillus brevis* (1). Isolated bacteria, then, were fermented in the 500 mL flask and followed with laccase activity analysis with adding guaiacol as substrate and read using spectrophotometer, resulted in 64 U/L, 54 U/L, and 202 U/L, respectively. In conclusion, Laccase enzyme, produced from food grade endogenous sago species, might likely improve the quality of sago starch, which leads to increase the broader utilization of sago starch.

Keywords: Laccase; bioleaching; *Lactobacillus plantarum*; and sago starch quality

## 1. Introduction

Indonesia's sago industry produces around 400 thousand tons per year, which is the second largest producer in the world, after Malaysia. The greatest sago plants can be found in Papua and West Papua (1). Although Indonesia produces the largest sago starch, its properties show a weakness in the term of color, which tends to brown and smells less pleasant. This characteristic might be caused by the presence of polyphenols in the extracted sago starch. On the other hand, generally sago starch quality criteria are determined by color, viscosity, and particle size (2). Some phenol compounds are not allowed in food standards specifically for food and health industry. Because of that, processing of sago starch that meets food safety standards (food grade) was necessarily required.

Polyphenol compounds in sago are formed because the soil, in which sago plants grow, is acidic containing metal elements such as Manganese (Mn), Iron (Fe), Magnesium (Mg), and water containing Sulphur, which are responsible for forming oxidation reactions. A study stated that soil, which is acid and high sulfur content of water, corresponds to high levels of ash and phenolic compounds causing the color of sago starch tended to pink and brown (3). During this time, bleaching of sago starch is done by adding chemicals which are feared to cause health and environmental problems. To improve the quality of sago starch, one approach can be used by adding laccase, an enzyme that was previously widely used in the pulp and textile industry as a dye-decolorization or purification agent. Laccase has the ability to act as a bioleaching through the mechanism of oxide-reductase polyphenol and amine.

The use of laccase in the food industry, especially beverages and baking bread, has been reported by many researchers (4-7). Not only above mentioned industry, but Laccase is also used in the wine and juice industry (8-10). Laccase can prevent the formation of polyphenol protein in the preparation of making fruit juices, thus, increasing the stability of juice (9). It is also used to improve the odor quality of plant oils by removing dissolved oxygen to prevent oxidation (4, 5). In addition, it has another properties, which is to enhance color quality when added to tea-based products (11). Furthermore, Laccase is also applied in pre-treatment of chocolate, before the roasting process, aiming to develop aroma and taste quality of chocolate-based products (4, 5, 12). Another example, which was written in the patent literature, is Laccase can reduce unpleasant odors. Therefore, despite its effect on aroma, it has been employed in the chewing gum industry to strengthen cross-links of beet sugar pectin through a double bond reaction to form a gel in the food ingredients. Moreover, review papers highlighted that laccase can have a new functional properties such as improving quality, or reducing production costs due to the reaction between laccase and carbohydrates, unsaturated fatty acids, phenol compounds and protein contained thiol in foods (4, 5). Lastly, the use of laccase to improve the quality of the dough- and bread-formed in the bread industry had been patented in America with US patent number 6,296,883 B1 dated October 2, 2001. Thus, it is implied that Laccase improves the quality and appearance of colors, removes polyphenol compounds and absorbs colors , which was more potent based on its physio-chemical properties.

Laccase, which is widely produced today, is derived from fungi such as *pleurotus, tremetes, agaricus, pycnoporus, trichoderma* (4, 5, 13). In addition, laccase is also produced from the bacteria, which were from *Brevibacillus* (14), *P. putida* (15), *Halo-bacillus* (16), *B. cereus* (17, 18), *B. subtilis* (19), *Thermus* sp 2.9 (20). Recent research showed that laccase was existed in *Lactobacillus plantarum* JLAU103, showing an enzyme with 12.4 KDa molecular weight, namely exopolysaccharide EPS 103 that had a strong scavenging ability against hydroxyl radicals, ABTS, and DPPH of 80.4%, 65.5%, and 60.5%, respectively (21). In addition, Laccase also found in recombinant *Lactobacillus plantarum* with a molecular weight of ~62.5 kDa and could oxidizes biogenic amines, mainly tyramine, which potentially could be used for eliminating toxic compounds present in fermented food and beverages (22).

In this study, thus, isolation and screening of laccase candidate from sago bacteria, was carried out through conventional method and followed by biochemical and molecular analysis. Laccase as bioleaching in sago starch is expected to enhance quality by improving color and texture as well as eliminating unpleasant odors.

## 2. Materials and Methods

### 2.1. Materials

Materials used in this study were sago waste from sago refineries in Bogor, wet and dry sago from sago refineries in Meranti, Selat Panjang, Riau. Nutrient agar (Merck), MRS agar (Merck), MRS broth (Merck), Glycerol (Merck), Molecular Material $Na_3P_5O_{10}$ (Merck), Mag genome extractor (TOYOBO), Tris HCl, EDTA, $MgCl_2$, $(NH_4)$ 2 $SO_4$, Tag Polymerase (Fermentas), TAE, acrylamide gel, and Urea (Merck) were used as microbiological reagents. In addition, 3% KOH solution, hucker's ammonium oxalate crystal violet (Gram A) solution, mordan lugol iodine solution (Gram B) and decolorizer (Gram C), safranin (Gram D), MRSA media, SIM media, Simons Citrat media, Urea agar media, TSIA media, $H_2O_2$ 3%, test bacterial isolates, sterile aquadest were used in biochemistry analysis.

IOP Conf. Series: Earth and Environmental Science **406** (2019) 012014 doi:10.1088/1755-1315/406/1/012014

# Nutrient Concentration and Population of Macrozoobenthos in Ciletuh Bay, Sukabumi District, West Java

**Delilla Suhanda[1], Yuniarti MS[2,3], Yudi Nurul Ihsan[2], and Syawaludin A. Harahap[2]**

[1]Marine Science Study Program, Faculty of Fisheries and Marine Sciences, Universitas Padjadjaran, Jl. Raya Bandung-Sumedang KM. 21, Jatinangor, West Java, Indonesia.

[2]Departemen of Marine Sciences, Fisheries and Marine Sciences Faculty, Universitas Padjadjaran, Jl. Raya Bandung-Sumedang KM. 21, Jatinangor, West Java, Indonesia.

[3]The Center for Conservation and Management of Maritime Areas of the Faculty of Fisheries and Marine Sciences, Universitas Padjadjaran, Jl. Raya Bandung-Sumedang KM. 21, Jatinangor, West Java, Indonesia.

Coresponden Authors: delillasuhanda@gmail.com

**Abstract.** Relationship between nutrient concentration and macrozoobenthos can indicate the level of fertility or pollution in the sea waters. The purpose of this research is to obtain nutrient concentrations in pore water, macrozoobenthos community structures, and the relationship between nutrient concentration and macrozoobenthos in pore water Ciletuh Bay. The research are using survey, purposive sampling and descriptive analysis method. The parameters observed in this research are nutrients including ammonia, phosphate, nitrate, nitrite, total N, and Total Organic Carbon (TOC). Analysis of population macrozoobenthos was determined by knowing abundance index and dominance index. From the results of the study, it has been found that two types of macrozoobenthos is *Donax* sp. and *Nassarius* sp which lived immersed, and its dominance influenced by sand sediment fraction. The higher concentration ammonia, nitrate, nitrite, phosphate, N-total, and TOC which is 0.028, 0.708, 1.663, 0.075, 3.066 and 1.4 $mgL^{-1}$. The factor that most influences the presence of macrozoobenthos is the type of sediment and the closest concentration of nutrient is phosphate.

**Keywords:** Ciletuh Bay, nutrient, population macrozoobenthos, pore water, sediment.

IOP Conf. Series: Earth and Environmental Science **406** (2019) 012014 doi:10.1088/1755-1315/406/1/012014

## 1. Introduction

Ciletuh Bay is part of Ciletuh National Geopark in West Java Province. Geographically, It is located on the southern coast of Pelabuhan Ratu Bay, which is the largest bay on the South Coast of Java Island and directly facing the Indian Ocean. Currently, Ciletuh Bay was recognized by UNESCO in 2015 as national Geopark in Indonesia and as a part of the Global Geoparks network in April 2018. Ciletuh Bay is a place that empties into Cimarinjung River and Ciwaru River which affect the physical and chemical conditions of Ciletuh Bay.

In addition to several rivers the changed of a physical and chemical condition of Ciletuh Bay generate sediment quality, there are also human activities such as fishing schemes, shrimp farms and uninhabited islands. Composition and concentration of nutrient will greatly affect the survival of an organism. One of the organisms that need nutrients and impacted by sediment and water quality declamation directly in coastal or sea ecosystem is macrozoobenthos (Asep *et al.*, 2018). The existence of macrozoobenthos is the one of the most important organism parts of food chain in water ecosystem as food for demersal fish and some mamals. Macrozoobenthos also can be used as an indicator of the coastal and sea ecosystem (Sahidin and Yusli, 2016; Putro, 2014).

Macrozoobenthos is an organism whose life relatively long-living at the bottom of the water and measuring 3-5 milimeters. Based on its existence, macrozoobenthos can be divided into two category, called macrozoobenthos epifauna and macrozoobenthos infauna. Macrozoobenthos infauna belongs to organisms living in substrate sediment which has the property to make holes and deposit eater tent to fade in the sediment area that contains organic matter in pore water, they modify chemical components in sediment and decompose of nutrient recycling, organic matter and energy transfer to other links within the food (Gaudencio and Cabral, 2007; Norling *et al.*, 2007; Putro, 2014). The presence of makrozoobenthos is strongly influence by biotic and abiotic factors. Factor of biotic component is food for makrozoobenthos and factor abiotic component is affected by physical-chemical water include: temperature, dissolved oxygen (DO), depth, salinity, substrate and content of nitrogen (N) (Allard, 1987; APHA, 2005; Shou *et al.*, 2009).

In pore water, nutrient can changes influence organism that living in sediment. Pore water comes from surface of the sediment or water column which carries a concentration of acidity and nutrients. Pore water is layers of sediment are filled with water or air (Yuniarti et al.,

2018). Community structure in each organism has its own role. Ecological dominance is a species that plays a major role in controlling energy flows and the environment.

Approachment to determine water condition would be conducted with two methods, using biota response to environmental pressure, describe the environment condition and water quality. Biological method could determining and describes waters condition in long-term rather than use physical-chemical methods. Due to biological method would analyze biota response in pressure water (Junshum *et al.*, 2008). This review paper was prepared as part of a nutrient concentration and population of macrozoobenthos which is in pore water in Ciletuh Bay, District of Sukabumi West Java. Based on foregoing, it should be examined nutrient concentration in pore water by looking at the impact to macrozoobenthos community structures in Ciletuh Bay.

## 2. Materials and methods

### 2.1. Study Area and Laboratory Analysis

The study area, in Ciletuh Bay, was conducted at five station in the monts of May to September. Five station are located around coastal of Ciletuh Bay, were selected as the sampling station. Area sampling of each station was be recorded by using a model 60csx Garmin Global Positioning System (GPS), the selection of sampling station was determined after considering characteristic and factors each station. Sampling was conducted on May 27, 2018. Three replicate samples macrozoobenthos and water quality were taken at each sampling station. Macrozoobenthos samples were sieved a surber mesh (40 cm x 20 cm), macrozoobenthos trapped on the sieves were fixed in 5% formalin. Sediment samples were taken by using piston core then putted sediment into zipper plastic. Pore water were taken by piston core then for sieved the water using Millipore sieve and putted into dark polythene bottle and collected in icebox to be brought to the laboratory. Water quality parameters (physical and chemical parameters) such as temperature, salinity, dissolved oxygen, pH, visibility and depth the sampling station were collected and measured in-situ at proper depth. A measurement temperature was recorded by using alcohol standard thermometer, salinity by using a refractometer Atago 2483 Master-S28M, dissolved oxygen by using DO meter Lutron DO-5510, pH by using PH meter Lutron PH-201, brightness by using a secchi disk and depth of waters with aid of scale rope, meanwhile the rest of the parameters was analyzed in laboratory.

Samples macrozoobenthos, sediment and sea water observed in laboratory. Macrozoobenthos identified by using magnifier (Lozarralde and Pittaluga, 2010) identifying macrozoobenthos morphologically to genus level was conducted by using standard macrozoobenthos identification books (Bunjamin, 2005; Agus *et al.*, 2008). Sediment particle size (sand, mud and clay) were analyzed using the wet sieving method. The water nutrient samples to analyze the nitrates, nitrites, ammonia, phosphate, Total Organic Carbon, and Total Nitrogen with the aid of the spectrophotometer (Asep *et al.*, 2018).

## 2.2. Field Sampling

This research was conducted in the months of May to September and were taken samples on May 27, 2018 which included of determining the research station, taking samples, processing sediment and water samples, identifying macrozoobenthos and analyzing seawater samples. Based on determination of characteristic, station 1 located at Kunti's Island at coordinate 7,186861°S, 106.431861°E; station 2 located at the shrimp farming at coordinate 7,189389°S, 106,443500°E; station 3 located at Mandra's Island at coordinate 7,182806°S, 106,448667°E; station 4 located at Ciwaru River which will flow toward the Estuary Ciletuh Bay at coordinate 7,182845°S, 106,457651°E; station 5 located at Cimarinjung River which will flow toward the Estuary Ciletuh Bay at coordinate 7,171157°S, 106,464503°E. The scope of the research can be seen in Figure 1.

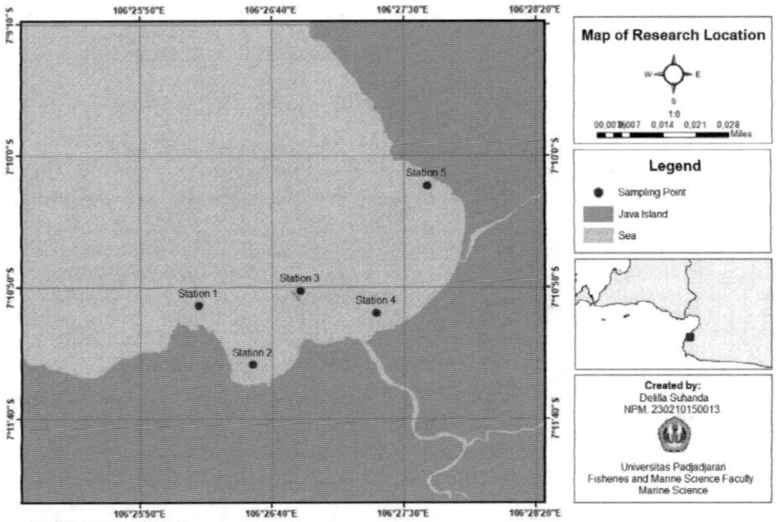


Figure 1. Map of Research Location

IOP Conf. Series: Earth and Environmental Science **406** (2019) 012014 doi:10.1088/1755-1315/406/1/012014

### 2.3. Data Processing

#### *2.3.1. Abundance Index*

Abundance index for each samples mathematically can be described as follows (Brower *et* al., 1990) :

$$DMZ = \frac{n1}{A}$$

#### *2.3.2. Dominance Index*

Species Dominance (*C*) of each samples was determined by the Simpson Index of Dominance (Brower *et al.*, 1990).

$$D = \frac{\sum ni\ (ni - 1)}{N\ (N - 1)}$$

$$C = 1 - D$$

## 3. Results and Discussion

### 3.1. Water Quality Parameters

The results of water quality in every sample was conducted by measuring the parameters of water quality in-situ. The results of measurements of water quality parameters observed for each station are presented in table 1.

**Table 1.** Parameters of Water Quality

| Station | salinity (°/₀₀) | Temperature (°C) | pH | DO (mgL$^{-1}$) | Brightness (%) | Depth (cm) |
|---|---|---|---|---|---|---|
| 1 | 15 | 28 | 7.34 | 9.5 | 41 | 100 |
| 2 | 26 | 29 | 7.13 | 8.7 | 62.5 | 80 |
| 3 | 33 | 29 | 7.1 | 8.3 | 37.5 | 40 |
| 4 | 31 | 29 | 7.13 | 7.7 | 87.5 | 40 |
| 5 | 34 | 30 | 7.07 | 8.7 | 20.83 | 120 |

As viewed from table 1, it is known that some parameters for measuring water quality. Salinity in Ciletuh bay ranges from 15 – 34 at station 1 shows that the waters is classified as brackish (Widi *et al.*, 2007). The temperature is a parameter that realty affects distribution, abundance and mortality of macrozoobenthos. Based on the temperature value obtained during the observation shows that is strongly influenced by the intensity of sunlight of each station. Variation of temperature value can influenced by nature processes such as biochemistry, through microorganism that can produce a heat and microbiology (Wenno, 1981). Moreover according to Officer (1976) expressed that temperature condition in sea water can be effected by atmosphere condition, weather and intensity of sunlight entering the waters. Temperature effect toward the metabolism of aquatic organisms so that needs dissolved oxygen increase (Nybakken, 1988).

Dissolved oxygen is needed by organisms that live in water (Evans, 1985). The dissolved oxygen conditions at each station are good because the dissolved oxygen level is not less than 2 $mgL^{-1}$ which is sufficient value for Gastropods and Bivalves. According to Bambang and Heron (2009) study of current patterns in Pelabuhan Ratu, adjacent to Ciletuh Bay, the pattern of current movements in the region is strongly influenced by tides. When the highest tide occurs, the water will move into the bay with maximum speed. The potential of Hydrogen is one of chemical parameter which is quite important for monitoring the stability of the waters. The organism of aquatic has certain limitations with pH values that are varies depending on sea water temperature, dissolve oxygen concentration and there are anions and cations (Pescod, 1978). Generally, pH in sea water range between 4 -9 (Marojahan, 2009) at each station shows a normal number and is suitable for marine biota to live in these waters.

The brightness level of a waters can indicated of the number of suspended particles or normal weather can give a hint in these waters has a low brightness (Baigo *et al.*, 2018). Based on field observations, in Ciletuh Bay has the level of brightness range between 20,83 – 87,5 % at range depth between 40 – 120 meters. The lowest level of brightness is at station 5 due to by influence of the Cimarinjung River flow. The river flow carries dissolved sediment particles, organic and inorganic materials through run off flow from the land which affects the brightness in station 5, so that the level of turbidity of the waters is high (Baigo *et al.*, 2018; Harpasis, 2004).

### 3.2. Grain Size

The Analysis of sediment samples carried out by the sieving method using a Sieve shaker were obtained by the percentage and type of sediment texture at each station in Ciletuh Bay, District of Sukabumi which was processed using KUMMOD SEL software. The result of the analysis sediment samples Ciletuh Bay showed two fractions sediments in the area are gravel and sand have different weight percentages at each point and take period.

**Table 2.** Types of sediments

| Station | Sediment Fraction | | Sediment Type |
|---|---|---|---|
| | Gravel | Sand | |
| **1** | 18.6 | 81.4 | Gravel Sand |
| **2** | 0 | 100 | Sand |
| **3** | 4.1 | 95.9 | Sand with little Gravel |
| **4** | 0.2 | 99.8 | Sand with little Gravel |
| **5** | 0.4 | 99.6 | Sand with little Gravel |

Based on the results of the sediment texture analysis at Ciletuh Bay, District of Sukabumi, the percentage value of the gravel fraction from station 1 to station 5 is quite far, 0.2 - 18.6%. The percentage of sand fraction is 81.4 - 100%. Based on these data it can be seen that the percentage of sand fraction is the largest sediment fraction when compared with gravel, clay and mud, so the result it can be seen that the type of sediment texture in Ciletuh Bay is sand dominance. The gravel fractions have a range of grain sizes from 2.0 to 10.00 with the type of grit particles *(Fine gravel).*The sediment substrate which is dominated by the sand fraction can be seen based on the research location in the coastal area of Ciletuh Bay. According to Hutabarat (1984) stated that sediments transported from land to sea by rivers in the ocean, the bigger grain sized tend to sink faster and settle.

## 3.3. Makrozoobenthos

The results of sampling and identification were obtained, there are *Nassarius* sp. and *Donax* sp. which dominate the waters of Ciletuh Bay. Ciletuh Bay as a whole scale, there are two species of macrozoobenthos and its dominanced by the species *Donax* sp. Station 1 is dominance by macrozoobenthos *Nassarius* sp. with 2 individuals while *Donax* sp. are not found at station 1. However, stations 2 to 5 are dominated by *Donax* sp. and there are no other species of macrozoobenthos at the station. At station 4 the number of species of *Donax* sp. was very abundant, reaching 45 individuals. This is due to its habit which cannot move actively and settles down somewhere (Hilda, 2011). These two types of individuals are Gastropods and Bivalves that are capable of extreme salinity.

**Table 3.** Structure community of Macrozoobenthos

| Station | Species | Abudance Index (ind/m3) | Dominance Index |
|---|---|---|---|
| 1 | Nassarius sp. | 250 | 1 |
| 2 | Donax sp. | 2250 | 1 |
| 3 | Donax sp. | 3125 | 1 |
| 4 | Donax sp. | 5625 | 1 |
| 5 | Donax sp. | 750 | 1 |

These shells live immersed in sand or sometimes in gravel. According to Yunitawati, *et al.* (2012), the main factor that greatly influenced the community structure of macrozoobenthos was coarse substrate. If the substrate changes, the macrozoobenthos structure will also change. This is evidenced by the abundance of macrozoobenthos types of *Donax* sp. at each station that has a substrate or type of sediment dominated by a sand fraction.

3.4. Nutrient Concentration

Nutrient conditions will be greatly influenced by the natural processes of a waters or even the process of entry of nutrients that can affect nutrients. Nutrient concentration will greatly affect

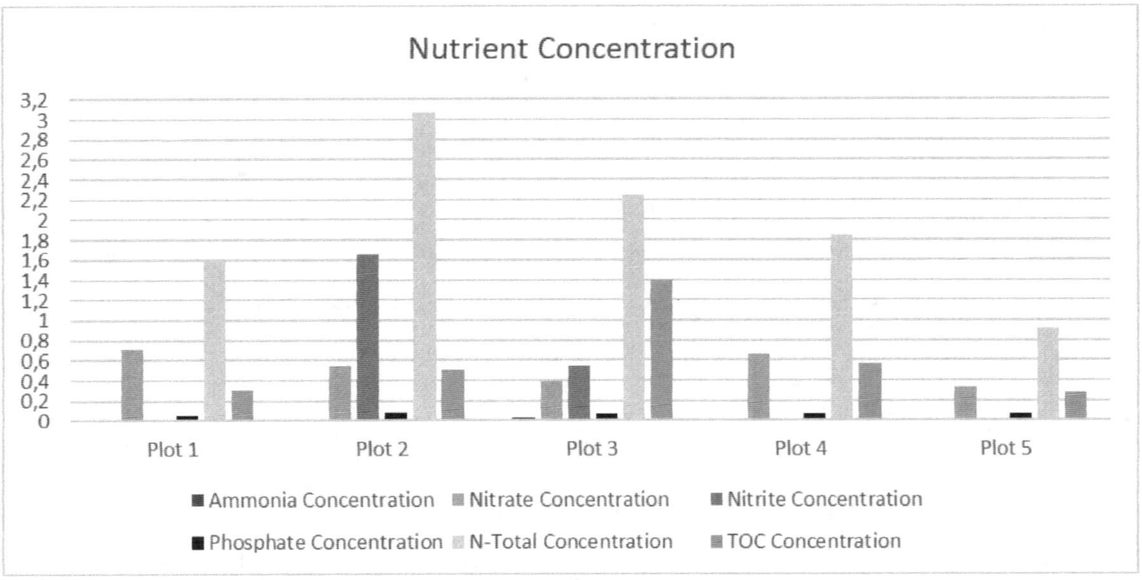

the nutrients needed by macrozoobenthos as nutrients for life

**Figure 2.** Nutrients in Water Pore

The amount of deposited organic contents can be influenced by sediment texture. The finer the texture of the sediment substrate in the water, more deposition of organic matter will be (Harpasis, 2004). The concentration of Ammonium ($NH_4^+$) compounds in the waters of Ciletuh Bay at each station ranges from 0.0099 - 0.028 $mgL^{-1}$ (Figure 2). Ammonia comes from organic nitrogen elements which are processed by heterotrophic organisms, namely organisms that need nutrients. In addition, ammonia comes from human activities such as domestic waste and industrial waste. Concentration ammonia in Ciletuh Bay very influenced by decomposition process component organic with some bacteria organism. The concentration of Nitrate ($NO_3^-$) compounds in the waters of Ciletuh Bay at each station ranges from 0.327 - 0.708 $mgL^{-1}$ (Figure 2). Nitrate is the main form of nitrogen in the waters and is the main nutrient for plant growth and algae. Nitrates are stable and very soluble in water (Bahari, 2006). Nitrate in Ciletuh Bay have the highest concentration at station 1, at station 1 have a ecosystem of coral reef and algae that influence by decomposition proses with bacteria ammonification and condition oceanography at station 1. The highest concentration of nitrite concentration is at station 2 which is equal to 1.663 $mgL^{-1}$. In station 2 there is shrimp culture, that dumps its waste into the water, component cultivated liquid waste was dominated by nitrite, total nitrogen, and phosphate (Mat Fahrur *et al.*, 2016). The highest total N-concentration is found at station 2, this can be caused by the activity of shrimp ponds at Station 2. Shrimp ponds greatly affect the nitrogen cycle in the waters and total nitogen in seawater was combined by ammonia, nitrite and nitrate concentration. The highest concentration of total nitrogen was compound by nitrite.

The concentration of Phosphate compounds ($PO_4^+$) in the waters of Ciletuh Bay at each station ranges 0.056 - 0.075 $mgL^{-1}$ (Figure. 2). The highest phosphate concentration is at station 2 which is in Shrimp Pond. This can be caused by the high phosphate diffusion process from sediment particles. Sediment is also a major place in phosphorus storage in cycles that occur in the ocean.

The carbon content can affect the level of ocean acidification, this impact can affect the supply of dissolved oxygen concentrations needed by marine biota as a process of respiration. The observations of Total Organic Carbon (TOC) concentrations ranged from

0.281 - 1, 4 $mgL^{-1}$. The highest TOC concentration is found at station 3, this can be caused by the activity of macrozoobenthos at station 3 and decaying vegetation from Mandra Island then settles and accumulates in pore water at station 3.

The correlation between the abundance index and each nutrient measured showed that almost all nutrients were negative. The regression results between abundance and ammonia concentration in pore water have a value of r = 0.1815, abundance index with nitrate concentration r = 0.0358, abundance index with nitrite concentration r = 0.0005, abundance index with phosphate concentration r = 0.7215, abundance index with N-Total concentration r = 0.1066, abundance index with TOC concentration r = 0.1768. Values show that the abundance value with the concentration value of ammonia, nitrate, nitrite, phosphate, N-Total and TOC is directly proportional, this correlation value shows the relationship between abundance index and nutrient concentration that closest nutrient concentration is phosphate. The nutrient concentration in the pore water and water column is interconnected from the activity of marine biota contained in it. If the higher the concentration of nutrient, the more abundance of the ecosystem will be.

**Acknowledgement**

This Research was totally supported by Universitas Padjadjaran, Indonesia (Project HIU-Hibah Internal Unpad). We thank Riska Zandhi, Fani Wulan Sari, Hazman Hiwari, and Ridlo Setiadeswan for sampling in Ciletuh Bay. Many thanks to reviewers for their comments that greatly improved the quality of the manuscript.

**Conclusion**

1. All nutrient parameters have relationship with macrozoobenthos, nutrient concentration that closest with abudance macrozoobenthos is phosphate and nutrient concentration that least influential with abudance macrozoobenthos is nitrate.
2. Macrozoobenthos in Ciletuh Bay is *Donax* sp. and *Nassarius* sp. in Ciletuh Bay dominated by macrozoobenthos *Donax* sp.
3. sediment in Ciletuh Bay is dominated by sand fraction.

**Citation and References**

[1] Allard M, Moreau G. (1987). Effects of experimental acidification on a lotic macroinvertebrate community. Hydrobiologia, 144(1):37-49

[2] Asep S. Zahidah. Heti H. Yusli W. Isdradjad S. Ruhyat P. (2018). Macrozoobenthos as bioindicator of ecological status in Tnajung Pasir Coastal, Tangerang District, Banten Province, Indonesia.

[3] Bahri and Andi Faizal. 2006. Analysis of Concentration Nitrate and Phosphate in Mangrove Sedimens utilized in the District of Mallusetasi, Barru Regency. Case Study of Utilization of Mangrove Ecosystems and Coastals Areas by Bungoro Community, Pangke Regency. Association of Environmental Consevators: Makassar.

[4] Baigo, H. Rosye H.R.T,. Suwito,. Hendra K.M,. Alianto. (2018). Sea Waters Quality Study and Pollution Index Based on Physics-Chemical Parameters in the Depapre District Waters, Jayapura. Jayapura: Journal Environment Science Volume 16 Issue 1 (2018) : 35 - 43.

[5] Brower, J. Jernold, Z., Von Ende, C. (1990). Field and Laboratory Method for General Ecology. Third Edition. USA: W. M. C. Brown Publishers.

[6] Bunjamin, D. (1988). Recent and Fossil Indonesian Shells. Conch Books.

[7] Gaud´encio, M.J., Cabral, H.N., (2007). Trophic structure of macrobenthos in the Tagus estuary and adjacent coastal shelf. Hydrobiologia 587, 241–251.

[8] Harpasis, S.S. (2004). Chemical Characteristic and Fertility Pelabuhan Ratu Bay Waters at West and East Monsoon. Journal Waters Sciences and Fisheries of Indonesia Volume 11, Number 2 : 93 – 100.

[9] Hutabarat, S. S.M. Evans. (1984). Introduction Oseanografi. UI-Press. Jakarta

[10] Junshum P, Choonluchanon S, Traichaiyaporn S. (2008). Biological indies for classification of water quality around Mae Moh power plant, Thailand. Mj Intl J Sci Tech 2 (1): 24-36.

[11] Lizarralde Z, Pittaluga S. (2010). Distribution and Temporal Variation of The Benthic Fauna In the Tidal Flat Of the Rio Gallegos Estuary, Patagonia, Argentina. *Thalassas, An International Journal of Marine Sciences* 27:9-20.

[12] Marojahan, S. (2009). Relationship of Chemical, Physics Environmental Factors of Plankton Distribution in East Belitung Waters, Bangka Belitung. Jakarta. LIPI: 31-45

pp.

[13] Mat Fahrur, Makmur Makmur, and Muhammad Chaidir Undu. 2015. Characteristics of Wastewater Culture Shrimp Vaname Superintensive. Forum of Innovation Technology Aquaculture.

[14] Norling, K., Rosenberg, R., Hulth, S., Gr´emare, A., Bonsdorff, E., (2007). Importance of functional biodiversity and speciesspecific traits of benthic fauna for ecosystem functions in marine sediment. Marine Ecology Progress Series 332, 11–23.

[15] Nybakken, W.J. (1988). Marine Biology. An Ecological Approach. Gramedia, Jakarta: 459

[16] Odum, E.P., (1993). The Basics of Ecology. Third Edition. Yogjakarta: Gadjah mada University Press.

[17] Officer, C.B. (1976). Physical Oseanography of Estuaries and Associated Coastal Waters. John Willey and Sons. New York: 465 pp.

[18] Putro, S. P. (2014). Sampling Method of Macrobenthos Research and its Aplications. Graha Ilmu. Yogyakarta. 1–37.

[19] R. Bambang A,M., Heron Surbakti. (2009). Simulation of Two Dimensional Flow Patterns in Pelabuhan Ratu Bay in September, 2004.

[20] Sahidin A, Yusli W. (2016). Spatial distribution of Polychaete at Tangerang coastal water, Banten Province. J Fish Mar 6 (2): 83-94. [Indonesian]

[21] Shannon, C.E., Weaver, W., (1963). *The Mathematical theory of communication*. University of Illinois Press, Urbana, Illinois.

[22] Water Environmental Federation and American Public Health Association (APHA). (2005). Standard methods for the examination of water and wastewater. Washington DC: APHA

[23] Wenno, L.F. 1981. Research report: Nature of Oceanology Shallow Waters Maluku. Maluku Aquatic Marine Resources Research and Development Project (1980 – 1981). LON-LIPI, SPA, Ambon: 185 pp.

[24] Widi A. Adil J. and Muhammad A. Brackish Water Desalination Using Surfactant Modified Zeolite. LIPI Vol.6 Number 1. May 2007.

[25] Yuniarti. Y. N. Ihsan, C. Asdak, Y. Dhahiyat, M. K.A. Kamarudin, M. B. Gasim, A. F. Ireana Yusra and H. Juahir. (2018). Impact Sedimentation To Community Structure

Macrozoobenthos In Segara Anakan Lagoon. Journal of Fundamental and Applied Sciences.

[26] Yunitawati. Sunarto. Zahidah, H. (2012). Relationship between Substrate to Structure Community of Macrozoobenthos at Cantigi River, District of Indramayu. Sumedang: Journal Fisheries and Marine Science Volume 3, Number 3 2012 : 221 – 227.

# Optimization of temperature and time of extraction of kecombrang stem and leaf (*etlingera elatior*) based on the quality of product bioactive components

**Farah Fatimah, Natalia Christin Larasati, Rumpoko Wicaksono, Rifda Naufalin*)**

Food Science and Technology Study Program, Faculty of Agriculture, Jenderal Soedirman University, Purwokerto 53122, Indonesia
*) Corresponding author, Email: rifda.naufalin@unsoed.ac.id

***Abstract.*** Foodstuffs are perishable due to the nutritional component contained therein. Therefore it is necessary to prevent food damage, one of which is the addition of preservatives in food. One of the natural ingredients that has the potential as a natural preservative is kecombrang plant. Kecombrang (*Etlingera elatior*) is known to have bioactive compounds including polyphenols, alkaloids, and flavonoids. Extraction is influenced by several factors including temperature and time. These factors need to be controlled so that the extract produced has the best quality. This study aims to determine the optimum temperature and time extraction of kecombrang stems and leaves on the quality of the bioactive components of the product produced. This study uses the Response Surface Methodology (RSM) method with the Design Expert v10 program. The experimental design used is Central Composit Design (CCD). The factors that were tried in this study were temperature and liquid extraction time. The temperature consists of three levels, namely 40°C, 50°C, and 60°C, and time consists of three levels namely 3, 4 and 5 hours. The results showed that the optimum temperature and time in the extraction of kecombrang stem using a temperature of 40°C with a duration of 5 hours produced a pH value of 3.85, a total flavonoid of 239.92 mg QE / 100 g, a total phenol of 111.41 mg TAE / 100 g and optimum temperature and time in kecombrang leaf extraction using a temperature of 51°C with a duration of 3 hours produced a pH value of 4.35, a total phenol of 327.13 mg TAE / 100 g and a total flavonoid of 24.911 mg QE / 100 g. Qualitative test results indicate that positive liquid extracts contain alkaloids, flavonoids, glycosides, saponins, phenolics and triterpenoids.

*Keywords: kecombrang, phenol, flavonoids, optimization, extraction*

## 1. Introduction

The current development is demanding food and beverage products that are all practical, durable and have an attractive appearance. The addition of preservatives is a solution made by the food industry so that the final product has good quality with a longer shelf life. The use of synthetic preservatives and antioxidants at this time is not recommended because it is suspected to cause degenerative diseases and even cancer [1]. Therefore it is necessary to develop alternative preservatives sourced from natural ingredients, one of which is kecombrang.

Kecombrang is a Zingiberaceae plant group that has long been known and used by the community as a nutritious food ingredient as a food preservative. Kecombrang stem section itself is a part that is often used as food additives in the community. According to research [2,3] kecombrang stem contains a number of bioactive substances such as alkaloids, saponins, phenolics, flavonoids, triterpenoids, steroids and glycosides. Kecombrang stem section is more able to inhibit oxidation when compared to other leaves and rhizomes. This shows that kecombrang stem has the potential as a bioactive compound that functions as an antioxidant, antimicrobia, antimicrobial coating for food product [4, 5, 6, 7].

Extraction process of bioactive compounds in kecombrang plants needs to be carried out to test its activity. Extraction is the activity of withdrawing soluble chemicals so that they are separated from insoluble substances by liquid solvents [8]. Extraction is influenced by several factors. There are many factors that influence the extraction process including temperature and time. These factors need to be controlled so that the extract produced can have the best quality. If the extraction process uses temperatures that are too high for too long, it is feared that the bioactive component in the kecombrang will suffer damage and can reduce the antioxidant activity of the extract produced. Previous studies have found the best way to extract kecombrang plants, but there is no optimization for the temperature and extraction time [9]. Therefore this research was conducted, namely to optimize the temperature and extraction time with the Response Surface Methodology (RSM) method.

The objectives of this study are: 1) knowing the optimal combination of treatment between temperature and extraction time to produce liquid extracts of kecombrang stems and leaves with good quality; 2) determine the effect between temperature and extraction time on the quality of liquid extracts of kecombrang stem and leaves produced.

## 2. RESEARCH METHODOLOGY

### 2.1 Experimental design

All treatments consisted of 13 processes in which each process condition followed the central composite design (Table 1) obtained from the Design Expert application ver 10. Temperature and time extraction estimates in making kecombrang stem extract using Response Surface Methodology (RSM) with central composite design (CCD composite design). The factors tested in this study were temperature and duration of extraction with 3 levels, are:

$X_1$ = The extraction temperature consists of 40, 50 and 60 (°C).

$X_2$ = The length of time consists of 3, 4 and 5 (hours).

Tabel 1. The combination of liquid extraction treatment of kecombrang stem and leaves

| No. | Stem | | Leaves | |
|---|---|---|---|---|
| | $X_1$ (Temperature) | $X_2$ (Time) | $X_1$ (Temperature) | $X_2$ (Time) |
| 1 | 64.1421 | 4 | 50 | 5.41 |
| 2 | 40 | 5 | 50 | 4 |
| 3 | 50 | 2.58579 | 50 | 2.59 |
| 4 | 40 | 3 | 35.86 | 4 |
| 5 | 50 | 5.41421 | 40 | 5 |
| 6 | 50 | 4 | 60 | 3 |
| 7 | 50 | 4 | 60 | 5 |
| 8 | 50 | 4 | 50 | 4 |
| 9 | 60 | 5 | 50 | 4 |
| 10 | 35.8579 | 4 | 50 | 4 |
| 11 | 60 | 3 | 40 | 3 |
| 12 | 50 | 4 | 64.14 | 4 |
| 13 | 50 | 4 | 50 | 4 |

### 2.2 Kecombrang preparation and powder

Young kecombrang stems and leaf are sorted. After that the kecombrang stems are cut into small pieces which are then dried using a cabinet dryer for 6 hours at 60°C. The simplicial kecombrang stem is size reduction using a disc mill until a homogeneous kecombrang stem powder is obtained. In kecombrang leaves, young kecombrang leaves are sorted. After that kecombrang leaves are cut into small pieces which are then dried using a cabinet dryer for 2 hours at 60°C. The kecombrang leaves are size reduction using a disc mill until a homogeneous kecombrang leaf powder is obtained [7, 9].

### 2.3 Kecombrang stem and leaf extraction

The extraction method used is the maceration method. Homogeneous kecombrang powder is put into the extractor, then added distilled water with a ratio of 1:14 through the extractor closing pipe. The extraction process is carried out in accordance with the treatment produced by the Expert Design application ver.10. Kecombrang liquid extract is filtered using a filter cloth and a press to separate the extract from the pulp. Kecombrang liquid extracts are dried again using a cabinet dryer for approximately 2-3 hours with a temperature of 40-50°C to dry, then extracted again with the same water ratio. The resulting kecombrang liquid extract slurry is refined using a filter cloth and a press to get the second kecombrang stem extract. The first and second extraction results were mixed until homogeneous to obtain liquid extracts of kecombrang stem and leaves.

### 2.4 Qualitative test

The variables observed in the study consisted of qualitative and quantitative variables. Qualitative variables include phytochemical tests including alkaloids, phenols, flavonoids, glycosides, saponins, steroids and tannins [10].

### 2.5 Quantitative test of total phenol

The total phenol test uses the Singelton method with Folin-ciocalteu reagents with the standard used tannic acid. A sample of 400 µL was added with 1.5 mL of Folin-Ciocalteu then allowed to stand for 5 minutes at room temperature. After that added 1.5 mL of NaHCO30,556 M then shaken and allowed to stand in a dark room for 90 minutes. After that the absorbance of the solution was measured using a spectrophotometer at a wavelength of 725 nm [11].

### 2.6 Quantitative test of total flavonoid

1000 µL samples were added with 1 mL of 2% AlCl 3 (2 grams of AlCl 3 in a 5% glacial acetic acid solution) and 1 mL of 120 Mm potassium acetate solution (1,176 grams of potassium acetate in 100 mL of distilled water). After that, it was incubated for 1 hour at room temperature and the absorbance was measured using a spectrophotometer at a wavelength of 435 nm [12].

### 2.7 Analysis of statistic

Data analysis was performed with Response Surface Methodology (RSM) which is a combination of statistical and mathematical techniques to find optimum conditions. Furthermore, the validation results are then tested independently to test whether there is a difference between the predicted results from the Design Expert application ver.10 and the results of the optimization treatment.

## 3. RESULT AND DISCUSSIONS

### 3.1 Qualitative test

The qualitative variables examined in this study include alkaloids, phenols, flavonoids, glycosides, saponins, steroids and tannins. The results of testing the qualitative variables of kecombrang stem liquid extract are presented in Table 2 and the results of testing the qualitative variables of kecombrang leaf liquid extract are presented in Table 3.

Table 2.The results of the qualitative test were kecombrang stem liquid extract.

| Treatment | Alkaloids | Phenol | Flavonoids | Glycosides | Saponin | Triterpenoid | Tannin |
|---|---|---|---|---|---|---|---|
| 64°C, 4 hours | ++ | +++ | + | + | + | - | +++ |
| 40°C, 5 hours | + | +++ | ++ | + | + | + | ++ |
| 50°C, 2.5 hours | ++ | ++ | + | + | + | - | + |
| 40°C, 3 hours | ++ | +++ | + | - | + | + | + |
| 50°C, 5.5 hours | +++ | ++ | ++ | + | ++ | - | ++ |
| 50°C, 4 hours | +++ | +++ | ++ | + | ++ | + | +++ |
| 50°C, 4 hours | ++ | +++ | ++ | - | + | + | +++ |
| 50°C, 4 hours | +++ | +++ | + | + | + | + | +++ |
| 60°C, 5 hours | ++ | +++ | + | - | ++ | - | +++ |
| 36°C, 4 hours | ++ | +++ | + | + | + | - | +++ |
| 60°C, 3 hours | ++ | +++ | ++ | - | ++ | + | ++ |
| 50°C, 4 hours | +++ | ++ | ++ | - | ++ | + | +++ |
| 50°C, 4 hours | +++ | +++ | + | + | + | + | +++ |

Note : +++ : strong; ++ : father strong; + : a little strong; - : over load.

Table 3. The results of the qualitative test were kecombrang leaf extract.

| Treatment | Alkaloids | Phenol | Flavonoids | Glycosides | Saponin | Steroids | Tannin |
|---|---|---|---|---|---|---|---|
| 50°C, 5.4 hours | +++ | +++ | ++++ | - | ++ | +++ | +++ |
| 50°C, 4 hours | ++++ | +++ | +++ | - | +++ | +++ | +++ |
| 50°C, 2.6 hours | +++ | ++++ | ++++ | - | ++ | +++ | +++ |
| 36°C, 4 hours | +++ | +++ | ++ | - | +++ | ++ | +++ |
| 40°C, 5 hours | ++++ | ++++ | +++ | - | +++ | ++ | +++ |
| 60°C, 3 hours | ++ | ++++ | +++ | - | ++ | + | +++ |
| 60°C, 5 hours | ++ | ++++ | ++++ | - | + | ++ | +++ |
| 50°C, 4 hours | +++ | +++ | +++ | - | ++ | +++ | +++ |
| 50°C, 4 hours | ++ | +++ | +++ | - | +++ | ++ | ++ |
| 50°C, 4 hours | +++ | +++ | ++++ | - | ++ | +++ | ++ |
| 40°C, 3 hours | ++ | ++++ | +++ | - | ++ | +++ | ++ |
| 64°C, 4 hours | ++ | ++++ | ++++ | - | ++ | +++ | +++ |
| 50°C, 4 hours | +++ | +++ | +++ | - | + | + | ++ |

Note : ++++ : very strong, +++ : strong; ++ : rather strong ; + : a little strong;
- : over load.

### 3.1.1. Alkaloid

The test results showed that all positive liquid extract treatments contained alkaloids which were characterized by red-orange deposits. These results are consistent with the literature which states that the positive kecombrang stem extract contains alkaloid compounds [2]. Sedimentation in alkaloid testing is caused by ligand changes, where nitrogen atoms in alkaloid compounds have free electron pairs so they can replace iodo ions in dragendorf and Wagner reagents [13].

### 3.1.2. Phenol

The phenol test results showed a positive value marked by the formation of green-blue-black color in liquid extracts of kecombrang stems and leaves. This is in accordance with [14] which states that the kecombrang stem and leaf extract contains phenol compounds. Discoloration during phenol testing was caused by the breakdown of hydroxy groups in phenol compounds and reacting with $Fe^{3+}$ in iron salts [15]. According to [10] the color formed is derived from iron phenolics, and this reaction can only take place in a weak or neutral acid atmosphere. The presence of high phenol components have an important role as antioxidants that can reduce free radicals [16].

### 3.1.3. Flavonoid

Flavonoid test is carried out in stages according to research [17]. A total of 2 mL of sample was put into a test tube and then added with magnesium powder, then added with amyl alcohol and concentrated hydrochloric acid. The formation of red, yellow or orange in amyl alcohol indicates the presence of flavonoid compounds in liquid extracts. In the flavonoid test there is a change in the color of amyl alcohol to pink which indicates that there are flavonoid compounds in liquid extracts of kecombrang stem and leaves. The results of this test are in accordance with the literature which states that liquid extracts of kecombrang stems and leaves contain flavonoid compounds [14]. According to [18] addition of HCl and magnesium powder in the flavonoid test will reduce the benzopiron nucleus contained in the flavonoid compound so that the orange-red color is formed in the solution.

### 3.1.4. glycoside

The results of glycoside testing on liquid extracts of kecombrang stem showed that as many as 8 positive treatments contained glycoside compounds and 5 negative treatments contained glycosides which were marked by a blue or green discoloration when the sample was added with anhydrous acetic acid and sulfuric acid. This is consistent with the literature showing that kecombrang stem extracts contain glycoside compounds [14]. The results of glycoside testing on liquid extract of kecombrang leaves showed that as many as 13 negative treatments contained glycosides. Kecombrang stem liquid extracts contained 5 samples which showed negative results on glycoside testing. This is because testing of glycosides using anhydrous acetic acid and concentrated sulfuric acid can only test non-sugar glycoside compounds and cannot detect the presence of sugar glycoside compounds [19], so they do not show positive results at the time of testing.

### 3.1.5. Saponin

Saponin compounds in the study of liquid extracts of kecombrang stems and leaves are known by the test method according to [20] namely the addition of 1 drop of HCl 2 N. The extract was dropped with HCl then shaken for 10 seconds then allowed to stand for 10 minutes. The formation of a stable foam indicates the presence of saponin compounds in the extract. The test results showed that the liquid extract of the positive kecombrang stem and leaves contained saponin compounds. This is characterized by the formation of a stable foam on the surface of kecombrang extract.

According to research by [21] the emergence of foam in the saponin test also showed that the presence of glycoside compounds which have the ability to form froth in water hydrolyzed into glucose and other compounds. This is what causes the formation of micelles when shaken with water. In the micelle structure the polar group will face outward and the non-polar group will lead inward, this micelle state which looks like a foam [22].

### 3.1.6. Triterpenoid

Triterpenoid testing is done by adding kecombrang stem extract with 3 drops of concentrated hydrochloric acid and 1 drop of sulfuric acid, red color changes indicate the presence of triterpenoid compounds in kecombrang stem extract. Triterpenoid test results showed that from 13 total treatments, there were 8 treatments that showed that the positive kecombrang stem extract contained triterpenoid compounds, while 5 treatments showed negative. The existence of triterpenoid compounds in kecombrang stem extracts is in accordance with [14] which states that kecombrang stem extracts contain triterpenoids. This is supported by the research of [23] triterpenoid compounds can also be bound to sugar groups so that they can be extracted with solvents that are semi-polar and polar.

Negative results in testing due to triterpenoid compounds are compounds that are non-polar and cannot be extracted completely with polar water solvents such as water [24]. This is supported by the research of [23] which states that triterpenoids are composed of C30 hydrocarbon chains which causes triterpenoid properties to tend to be non-polar and easier to extract using non-polar solvents.

### 3.1.7.Steroid

Based on the results of the qualitative analysis of the steroid compounds in Table 3 it can be seen that the liquid extract of kecombrang leaves which are extracted using water is then analyzed using the Lieberman Burchard reagent. There are 2 possible changes in color types, namely the formation of green for steroids and red or violet for triterpenoids. In this study, the resulting color was only green as positive as +++ with strong positive note on liquid extract of kecombrang leaves containing steroids.

Changes in color to red, pink or violet indicate positive containing terpenoids and if the color changes to green or purple then the sample is tested positive for steroids. Steroids act as antibacterial by damaging the lipid membrane, so that the liposomes leak.

### 3.1.8.Tanin

Tannin compounds were detected by changing the color of the extract to blue-green when added with 1% $FeCl_3$ [20]. In tannin testing, all samples of liquid extracts of positive kecombrang stem and leaf contain tannin compounds which are marked by changing the color of the sample to blue green. This is not in accordance with the literature which states that tannins are compounds that are not contained in kecombrang stem extracts [14].

The main properties of tannins depend on the phenolic-OH group contained in tannins. Color reactions occur when combined with iron salt. Iron salt ($FeCl_3$) will give a green or blackish blue color when reacted with tannin. But this test is not good because in addition to tannins that can provide a color reaction, other substances can also provide the same color reaction [15]. Other compounds that can form dark green-blue deposits are polyphenolic compounds and tannins.

### 3.2. Analisis Result of Qualitative Data

This study uses the expert ver.10 design program for processing statistical data. The response results from the central composite design of kecombrang stem liquid extract are presented in Table 4 and the response results from the central composite design of kecombrang leaf liquid extract are presented in Table 5

Tabel 4. Response data from the central composite design of kecombrang stem liquid extract

| No. | Variable | | Response | | |
|---|---|---|---|---|---|
| | $X_1$ (Temperature) | $X_2$ (Time) | Total Phenol (mg TAE/100 g) | Total Flavonoids (mg QE/100 g) | pH |
| 1 | 64.1421 | 4 | 23.02 | 9.93 | 3.8 |
| 2 | 40 | 5 | 24.96 | 11.79 | 3.7 |
| 3 | 50 | 2.58579 | 23.16 | 7.34 | 3.75 |
| 4 | 40 | 3 | 28.11 | 9.59 | 3.8 |
| 5 | 50 | 5.41421 | 26.75 | 10.69 | 3.8 |
| 6 | 50 | 4 | 32.18 | 9.30 | 3.9 |
| 7 | 50 | 4 | 29.51 | 7.26 | 3.6 |
| 8 | 50 | 4 | 37.65 | 7.84 | 3.6 |
| 9 | 60 | 5 | 30.62 | 9.64 | 4.0 |
| 10 | 35.8579 | 4 | 34.84 | 10.61 | 3.9 |
| 11 | 60 | 3 | 35.76 | 7.79 | 3.8 |
| 12 | 50 | 4 | 32.95 | 7.31 | 3.8 |
| 13 | 50 | 4 | 37.99 | 7.42 | 3.8 |

Tabel 5. Response data from the central composite design of kecombrang leaf liquid extract

| No. | Variable | | Response | | |
|---|---|---|---|---|---|
| | $X_1$ (Temperature) | $X_2$ (Time) | Total Phenol (mg TAE/100 g) | Total Flavonoids (mg QE/100 g) | pH |
| 1 | 50 | 5.41 | 400.29 | 20.01 | 4.35 |
| 2 | 50 | 4 | 339.24 | 20.64 | 4.40 |
| 3 | 50 | 2.59 | 516.09 | 29.17 | 4.40 |
| 4 | 35.86 | 4 | 331.49 | 12.71 | 4.05 |
| 5 | 40 | 5 | 126.55 | 16.14 | 4.25 |
| 6 | 60 | 3 | 391.09 | 17.81 | 4.20 |
| 7 | 60 | 5 | 224.42 | 19.12 | 4.30 |
| 8 | 50 | 4 | 186.14 | 17.65 | 4.35 |
| 9 | 50 | 4 | 221.15 | 18.47 | 4.40 |
| 10 | 50 | 4 | 230.23 | 16.61 | 4.40 |
| 11 | 40 | 3 | 226.84 | 15.06 | 4.50 |
| 12 | 64.14 | 4 | 365.89 | 23.31 | 4.40 |
| 13 | 50 | 4 | 289.34 | 16.50 | 4.30 |

The Expert Design Program will provide a regression model that matches the measurement results of each response. The model obtained can be used to predict responses. The responses analyzed were total phenol levels, total flavonoids and pH.

### 3.2.1 Total phenol

The total phenol test was carried out by reacting the sample with 10% Focal Ciocalteu and NaHCO3 and then absorbed the sample at 725 nm wavelength. When the phenolic compounds contained in the sample are reacted with the folin-ciocalteu reagent, the sample color changes from yellow to blue. The more concentrated blue color formed, the higher the phenolic compounds contained in the sample [25].

Based on RSM calculations, the temperature and time treatments do not have a significant effect on total phenols. Two-dimensional contour graph of the effect of temperature and time of extraction of kecombrang stem is presented in Figure 1.

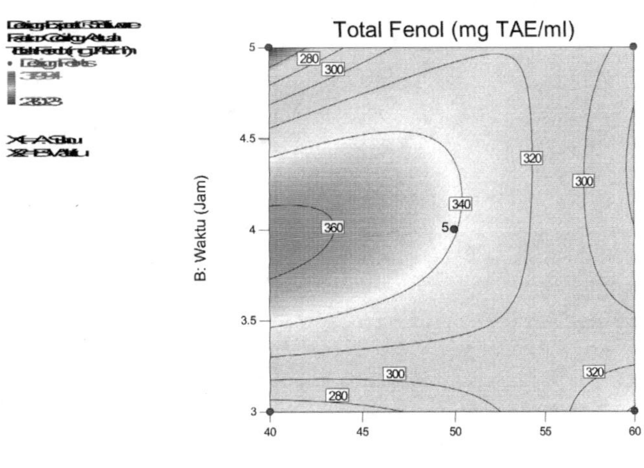


Figure 1. Two-dimensional contour graph of total response of phenol liquid extract of kecombrang stem.

The result of total phenol liquid extract of kecombrang stem provides a determination of the level of significance and suitability of the model carried out by ANOVA analysis on Design Expert ver.10. The type of regression chosen for the analysis of total phenols is a cubic model with a coefficient of determination (R2) of total phenols of 0.7812. [26] states that a value of R2 close to 1 indicates a high degree of correlation between observation and predictive value. Based on the lack of fit tests the results obtained are 0.3233 (32.33%) which indicates this model is not significant. The model will be considered appropriate if the lack of fit tests of the model are insignificant [27].

One of the possible causes of the temperature factor and duration of extraction being not affected in the liquid extraction process of kecombrang stem and leaves is due to the use of distilled aquades which are less appropriate in the extraction process. Phenols are polar compounds that have high solubility when extracted with polar solvents such as ethanol and distilled water. In the research of [28] polar solvents such as ethanol were able to extract phenols higher than other solvents. Based on these studies it is suspected that the phenol component contained in kecombrang has a polarity that approaches the ethanol polarity so that the use of ethanol solvents is more effective to dissolve phenol compounds in kecombrang. The solubility of phenol the highest is not always present in polar extracts, but depends on the structure of the phenol compound itself. The equation of the regression model obtained is as follows:

$$Y = 340{,}60 - 41.79(A) + 12{,}67(B) - 4{,}97(AB) - 38{,}20(A^2) - 38{,}20(B^2) - 33{,}39(A^2B) + 75{,}10(AB^2).$$

Note :

Y = Respons value of total phenol (mg TAE/mL)
A = temperature (°C)
B = time (hours)

The ANOVA model used in the total phenol response of kecombrang leaf liquid extract was a quartic model with R2 as large as 0.8924. P-values in the ANOVA model, temperature, time and interactions between temperature and time on the total phenol response were 0.0928, respectively; 0.7091; 0.2481 and 0.6136. Especially the p-value in the ANOVA total phenol model is greater than $p < 5\%$ which means that the model is not significant or does not significantly affect the response. The ANOVA total phenol model of the quartic model produces a two-dimensional color contour graph like in Figure 2.

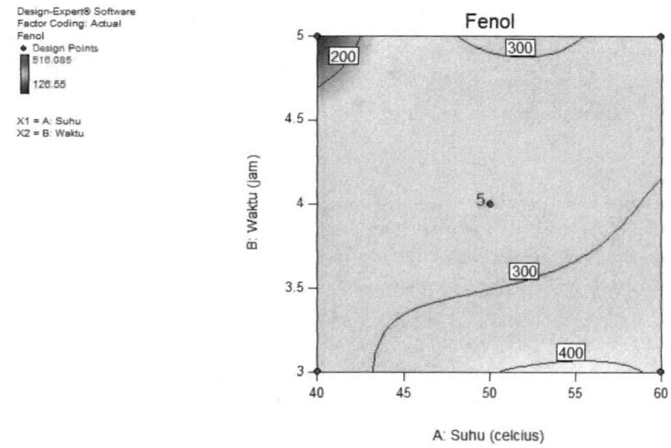


Figure 2. Two-dimensional contour graph of total response of phenol liquid extract of kecombrang leaves.

The results of the analysis of total phenol responses by using a two-dimensional color contour graph can be seen in Figure 2. The response shows that the temperature and time of extraction did not have a significant effect on the total phenol content of liquid extract of kecombrang leaf powder. The value of total phenol content of kecombrang leaf extract ranged from 126,550 - 516,085 mg TAE /

100 g. The highest value of total phenol in the treatment combination was obtained from a treatment temperature of 50°C with a 2.6-hour duration of 516.085 mg TAE / 100 g. The lowest total phenol value in the treatment combination was obtained from the treatment temperature of 40°C with a duration of 5 hours of 126,550 mg TAE / 100 g.

The value of total phenol content in this study was lower when compared with the results of research conducted by [28], liquid extract of kecombrang leaves with ethanol solvent resulted in total phenol content values ranging from 1338.06 - 8636.15 mg TAE / 100 g . This difference in total phenol content can be caused by differences in the types of solvents used. Phenol is a polar compound so that its solubility is highest in polar solvents. Polar (ethanol) solvents are able to extract higher phenols so that the total value of the ethanol fraction is higher [29]. The bioactive component of kecombrang flower extract varies according to its polarity. The phytochemical components of hexane extract consist of steroids, triterpenoids, alkaloids, and glucosides. Phytochemical components of ethyl acetate extract are steroids, terpenoids, alkaloids, flavonoids, and glycosides, while ethanol extracts produce phenolic components, terpenoids, alkaloids, saponins, and glycosides [14].

According to [29], phenol is a polar compound so that its solubility is highest in polar solvents. Polar (ethanol) solvents are able to extract higher phenols so that the total phenol fraction value of ethanol is higher. The chlorophyll content also influences the high total phenol produced by kecombrang leaves, although it is not a phenol compound, hydrophilic chlorophyll is extracted in a polar ethanol solution and is read when measuring [28].

### 3.2.2. Total Flavonoid

The regression model chosen for the analysis of total flavonoid liquid extract of kecombrang stem is a quadratic model with a coefficient of determination (R2) of total flavonoids of 0.8578 which is close to 1. Based on the lack of fit tests, a result of 0.6808 (68.08%) is obtained, which indicates insignificant results. The equation of the regression model obtained is as follows:

$$Y = 78{,}32 - 6{,}14(A) + 11{,}00(B) - 0{,}85(AB) - 12{,}37(A^2) - 6{,}09(B^2)$$

Note :

Y = Respons value of total flavonoid (mg QE/mL)
A = temperature (°C)
B = time (hours)

RSM calculation shows the influence of temperature factor and extraction time of liquid extract of kecombrang stem on total flavonoids. The regression model equation shows that the increase in total flavonoid extract response is inversely proportional to the increase in time (B) but directly proportional to the increase in temperature (A). The longer the extraction time used, the total value of the resulting flavonoids will be minimum, whereas the higher the extraction temperature, the total value of the resulting flavonoids will increase and at a certain point will be constant. This is shown by the two-dimensional graph in Figure 3.

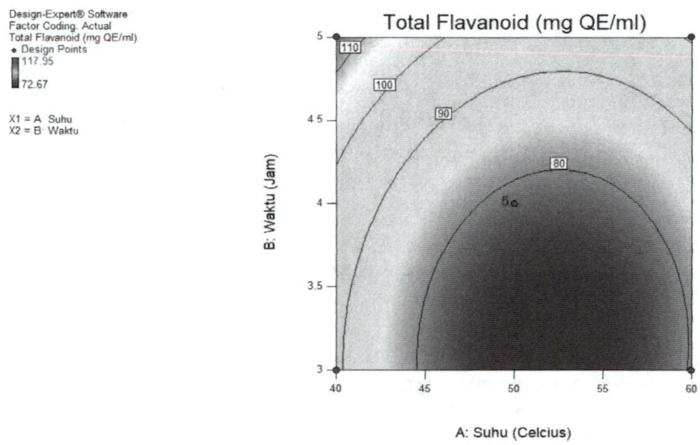


Figure 3. Two-dimensional contour graph of total response of flavonoid liquid extract of kecombrang stem.

Determination of total flavonoid levels of kecombrang stem liquid extract in this study used quercentin as a standard solution. Quercentin is a class of flavonoids that are often found in plants that are known to have biological activities as antioxidants [30]. The highest total flavonoid value was found in the treatment temperature of 40°C with 5 hours extraction time. The use of high temperatures can increase the levels of flavonoids obtained. This is consistent with the research of [31] which states that high temperatures are able to release biocative compounds that are bound in cells due to damage to cell elements. This causes more and more compounds to be extracted.

Increased straight-line extraction time with increased flavonoid content. This is consistent with the research of [32] which states the increase in extraction time is directly proportional to the increase in the total value of flavonoids. Within a certain time the concentration of flavonoids in the solvent will not increase due to equilibrium in the extraction has been reached. So that at this point the increase in extraction time does not affect the solute [31].

At a treatment temperature of 60°C within 5 hours a decrease in the total value of flavonoids. This is because flavonoids have been damaged. Bioactive compounds such as flavonoids are generally thermosensitive and not resistant to exposure to high temperatures above 50°C, so that they undergo structural changes and produce extracts with low bioactive content [33].

The ANOVA model used in the total flavonoid response of kecombrang leaf liquid extract was a quartic model with R2 as large as 0.9436. P-values in the ANOVA model, temperature, time and interaction between temperature and time on the total flavonoid response were 0.0284; 0.0115; 0.0187 and 0.9476. Especially the p-value in the ANOVA total flavonoid model is smaller than $p < 5\%$ which means that the model is significant or significantly influences the response. The ANOVA model of the total flavonoid quartic model produces a two-dimensional color contour graph like in Figure 4.

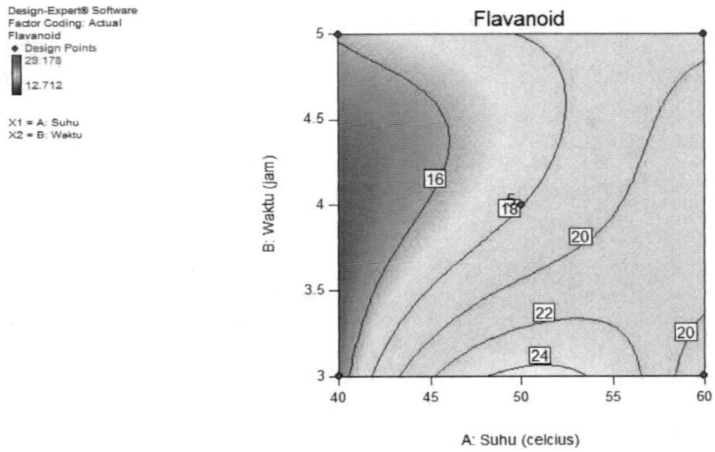


Figure 4. Two-dimensional contour graph of total flavonoid response of kecombrang leaf liquid extract.

The results of the analysis of the response of the flavonoid levels of the kecombrang leaf liquid extract can be seen in Figure 4. The results of the response of the flavonoid levels showed that the temperature and time of extraction had a significant effect on the flavonoid levels of the liquid extract of the kecombrang leaf powder. The value of flavonoids of kecombrang leaf liquid extracts ranged from 12,712 - 29,178 mg QE / 100 g. The highest value of flavonoid levels in the treatment combination was obtained from the treatment temperature of 50°C with a duration of 2.6 hours at 29.178 mg QE / 100 g. The lowest value of flavonoid levels in the treatment combination was obtained from a treatment temperature of 36°C with a duration of 4 hours amounting to 12.712 mg QE / 100 g. Graph of the effect of extraction temperature on the levels of liquid extracts of kecombrang leaf extract can be seen in Figure 5.

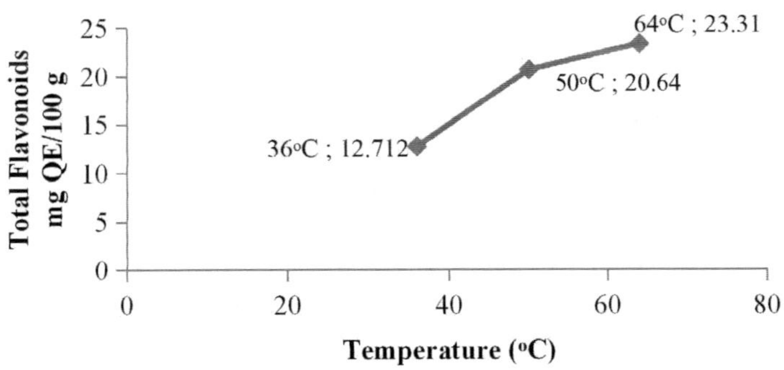


Figure 5. Graph of the effect of extraction temperature on the levels of liquid extract flavonoids kecombrang leaves at 4 hours

It is seen in Figure 5, at 4 hours with a temperature of 36°C to 64°C the levels of flavonoid extract extract tends to increase. This is in accordance with research conducted by [34] the increase in extraction temperature used, can make it easier for the solvent to attract chemicals contained in the material. According to [35] extraction temperatures that were too low caused not all active compounds to be extracted from the ingredients, so the active compounds produced were low. The graph of the effect of extraction time on flavonoid levels can be seen in Figure 6.

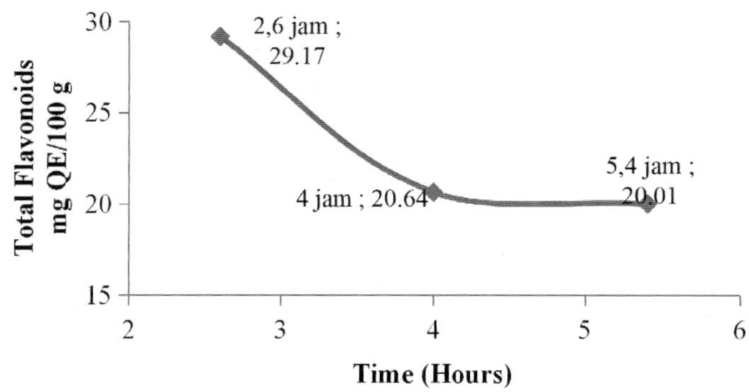


Figure 6. Graph of the effect of extraction time on flavonoid levels of kecombrang leaf liquid extract at 4 hours

Figure 6 shows a graph of the effect of extraction time at 50°C on the levels of flavonoids of kecombrang leaf liquid extract. At the time of 2.6 hours to 5.4 hours the level of flavonoids has decreased. Increasing the amount of time used, flavonoid levels tend to decrease. This is consistent with research conducted by [33] that the addition of the extraction time, the total flavonoids produced did not increase significantly, even decreased the total flavonoids. [35] states that the increase in temperature and extraction time used need to be considered. Because the extraction temperature is too high and the extraction time is too long and exceeds the optimum time limit can cause the loss of compounds in the solution due to evaporation.

### 3.2.3. pH.

The selected regression model for the variable pH of liquid extract of kecombrang stem is the cubic model. The coefficient of determination (R2) produced was 0.5204 (52.04%) with a lack of fit value of 0.9015 (90.15%) which showed insignificant results. The equation of the regression model obtained is as follows:

$$Y = 3.74 - 0.035(A) + 0.018(B) + 0.075(AB) + 0.058(A^2) + 0.021(B^2) + 7.322E\text{-}003(A^2B) + 0.11(AB^2).$$

Note :

Y = Respons pH
A = temperature (°C)
B = time (hours)

RSM calculation shows that the temperature and extraction time of the liquid extract of kecombrang stem does not significantly affect the pH value. A two-dimensional contour graph of the pH response is presented in Figure 7.

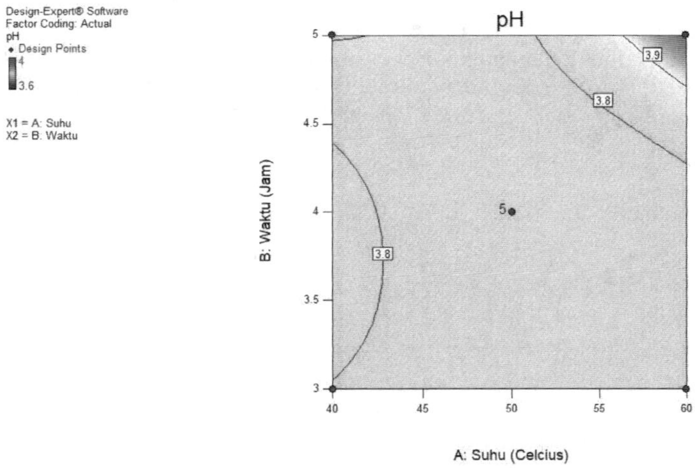


Figure 7. Two-dimensional contour graph of pH response of liquid extract of kecombrang stem.

The effect of temperature on the acidity of the extract, that the increase in temperature does not affect the pH of the extract produced. The increase in extraction time also did not affect the acidity of the extract. This is in accordance with [37] which states that the extraction time has no significant effect on pH. It is suspected that the type of solvent used in extraction can affect the acidity of the extract. In this research, distilled water is used as a solvent in the extraction process, where distilled water is not suitable for extracting phenol compounds in kecombrang stems. Phenolic compounds are acidic because of the nature of the $H^+$ group that easily breaks away from its main structure [38]. This $H^+$ group causes the atmosphere of the solution to become acidic and affects the pH of the solution.

The ANOVA model used in response to the pH value of kecombrang leaf liquid extract was a quartic model with R2 as large as 0.9493. P-values in the ANOVA model, temperature, time and the interaction between temperature and time in response to the pH value of 0.0232 respectively; 0.0052; 0.4734 and 0.0173. Especially the p-value in the ANOVA model pH value is smaller than $p < 5\%$ which means that the model is significant or significantly influences the response. ANOVA model the pH value of the quartic model produces a two-dimensional color contour graph like in Figure 8.

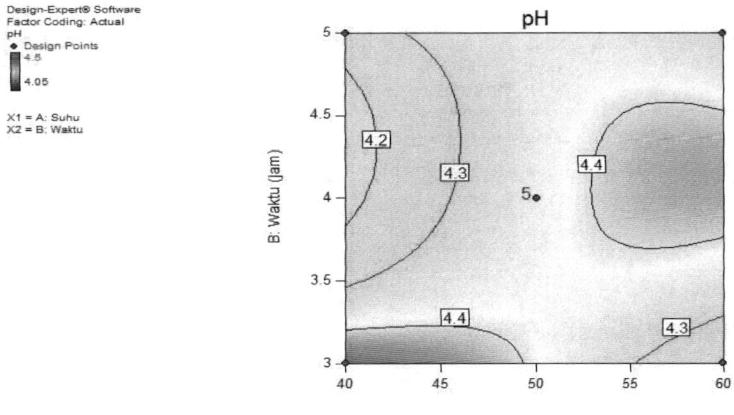


Figure 8. Two-dimensional contour graph of pH value of kecombrang leaf liquid extract

The results of the analysis of pH response responses can be seen in Figure 8. The response results of pH values indicate that the temperature and extraction time have a significant influence on the pH value of liquid extract of kecombrang leaf powder. The value of flavonoids of kecombrang leaf liquid extracts ranged from 4.05 - 4.5. This is consistent with the literature. According to [14], the pH value of foodstuffs and processed foods in general ranges between 4-7. The level of acidity (pH) indicates the ability of antimicrobial compounds in food to inhibit the growth of bacteria that is influenced by the stability of the food processing of the product. The graph of the effect of extraction temperature on the pH value can be seen in Fig 9.

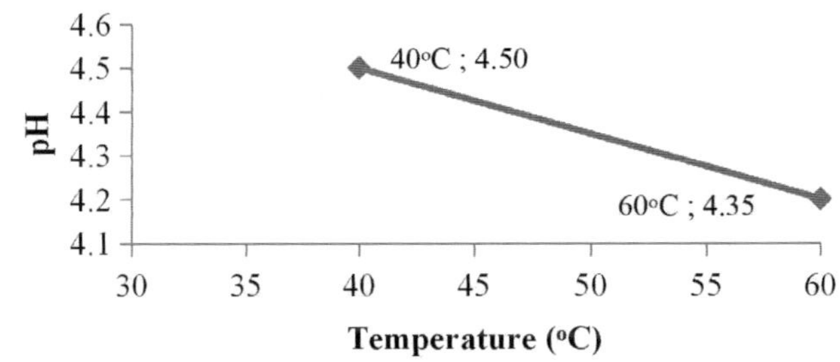


Figure 9. Graph of the effect of extraction temperature on the pH value of kecombrang leaf liquid extract at 3 hours

Figure 9 shows a graph of the effect of temperature at 3 hours on the pH value of the liquid extract. Addition of extraction temperature used, the pH value tends to be small or acidic. According to [2] pH will tend to decrease due to the addition of extraction time and temperature. This is due to the increasing number of elements in the material that decompose and form chemical compounds that are acidic. The graph of the effect of extraction time on the pH value can be seen in Figure 10.

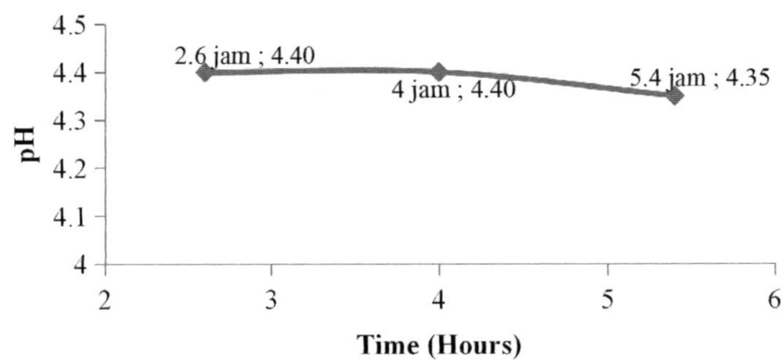


Figure 10. Graph of the effect of extraction temperature on the pH value of kecombrang leaf liquid extract at temperature 50°C

Figure 10 shows a graph of the influence of time at 50oC on the pH value of kecombrang leaf liquid extract. In addition to the extraction time used, the pH value tends to be small or acidic. According to [39] pH tends to decrease due to the addition of extraction time and temperature. This is due to the increasing number of elements in the material that decompose and form chemical compounds that are acidic.

### 3.3 Verification of Optimization Results

After conducting research with 13 treatments, RSM recommends several optimum formulas according to the desired optimum criteria. Optimization of kecombrang stem liquid extract, the expected response criteria are in range for pH and total phenol responses, and maximize for flavonoid criteria. Based on these response criteria RSM recommends eleven formulations. From the eleven optimum formulas recommended by RSM, the formula with temperature 40°C and extraction time of

5 hours which has a high desirability of 0.964 was selected, which is presented in Table 6. Optimization of liquid extract of kecombrang stem, the expected response criteria are in range for the pH response and total phenol, and maximize for flavonoid criteria. Based on these response criteria RSM recommends eleven formulations. From the eleven optimum formulas recommended by RSM, a formula with a temperature of 51°C and a extraction time of 3 hours which has a high desirability of 0.721 was chosen, which is presented in Table 7.

Table 6. Recommended RSM formula for liquid extract of kecombrang stem.

| No. | Temperature (°C) | Time (Hours) | Total Phenol | Total Favanoids | pH | Desireability |
|---|---|---|---|---|---|---|
| 1 | 40 | 5 | 23.51 | 11.47 | 3.694 | 0.964 |

Table 7. Recomended RSM formula for liquid extract of kecombrang leaf

| No. | Temperature (°C) | Time (Hours) | Total Phenol | Total Favanoids | pH | Desireability |
|---|---|---|---|---|---|---|
| 1 | 50.848 | 3 | 327.13 | 24.586 | 4.35 | 0.721 |
| 2 | 50.951 | 3 | 325.13 | 24.585 | 4.37 | 0.721 |
| 3 | 51.092 | 3 | 328.10 | 24.581 | 4.37 | 0.721 |
| 4 | 55.198 | 3 | 330.32 | 23.057 | 4.30 | 0.628 |
| 5 | 60 | 4.101 | 302.67 | 21.777 | 4.42 | 0.551 |
| 6 | 60 | 4.111 | 302.77 | 21.777 | 4.42 | 0.551 |
| 7 | 60 | 4.085 | 304.68 | 21.776 | 4.42 | 0.550 |
| 8 | 60 | 4.120 | 301.65 | 21.776 | 4.42 | 0.550 |

There are eleven optimum formulas of kecombrang stem liquid extract recommended by RSM, selected formula with a temperature of 40°C and a extraction time of 5 hours which has a high desirability of 0.964. The optimum formula produced by RSM for liquid extraction of kecombrang leaves is 8 formulations. The optimum formula was selected with a temperature of 51°C and extraction time of 3 hours which has a high desirability of 0.721. The desirability value is the value of the optimization objective function that shows the ability of the program to fulfill the desire based on the criteria set on the final product. The desirability value which is close to 1.00 indicates the ability of the program to produce the desired product more perfect [38].

The verification results of the optimum conditions recommended by RSM, obtained liquid extract of kecombrang stem with a total phenol value of 23.99 mg QE / 100 g, a total flavonoid value of 11.14 mg TAE / 100 g, and a pH value of 3.85. When compared with the predicted value by RSM (Table 8), the verification value ranges between 95% PI low and 95% PI high, so that the results of RSM optimization can be accepted.

Table 8. Verification results of kecombrang stem liquid extract

| Respon | Verification Value | Prediction Value | PI 95% Low | PI 95% High |
|---|---|---|---|---|
| Total Phenol | 23,99 | 23,50 | 10,25 | 36,75 |
| Total Flavonoids | 11,14 | 11,47 | 9,16 | 13,78 |
| pH | 3,85 | 3,69 | 3,27 | 4,12 |

The results of verification of optimum conditions of kecombrang leaf liquid extract with a total phenol value of 327.13 mg QE / 100 g, a total flavonoid value of 24,911 mg TAE / 100 g, and a pH value of 4.35. When compared with the predicted value by RSM (Table 9), the verification value ranges between 95% PI low and 95% PI high, so that the results of RSM optimization can be accepted.

Table 9. Verification results of kecombrang leaf liquid extract

| Respon | Verification Value | Prediction Value | PI 95% Low | PI 95% High |
|---|---|---|---|---|
| Total Phenol | 327.13 | 253.30 | 68.68 | 437.91 |
| Total Flavonoids | 24.911 | 24.586 | 18.98 | 30.20 |
| pH | 4.35 | 4.37 | 4.23 | 4.53 |

Furthermore, an independent t-test with a 95% confidence level is performed to validate the verification results with the predicted values. The test results obtained from all responses (total phenols, total flavonoids, and pH) is the absence of a significant difference between the results of verification with prediction results (F count> 0.05), so that the results of temperature optimization and liquid extraction time of stem and leaf kecombrang can be accepted because it is not significantly different from the results of RSM predictions.

## 4. CONCLUSION

The optimum formula of kecombrang stem liquid extract based on the response of total flavonoid value, total phenol and pH value is a temperature treatment of 40°C with extraction time of 5 hours with total flavonoid response value of 11.14 mg QE / 100 g, total phenol of 23.99 mg TAE / 100 g, and a pH value of 3.85. The optimum formula of kecombrang leaf liquid extract based on the response of total flavonoid value, total phenol and pH value is 51°C temperature treatment with extraction time of 3 hours with total flavonoid response value of 327.13 mg QE / 100 g, total phenol of 24,911 mg TAE / 100 g , and a pH value of 4.35. The total value of flavonoids, total phenols and pH values of liquid extracts of kecombrang stems and leaves did not differ greatly with RSM predictions. Kecombrang stem liquid extract contains bioactive compounds such as alkaloids, phenols, flavonoids, tannins, glycosides, saponins and triterpenoids. Kecombrang leaf liquid extract contains bioactive compounds such as alkaloids, phenols, flavonoids, saponins, tannins and steroids.

**Acknowledgements**

Authors wish to thanks to the DRPM Kemenristek Republic of Indonesia which has provided research funding through Applied Research Grant Scheme 2019 also to Institute for Research and Community Service of Jenderal Soedirman University for the support.

## REFERENCES

[1] Barus, P. 2009. Pemanfaatan Bahan Pengawet dan Antioksidan Alami Pada Industri Bahan Makanan.*Seminar pengangkatan guru besar FMIPA* USU. Medan.3 Oktober 2009.

[2] Naufalin, R., 2019. Natural preservation opportunities and challenges in improving food safety. In *AIP Conference Proceedings.* Vol. 2094, No. 1, p. 020032. AIP Publishing.

[3] Naufalin, R., Herastuti, S. R. 2017. Antibacterial activity of *Nicolaia speciosa* fruit extract. *International Food Research Journal* 24.1:379-385.

[4] Naufalin, R., Wicaksono, R., Arsil, P. And M.F. Salman. 2018. Antimicrobial coating on quality attributes of sausage during refrigerated storage. E3S Web of Conferences 47, 01002 (2018) https://doi.org/10.1051/e3sconf/20184701002

[5] Naufalin, R., and S. R. Herastuti. 2018. Antibacterial activity of kecombrang flower extract (*Nicolaia speciosa*) microencapsulation with food additive materials formulation. IOP Conf. Series: Earth and Environmental Science. 102 012035. doi :10.1088/1755-1315/102/1/012035.

[6] Latifasari, N., R. Naufalin, R. Wicaksono. 2019. Edible coating application of kecombrang leaves to reduce gourami sausage damage. IOP Conf. Series: Earth and Environmental Science 250 (2019) 012055: 1-7.

[7] Hanifah, M.R., R. Naufalin, R. Wicaksono. 2019. The effect of edible coating contained Kecombrang leaves concentrate on gourami fish fillet quality. IOP Conf. Series: Earth and Environmental Science 250 (2019) 012057: 18.

[8] Direktorat Jenderal POM. 2000. *Parameter Standar Umum Ekstrak Tumbuhan Obat: Cetakan Pertama.* Departemen Kesehatan RI.Jakarta.

[9] Naufalin, R., Wicaksono, R., Arsil, P. and Khusna, V.Z., 2019, April. Physicochemical Analysis Of Gouramy Fish Sausage With Kecombrang Edible Coating Addition. In *IOP Conference Series: Earth and Environmental Science* (Vol. 255, No. 1, p. 012039). IOP Publishing.

[10] Harborne. J. B. 1987. *Metode Fitokimia.* Institut Teknologi Bandung. Bandung.

[11] Singleton, V. L., Rossi, J. A. 1965. Colorimetry of total phenolics with phosphomolybdic-phosphotungstic acid reagents. *American Journal of Enology and Viticulture, 16:144-158*

[12] Aminah, D. N., Tomayahu, N., Abidin, Z. 2017. Penetapan kadar flavonoid total ekstrak etanol kulit buah alpukat *(Persea Americana Mill.)* dengan metode spektrofotometri Uv-Vis. *Jurnal Fitofarmaka Indonesia,* 4(2).

[13] Simaremare. E. S. 2014. Skrining fitokimia ekstrak etanol daun gatal (*Laportea decumana (*Roxb.)Wedd).*PHARMACHY.*11(1) 98-107.

[14] Naufalin. R. 2017. *Kecombrang Antimikroba dan Pemanfaatanya Sebagai Pengawet Makanan.*Universitas Jenderal Soedirman. Purwokerto.

[15] Susanti. N. M. P., Dewi, L. P. M. K., Manurung, H. S., & Wirasuta, M. A. G. 2017. Identifikasi senyawa golongan fenol dari ekstrak etanol daun sirih hijau (*Piper bettle* Linn.) dengan metode KLT-spektrofotodesitometri.*Jurnal Metamorfosa.* 6(1):108-113.

[16] Zuraida., Sulistiyani., Sajuthi, D., & Suparto, I. H. 2017. Fenol, flavonoid, dan aktifitas antioksidan pada ekstrak kulit batang pulai (R.Br).*Jurnal Penelitian Hasil Hutan.* 35(3):211-219.

[17] Noer, S. 2016. Uji kualitatif fitokimia daun *Ruta angustifolia.Jurnal Factor Exacta.*9(3):200-206.

[18] Agustina, W., Nurhamida, & Handayani, D. 2017. Skrining fitokimia dan aktifitas antioksidan beberapa fraksi dari kulit batang jarak (*Ricinus communis* L.) *ALOTROP Jurnal Pendidikan dan Ilmu Kimia.*1(2):117-122.

[19] Jannah, R., Husni, M. A., & Nursanty, R. 2017.Uji daya hambat ekstrak methanol daun sirsak (*Annona muricata Linn.)* terhadap bakteri *Streptococcus mutan. Jurnal Natural.* 17(1):23-30.

[20] Syafitri, N.E., Bintang, M., & Falah, S. 2014.Kandungan fitokimia, total fenol ekstrak buah harendong (*Melastoma affine* D. Don).*Current Biochemistry.1(3):105-115.*

[21] Pangesti, R, D., Cahyono. E., & Kusumo, E. 2017.Perbandingan daya antibakteri ekstrak dan minyak *piper betlel.*terhadap bakteri *Streptococcus mutans. Indo.Jurnal Chemical Science.*6(3):270-278.

[22] Sangi, M., Runtuwene, M. R. J., Simbala, H. E. I., & Makang, V. M. A. 2008.Analisis Fitokimia Tumbuhan Obat di Kabupaten Minahasa Utara.*Chemical Program.* 1(1): 47-53

[23] Astarina, N. W. G., Astuti, K. W., & Warditiani, N. K. 2013. Skrining fitokimia ekstrak methanol rimpang bangle (*Zingiber purpureum* Roxb.)*Jurnal Farmasi Udayana.* 1-7.

[24] Ergina., Nuryanti, S., & Puspitasari, I. D. 2014. Uji kualitatif senyawa metabolit sekunder pada daun palado (*Agave angustifolia)* yang diekstrak dengan pelarut air dan etanol.*Jurnal Akademik Kimia.* 3(3):165-172.

[25] Ismail, J., Runtuwene, M. R. J., & Fatimah, F. 2012. Penentuan total fenolik dan ujI aktivitas antioksidan Pada biji dan kulit buah pinang yaki (*Areca vestiaria giseke*). *Jurnal Ilmiah Sains.* 12(2):84-88.

[26] Zhao, Q., Kennedy, J.F., & Wang, X. 2011.Optimization of ultrasonic circulating extraction of polysaccharides from *Asparagus officinalis* using response surface methodology.*International Journal of BiologicalMacromolecules.*49:181–187.

[27] Bas, D., & Boyaci, I.S., 2007. Modeling and Optimization I: Usability of response surface methodology. *JournalofFood Engineering.* 78:836–845.

[28] Naufalin, R, & Rukmini, H.S. 2010. Potensi Antioksidan Hasil Ekstraksi Tanaman Kecombrang (*Nicolaia speciosa Horan)* Selama Penyimpanan.*Seminar Nasional Daya Saing Produk Pangan Berbasis Lokal.*Surakarta. 8 Juni 2010.

[29] Estiasih, T & D.A. Kurniawan. 2006. Aktivitas antioksidan ekstrak umbi akar gingseng jawa (*Talinum triangulare Willd.).Jurnal Teknologi dan Industri Pangan.* 17(3): 166-175.

[30] Kusuma, A. T., Adelah, A., & Najib, A. 2018.penentuan kadar flavonoid ekstrak etil asetat daun sukun (*Artocarpus altilis*).*Ad- Dawa'Journal Pharmacy Science.*1(1):25-31.

[31] Iriany., Septiawan, I., & Gustia., S. J. 2017. Model kinetika ekstraksi flavonoid dari bayam merah (*Alternanthera amoena voss*).*Jurnal Teknik Kimia USU.* 6(4):8-14.

[32] Dewi, S. R., Ulya, N., & Argo. B. D. 2018. Kandungan flavonoid dan aktivitas antioksidan ekstrak *Pleurotus ostreatus*.*Jurnal Rona Teknik pertanian.*11(1):1-11.

[33] Yuliantari, N. W. A., Widarta, I. W. R., & Permana, I. D. G. M. 2017. Pengaruh suhu dan waktu ekstraksi terhadap kandungan flavonoid dan aktivitas antioksidan daun sirsak (*annona muricata l.)*menggunakan ultrasonic. *Media Ilmiah Teknologi Pangan.*4(1):35-42.

[34] Tambun, R., H. P. Limbong, C. Pinem & E. Manurung. 2016. Pengaruh ukuran partikel, waktu dan suhu pada ekstraksi fenol dari lengkuas merah. *Jurnal Teknik Kimia.* 5(4): 53-56

[35] Ibrahim, A.M., Yunita & H.S. Feronika. 2015. Pengaruh suhu dan lama waktu ekstraksi terhadap sifat kimia dan fisik pada pembuatan minuman sari jahe merah dengan kombinasi penambahan madu sebagai pemanis. *Jurnal Pangan Dan Agroindustri.* 3(2):530-541.

[36] Firdiyani, F., Agustini, T. W., & Ma'aruf, W. F. 2015. Ekstraksi senyawa bioaktif sebagai antioksidan alami *Spirulina platensis* segar dengan pelarut yang berbeda. *JPHPI.*18(1):28-37.

[37] Akbar, A., R. Paindoman & P. Coniwanti. 2013. Pengaruh variabel waktu dan temperatur terhadap pembuatan asap cair dari limbah kayu pelawan (*Cyanometra cauliflora). Jurnal Teknik Kimia.* 19(1) : 1-8.

[38] Nurmiah, S., Syarief, R., Sukarno., Peranginangin, R., & Nurtama, B. 2012. Aplikasi *Responses Surface Methodology* pada optimalisasi kondisi proses pengolahan *Alkali treater cottonii (atc). JPB Kelautan dan Perikanan.*8(1):9-22.

IOP Conf. Series: Earth and Environmental Science **406** (2019) 012016 doi:10.1088/1755-1315/406/1/012016

# Phytotoxicity in Foxtail millet seed polluted batik wastewater and its reduction by ArbuscularMycorrhizal Fungi

**RatnaStia Dewi[1,*], MardiyahKurniasih[2], Uki Dwiputranto[1]**

[1]Microbiology Department, Faculty of Biology, UniversitasJenderalSoedirman, Jl. Dr. Soeparno 63, Purwokerto 53123, Indonesia

[2]Chemistry Department, Faculty of Mathematics and Natural Sciences, UniversitasJenderalSoedirman, Jl. Dr. Soeparno, Purwokerto 53123, Indonesia

ratna.dewi0509@unsoed.ac.id

**Abstract**. The aim of this study was to determine the effect of giving ArbuscularMycorrhizal Fungi (AMF), to the toxicity of Millet seeds (Foxtail millet) contaminated with batik waste water. This study used seeds contaminated with batik waste water with the addition of AMF and without AMF (negative control), as well as water-treated waste, Millet seeds that were not polluted with batik waste water with and without AMF added with distilled water (positive control). The results showed that AMF addition had an effect on reducing the toxicity of Millet seeds contaminated with batik waste water. Millet exposed to batik wastewater with the provision of AMF + water produced the best value of 57.78% germination, 42.22% inhibition, highest leaf length 1.13 cm, plumulae 1.89 cm, radics 1.58 cm. The provision of AMF is promising in preventing pollutants from entering plants as well as improving soil quality.

## 1. Introduction

ArbuscularMycorrhizal Fungi (AMF) is an association between certain fungi and plant roots by forming complex interwoven interactions. AMF is known as soil fungi because its habitat is in the soil and is in the rooting area of the plant (rhizosphere). Besides being referred to as soil fungi, AMF is also commonly said to be root fungi. The specialty of this fungus is its ability to help plants to absorb nutrients especially phosphorus nutrients or P [1]. Besides being useful in increasing nutrient absorption, AMF increases plant tolerance to degraded land conditions in the form of drought and the presence of heavy metals. This function is used as an alternative technology to assist growth, increase productivity and quality of plants grown on marginal lands [2].

AMF can protect host plants from the absorption of these toxic elements through the effects of filtration, complexation and accumulation. AMF can act as a biocontrol of heavy metal absorption, and can help plants avoid heavy metal poisoning [3]. AMF symbiosis also increases plant resistance to extreme drought and humidity, helping to accumulate substances that are poisonous to plants such as As, Cr, and Pb [4]. The AMF genus Glomus associated with plants has been proven effective in absorbing heavy metals, namely Cd, Zn, and Pb. The things above show that AMF inoculation is very important in the process of plant growth and absorption of heavy metals in polluted soils [5].Another section of your paper.

Batik waste contains 0.1385 mg / l total Crom, 2.0587 mg / l Iron, 0.2696 mg / l Copper, 54.7175 mg / l Zinc, 0.0063 mg / l Cadmium, 0.2349 mg / l Lead (Riwayati). The content of this metal can accumulate in the environment, for example on the ground. Batik waste from many home industries is discharged into the environment without processing. The metal content is bad for exposed plants and also for seeds. Seed germination bioassay research shows that when a dye solution is used, the results inhibit seed germination [6].

Millet ranks sixth as the most important grain and consumed by one third of the world's population. One of the main sources of energy providers, protein, vitamins and minerals, rich in B vitamins, especially niacin, B6 and folacin as well as essential amino acids such as isoleucine, leucine, phenylalanine and threonine and contain nitrileoside compounds that play an important role in inhibiting the development of cancer cells (anti-cancer), also reduce the risk of heart disease (artheriosclerosis, heart attack, stroke and hypertension). Millet thrives in high-temperature areas, limited water availability, without the application of fertilizers and other technological inputs, and in critical lands that are difficult to plant other grains such as wheat and rice [7]. batik waste removal experiments on millet seeds need to be done because this plant is considered good on critical land.

The provision of AMF is promising in preventing pollutants from entering plants as well as improving soil quality, however, toxicity is still not studied. This study aims to determine the effect of AMF administration on the phytotoxicity of F. millet seeds contaminated with batik waste water.

## 2. Methode

### *2.1. Material*

AMF used is a mixture of *Glomus* sp., *Glomusmanihotis*, *Glomusetunicatum*, *Gigaspora margarita*, *Acaulosporaspinosa* with a zeolite carrier containing 20-30 spores / gram. The seeds used in this study were Millet (*F. millet*).

### *2.2. Evaluation of Phytotoxicity*

Sterilized millet seeds are germinated in petri dishes that have been sterilized with sterile cotton [8], then stored at room temperature. seeds on cotton surfaces are sprayed with each treatment solution every 24 hours to reach 80%. The treatments used were seeds exposed to batik waste added with AMF and without AMF (negative control), waste given distilled water, Millet seeds were not polluted with batik waste water with and without AMF added with distilled water (positive control). The study was carried out at room temperature.Percentage of seed germination and inhibition were estimated, leaf length, shoot length (plumule) and root length (radicle) were measured after 9 days.

## 3. Result and discussion

Phytotoxicity is used to measure the level of plant poisoning caused by pollutants. Batik waste exposed to Millet seeds results in damage to seeds which can inhibit millet growth resulting in other variables being disrupted. AMF application reduces the toxic effects of batik waste.

The results showed that AMF administration affected the level of weed puzzle poisoning as indicated by the calculation of germination, inhibition, measurement of leaf length, plumulae, and radicle. The AMF + Water + waste treatment results in a lower level of poisoning against F. millet compared to the administration of only waste or AMF + waste. This is characterized by a higher level of germination, lower inhibition, longer leaves, higher plants and longer roots.

The percentage values of F. millet germination and inhibition in the treatment of Water, Effluent, AMF + Water, AMF + Effluent, AMF + Effluent + Water 88.89%, 21%, 100%, 52.22%, 57.78% and 11.11%, 79%, 0%, 47.78 %, 42.22% (Figure 1). The length of F. millet leaves was 1.36, 0, 1,029, 0.57,1.13 in the treatment of Water, Effluent, AMF + Water, AMF + Effluent, AMF + Effluent + Water (Figure 2). Shoot lengths (plumules) of F. millet with the treatment of Water, Effluent, AMF + Water, AMF + Effluent, AMF + Effluent + Water are 6.52, 0, 6.4, 0.78, 1.89 cm, respectively (Figure 3). Whereas for root lengths (radicles) are 4.3, 0, 2.45, 0, 1.58 cm (Figure 4).

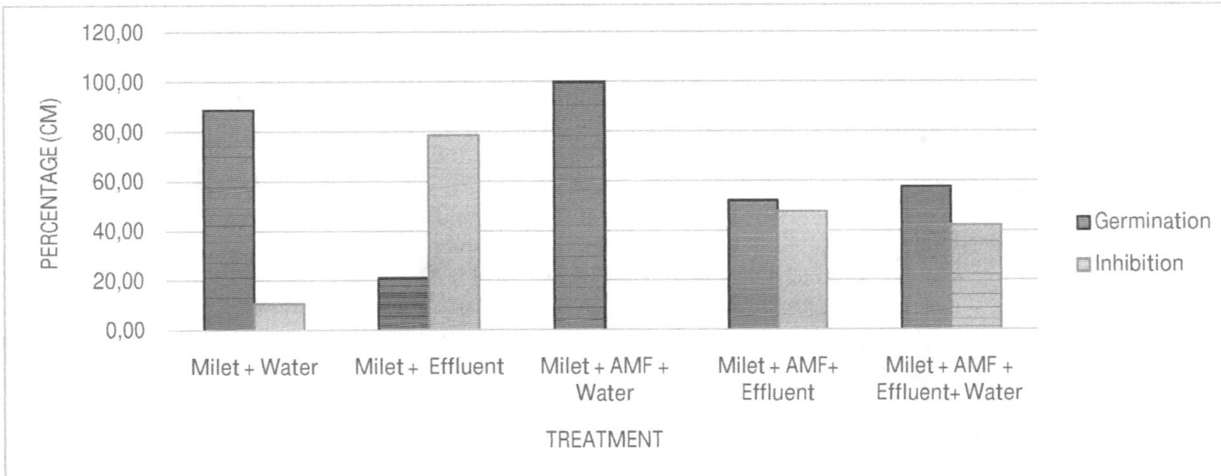


Figure 1. Effect of treatment on germination and inhibition of*F. millet.*

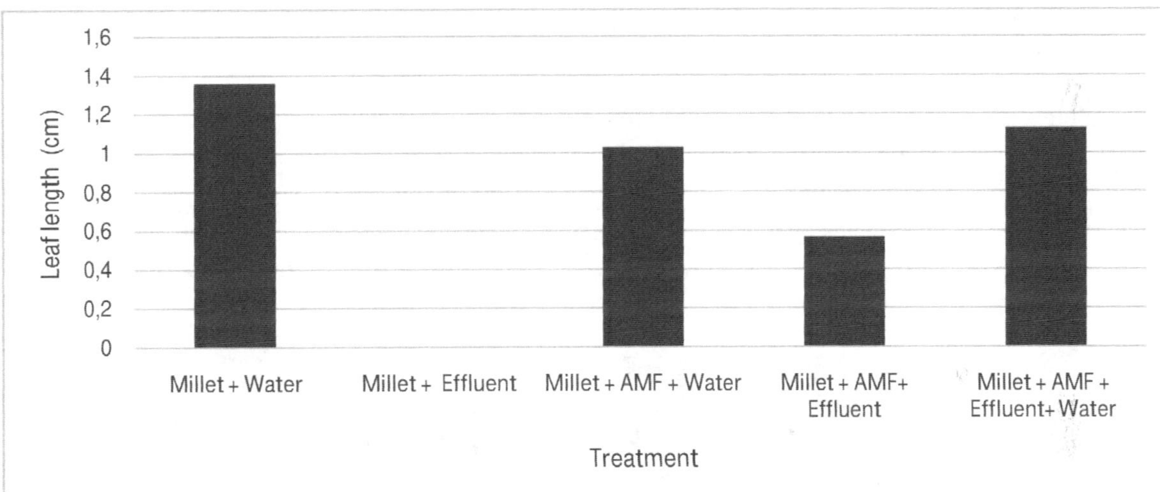


Figure 2. Effect of treatment on leaf length of *F. millet.*

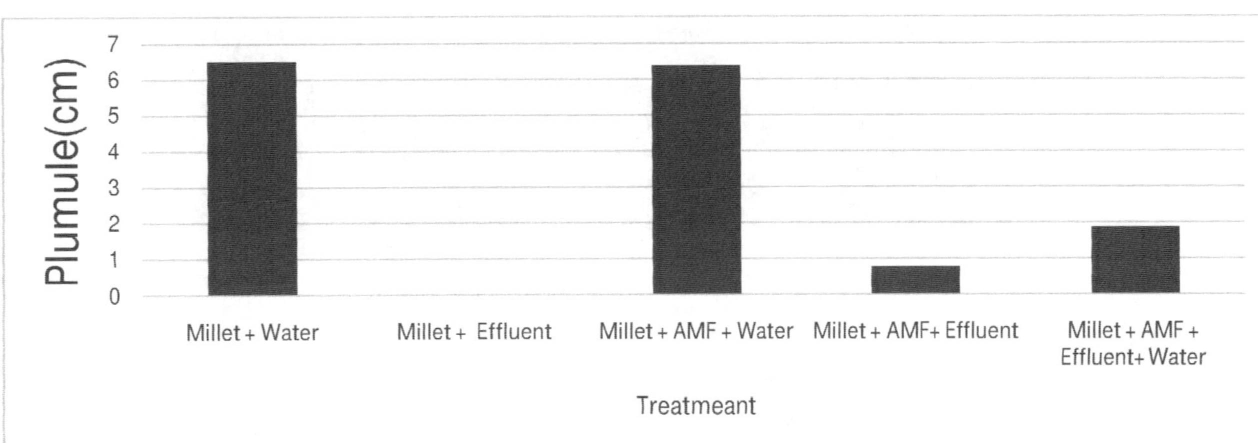


Figure 3. Effect of treatment on shoot lengths (plumule) of*F. millet.*

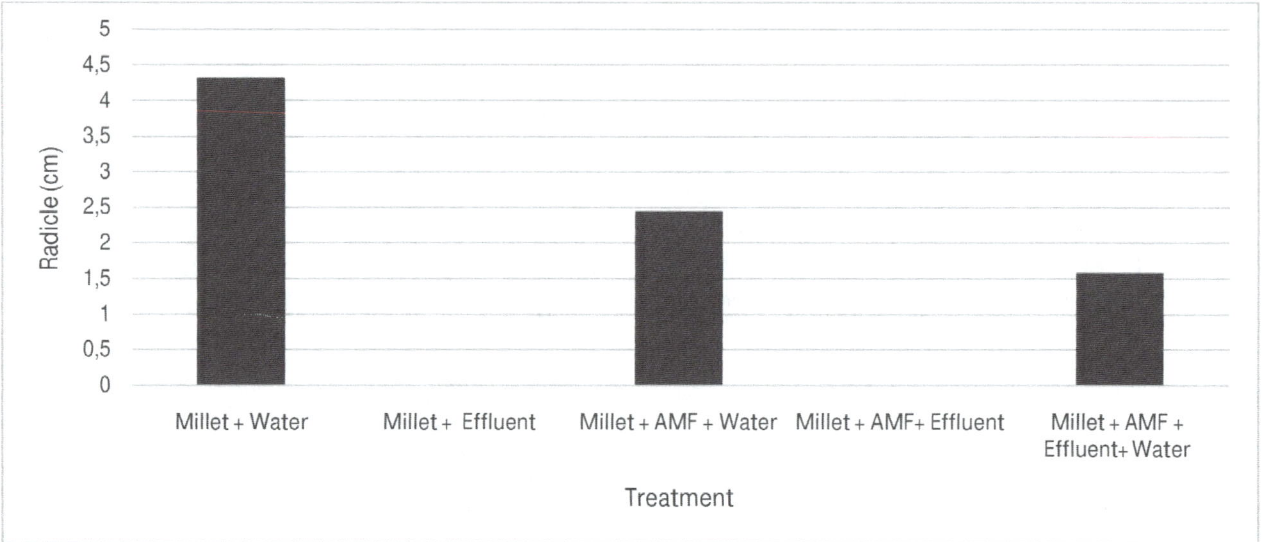


Figure 4. Effect of treatment on root lengths (radicle) of *F. millet.*

Millet seeds immersed in water showed 88.89% germination, leaf length & roots and plumulae were higher than those exposed to batik waste showed 21% germination as well as leaf length, roots and plumulae. Batik waste shows signs of toxicity compared to water treatment. whereas seeds treated with the addition of AMF exposed to batik show a higher value in leaf and root length. The results showed that seedlings treated by AMF reduced millet toxicity, which showed that AMF was efficient in reducing toxicity. These results are in line with previously reported studies that the toxicity of pollutants can be reduced using AMF (kedia, dias, money). the efficiency of reducing the toxicity of millet exposed to batik waste and other treatments can be seen visually in Figure 5.

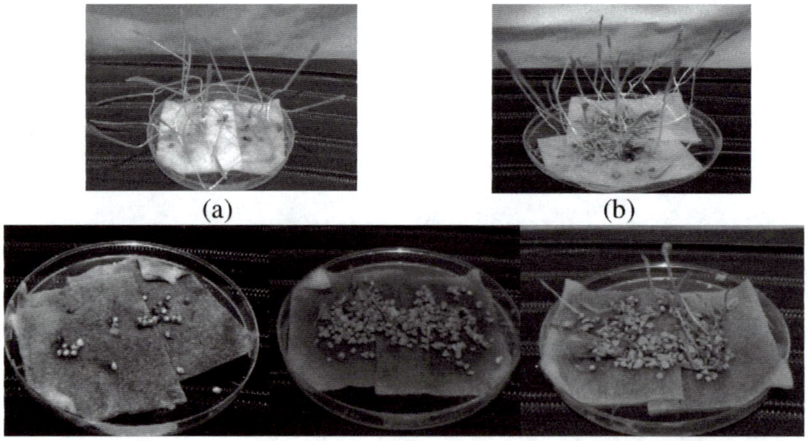

(a) (b)

(c) (d) (e)

Figure 5. *F. millet* on variations in the treatment of AMF and water.

Explanation :

(a) Millet + Water

(b) Millet + AMF + Water

(c) Millet + Effluent

(d) Millet + AMF+ Effluent

(e) Millet + AMF + Effluent+ Water

In addition to suppressing AMF toxicity, it plays a good role in the growth of proven seedlings that are exposed to water better in germination (100%) than without AMF (88.89%). By plants, (2) increasing tolerance to heavy metal contamination, drought, and root pathogens, and (3) providing access for plants to utilize nutrients that are not available to be available to plants [9].

AMF can dissolve heavy metals in batik waste. The secretion of secondary metabolites produced by AMF can be in the form of organic acids, causing the micro elements to be easily dissolved. The situation with acidic pH causes the heavy metals contained in the medium to become soluble and actively absorbed by plants [10]. AMF plays an important role in protecting plant roots from toxic elements, including heavy metals. The mechanism of protection against heavy metals and toxic elements by AMF can be through the effects of filtration, chemical deactivation, or the accumulation of these elements in hyphae fungi. Absorption of micro elements by AMF plants depends on several factors, namely the physical-chemical condition of the soil, soil fertility, pH, type of plants, and the concentration of micro elements in the soil [11]. This occurs because AMF is known to be able to bind the metal to carboxyl groups and pectak compounds (hemiseslulose) on the AMF contact matrix and host plant, on the polysaccharide sheath and hyphal cell wall [12]. AMF can bind metal ions in the cell walls of HIV and can protect plants from these metal ions. Heavy metals are stored in crystalloids in mycelium fungi and in root cortex cells in plants using AMF [13].

Figure 5c shows millet seedlings exposed to batik waste that cannot grow properly, Figure 5d and 5e show better growth of seedlings with the addition of AMF. At low concentrations of heavy metals does not affect plant growth but at high concentrations will cause plant damage [14]. AMF can increase plant tolerance to toxic metals by accumulation of metals in external hyphae thereby reducing its absorption into host plants. The use of AMF in bioremediation of polluted soils, in addition to the accumulation of these materials in hyphae, can also be through the mechanism of metal complexing by external hyphae secretions. This shows that there is a filtration mechanism, so that the toxic material is not absorbed by plants [15].

Plants gain a lot of benefits from being associated with AMF. AMF is able to replace 40% Nitrogen and 25% Potassium. In addition, AMF can also protect against soil infectious diseases and increase plant resistance to drought [16]. AMF can produce growth regulators, be able to form physical barriers and issue certain antibiotics to block the development of soil borne pathogens, and help plants competing with weeds [17].

**References**

[1] Syib'li. M. A. 2008. *Jati Mikoriza, Sebuah Upaya Mengembalikan Eksistensi Hutan dan Ekonomi Indonesia*. http://-www.kabarindonesia.com. Accessed on 28 februari 2009.

[2] Al-Karaki, G., McMichael, B. Z. A. K. J., & Zak, J. (2004). Field response of wheat to arbuscular mycorrhizal fungi and drought stress. *Mycorrhiza*, *14*(4), 263-269.

[3] Tisdall, J.M. 1991. Fungal hyphae and structural stability of soil. J. Soil. Res. 29:729-743. Australia

[4] Aisyah, L dan Hardiani, H. 2009. Fitoremediasi Tanah Terkontaminasi Logam Cu Limbah Padat Proses Deinking Industri Kertas Oleh Tanaman Bunga Matahari (Helianthus annuus) Dengan Penambahan Mikoriza. BS, Vol. 44, No. 1. Bandung.

[5] Adholeya dan Gaur, A. 2004. *Prospect of Arbuscular Mycorrizal fungi in Phytoremediation of Heavy Metal Contaminated Soils*. Centre for Mychorrhizal Research, The Energy and Resources Institute, Darbari Seth Block, Habitat Place, Lodhi road, New Delhi 110003, India.

[6] Rani, B., Kumar, V., Singh, J., Bisht, S., Teotia, P., Sharma, S., & Kela, R. (2014). Bioremediation of dyes by fungi isolated from contaminated dye effluent sites for bio-usability. *Brazilian Journal of Microbiology*, *45*(3), 1055-1063.

[7] Bhuja, 2009. Teknologi Budidaya Millet.Departemen Pertanian Balai Informasi Pertanian Provinsi Irian Jaya. Jayapura.

[8] Nouren, S., Bhatti, H. N., Iqbal, M., Bibi, I., Kamal, S., Sadaf, S., ... & Safa, Y. (2017). By-

product identification and phytotoxicity of biodegraded Direct Yellow 4 dye. Chemosphere, 169, 474-484.

[9] Gentili, F., & Jumpponen, A. (2006). of Bacterial and Fungal Biofertilizers. *Handbook of microbial biofertilizers*, 1.

[10] Sarwono, Hardjowigeno. 1995. *Ilmu Tanah*. Penerbit Akademia Pressindo. Jakarta

[11] Hardiani, H. 2009. Potensi Tanaman Dalam Mengakumulasi Logam Cu Pada Media Tanah Terkontaminasi Limbah Padat Industri Kertas. BS, Vol. 44, No. 1. Bandung.

[12] [14] Rossiana, N. 2003. Penurunan Kandungan logam Berat dan Pertumbuhan Tanaman Sengin (Paraserianthes falcataria L (Nielsen)) Bermikoriza Dalam Medium limbah Lumpur Minyak Hasil Ekstraksi. Universitas Padjajaran: Bandung.

[13] [15] Lakitan, B. 2001. *Dasar – Dasar Fisiologi Tumbuhan*. PT. Raja Grafindo Persada, Jakarta.

[14] [16] Dwidjoseputro, D. 1994. *Pengantar Fisiologi Tumbuhan*. PT Gramedia Pustaka Utama. Jakarta.

[15] [11] Donelly, PK. 1994. Potential *Use of Mycorrhizal Fungi as Bioremediation Agents*. American Chemical Society. USA. 94-97.

[16] Setiadi Y. 2006 *Pengembangan Cendawan Mikoriza Arbuskular untuk Merehabilitasi Lahan Marginal*. Prosiding Workshop Teknologi Baru Bekerja dengan Cendawan Mikoriza. Bogor.

[17] Halim. 2012. Peran Mikoriza Indegenous Terhadap Indeks Kompetesi Antara Tanaman Jagung (Zea May L). Dengan Gulma Agregatum conyzoides. Berkas Penelitian Agronomi. 1 : 86-92

# The effect of packaging type and storage temperature on the characteristics of cheese spread analogues from corn extract

**Nur Aini[1], Budi Sustriawan[1], Juni Sumarmono[2], V. Prihananto[1], Tanty Purwaning Atmajayanti[1]**

[1]Department of Food Science and Technology, Jenderal Soedirman University, Purwokerto, Central Java, Indonesia, 53123

[2]Department of Food Science and Technology, Jenderal Soedirman University, Purwokerto, Central Java, Indonesia, 53123

nur.aini@unsoed.ac.id

**Abstract.** Cheese spread analogues composed of corn milk have a limited shelf life due to their high water and fat content. Therefore, appropriate packaging and storage temperatures are required to maintain the nature of these products for long-term storage. The purpose of the present study was to 1) study the effect of packaging type and storage temperature on the chemical and sensory characteristics of corn milk-based cheese analogues; 2) determine the most effective packaging type for corn milk-based cheese analogues; 3) determine the optimal storage temperature to be applied for corn milk-based cheese analogue. The present study used a completely randomised split-plot design. The effects of various storage factors were assessed, including storage time (1, 2, 3 and 4 weeks) as the main plot, type of packaging (polypropylene, polyethylene terephthalate, and glass) as a subplot, and storage temperature (-5°C, 5°C, and 10°C) as a subplot. The studied cheese analogue variables included moisture, free fatty acid content, soluble protein content, fat content, pH and sensory characteristics. The results indicated that packaging type and storage temperature had a significant effect on analogue cheese during storage. The most effective packaging types applied for 4 weeks of analogue cheese storage were glass packaging with a moisture content of 69.60%, soluble protein of 6.91%, a fat content of 5.4%, free fatty acid content of 3.65% and pH of 4.6. The sensory characteristics included smooth texture, easy to spread, a bright, attractive and shiny appearance, pale yellow colour, and acidic smell. The optimum storage temperature for corn milk-based analogue cheese spread for 4 weeks was 5°C with a moisture content of 69.26%, dissolved protein content of 7.23%, fat content of 5.68%, free fatty acid content of 3.54% and pH 4.6. Sensory characteristics included smooth texture, easy to spread, a bright, attractive and shiny appearance, pale yellow colour, and slightly acidic smell smelling only slightly of cheese.

**Keywords:** cheese analogues, corn, glass packaging

## 1. Introduction

Cheese spread analogues can consist of milk or non-milk protein while substituting oil or milk fat for milk solids [1]. One ingredient that can be used to make cheese spread analogue is corn milk [2]. As

IOP Conf. Series: Earth and Environmental Science **406** (2019) 012017 doi:10.1088/1755-1315/406/1/012017

per cheese in general, cheese spread analogues have a high water content; thus, they can easily become damaged [3].

Packaging strongly influences changes in product quality during storage. Some types of packaging materials can be used to package corn milk-based cheese spread analogues, include polypropylene (PP), polyethylene terephthalate (PET), and glass. PP packaging is strong, transparent but not clear, has a light mass, waxy surface and is resistant to chemicals, heat and oil [4]. PET packaging has a light mass, is clear, transparent, strong, solvent resistant and impermeable to gas and water; however, it softens at 80°C [5]. Glass packaging has the advantages of not reacting with packaged materials, being resistant to acids, bases and the environment, being made transparent or dark, being odourless; however, glass has a heavy mass and is easily broken [6].

Storage temperature also affects the deterioration of product quality and must be stably maintained to prevent the loss of quality during storage [7]. The temperature that can be used to store cheese spread analogue is low, as low temperatures are intended to slow down the speed of metabolic reactions and inhibit chemical reactions as well as enzymatic or microbial growth.

The present study aims to 1) determine the most effective type of packaging for corn milk-based cheese spread analogue and to 2) determine the optimum storage temperature for corn milk-based cheese spread analogue.

## 2. Materials and methods

The main ingredients used in the present study included sweet corn sourced from Wage Market, Purwokerto, Central Java, Indonesia; whey protein isolate (Global Milk Specialties), Arabic gum, Tween 20, virgin coconut oil (Mutia, Yogyakarta) and other ingredients. The main equipment used included a blender, filter, PET plastic packaging, PP plastic packaging, glass packaging, a refrigerator, a freezer, and a set of analysers.

Cheese spread production was performed using a modified [2] method. Corn was steam blanched for 30 minutes and then ground with the addition of water (1: 2). Corn extract plus 15% whey protein isolate was then heated to 70°C for 15 seconds. The corn extract was then added by 7.5% VCO and 0.5% Tween 20, then stirred and allowed to stand for 5 minutes. The next step involved pressing for 15 minutes. The solids obtained were then combined with 6% sugar and 1% salt. The obtained cheese was then heated to 40°C for 15 seconds. The resulting analogue cheese was then packaged using polypropylene, polyethylene terephthalate and glass packaging and stored at -5, 5 and 10°C. Furthermore, an analysis of water content, fat content, free fatty acid (FFA) content, dissolved protein content, pH and organoleptic tests were performed every week for 4 weeks.

The study used a split-plot design involving a completely randomised design (CRD). The assessed factors included 1) storage time as the main plot (1 week, 2 weeks, 3 weeks and 4 weeks); 2) types of packaging as subplots (PP packaging, PET packaging, glass packaging); and 3) storage temperature as subplots (-5°C, 5°C, and 10°C). Data were analysed using the analysis of variance (ANOVA) test; if a significant effect was observed, Duncan's Multiple Test (DMRT) at a 5% confidence level was performed. The organoleptic test results were analysed using the Friedman test; if this test had a significant effect, it was followed by a multiple comparison test with a significant level of 5%.

## 3. Results and discussion

### *3.1. Moisture content*

The moisture content of cheese spread analogues increased over storage time, with the lowest moisture content being observed a storage time of 1 week (68.29%) and the highest at 4 weeks (69.48%). This increase in water content can be predicted using the equation $y = 0.397x + 67.905$ with $R^2 = 0.999$ (Fig. 1). According to the United States Department of Agriculture (2001), the water content of soft cheese is no more than 80%. When viewed from the prediction equation, the water content of cheese spread analogue can be stored for up to 30 weeks, with water content reaching 79%.

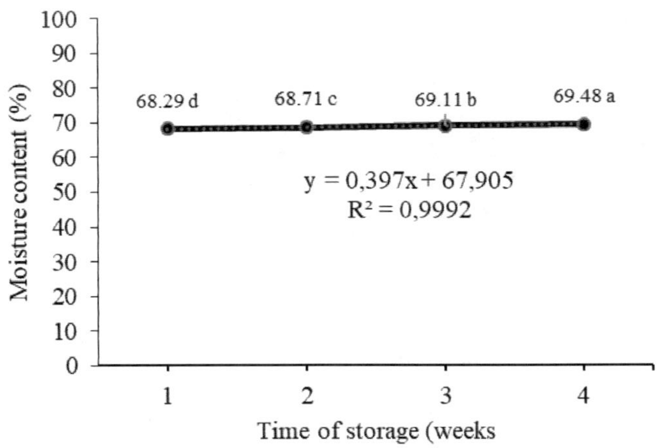


Figure 1. The moisture content of cheese spread analogue during storage

The increased moisture content of cheese spread analogues during storage is likely due to the product absorbing water from the environment. High humidity in storage rooms can cause the process of water vapour absorption from the air to the product, resulting in increased water content [8]. The increased water content during storage is in accordance with [9], who stated that the moisture content of cheese spread analogues tends to increase during storage. In food products, increased water content can cause damage and affect shelf life [4].

### *3.2. Dissolved protein content*

Packaging type, temperature and storage time significantly influence the level of dissolved protein. The dissolved protein content of cheese spread analogues in glass packaging (7.5%) was higher when compared to cheese spread analogues in PP (7.39%) and PET (7.26%) packaging (Fig. 2). This is due to PP and PET packaging having higher permeability, thus absorbing more air from the storage environment when compared to glass packaging. According to [10], glass packaging is impermeable, whereas polypropylene and polyethylene packaging have water vapour permeabilities of 6.8 $g/m^2/24$ hours and 1.3 $g/m^2/24$ hours, respectively. Air contact with cheese spread analogues triggers fat oxidation, causing the fat contained therein to become damaged. This causes cheese spread analogues to become acidic, thereby resulting in protein denaturation [11]. Acid or base reagents can break the intermolecular hydrogen bonds that cause protein coagulation [9]. If the protein is in long-term contact with acids or bases, it is very likely that the peptide bonds will be hydrolysed, which results in the primary protein structure to become completely damaged. Protein damage causes proteins to lose certain properties (e.g. solubility).

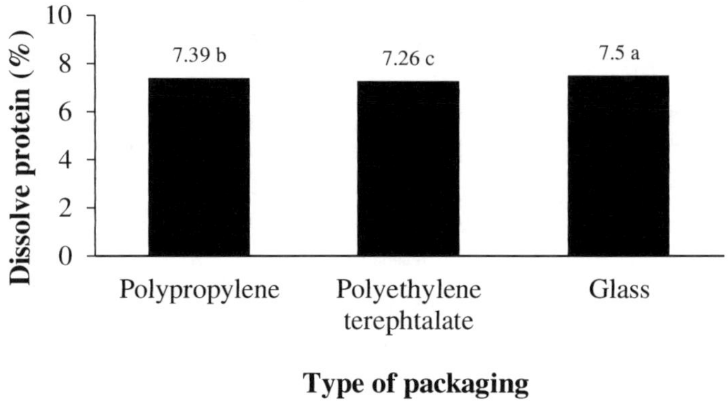


Figure 2. The dissolved protein content of cheese spread analogue stored in three types of packaging

The highest dissolved protein content was observed at a storage temperature of 5°C (7.73%), while the lowest value was observed at 10°C (7.17%) (Fig. 2). This result is due to the higher temperature of more rapid fat hydrolysis and the production of more free fatty acids; therefore, the protein contained in cheese spread analogue will be denatured due to the acidic environment. This is consistent with [9], who stated that the higher the storage temperature, the greater the number of molecules with greater kinetic energy, thus making the particles involved in the reaction move faster. The movement of particles involved in this reaction causes damage to fat and turns into free fatty acids so that the product becomes acidic and causes denatured proteins.

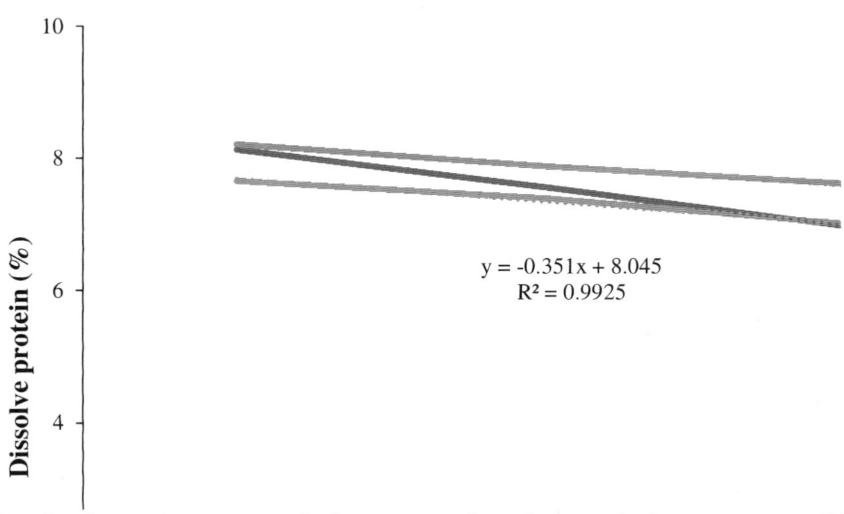


Figure. 3. The dissolved protein content of cheese spread analogues during storage at different temperatures

Cheese spread analogues exhibited decreased levels of dissolved protein during the storage process at all temperatures. This is because protein decomposition and hydrolysis occur during storage, resulting in the levels of dissolved protein decreasing over time. A decrease in the highest levels of dissolved protein (found at -5°C) can be observed from the sharp linear regression graph (Fig. 3). This is presumably because ice crystals will form at a temperature of -5°C, which can damage the protein membrane and cause protein solubility to decrease. According to [12], the freezing process causes damage to the function and structure of the membrane, which is caused by the formation of intracellular ice crystals that can damage protein cells. A decrease in the lowest dissolved protein content occurred at 5°C, where it still had a dissolved protein content of 7.23% at 4 weeks of storage time. This is in accordance with [13], who noted that soft cheeses should be stored at 5-10°C to inhibit and prevent cheese damage from microbial contamination, fat hydrolysis and protein denaturation, which can reduce cheese quality and make cheese storage low.

### *3.3. Fat content*

Storage temperature, storage time and packaging type significantly affected the fat content of cheese spread analogue. The fat content of cheese spread analogue decreased with increasing storage temperature, following a linear regression with the equation $y = -0.02x + 6.33$ (Fig. 4). Thus, higher temperatures increase the potential for fat hydrolysis. With higher storage temperature, the number of molecules having greater kinetic energy makes the particles involved in the reaction move faster. The movement of particles involved in this reaction causes damage to fat and turns it into free fatty acids, resulting in decreased fat content [14].

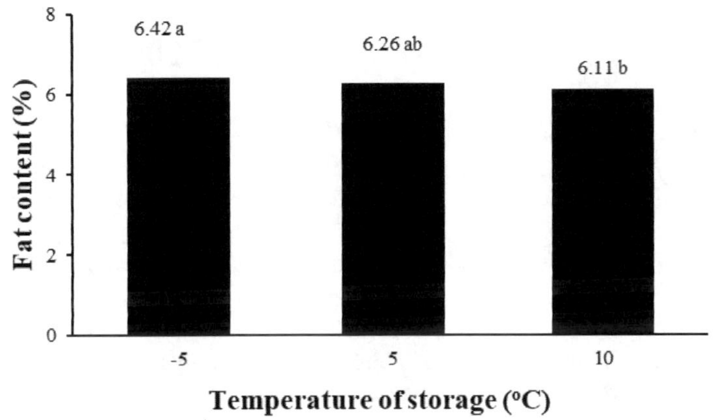


Figure 4. The fat content of cheese spread analogue at different storage temperatures

In all types of packaging, the fat content of cheese spread analogue decreased during the storage process. This is because, during the storage, the fat in cheese spread analogue undergoes hydrolysis, which causes the triglyceride molecular bonds to break and turn into free fatty acids. Moreover, the rupture of triglyceride molecular bonds causes the fat content to decrease and free fatty acid level to rise.

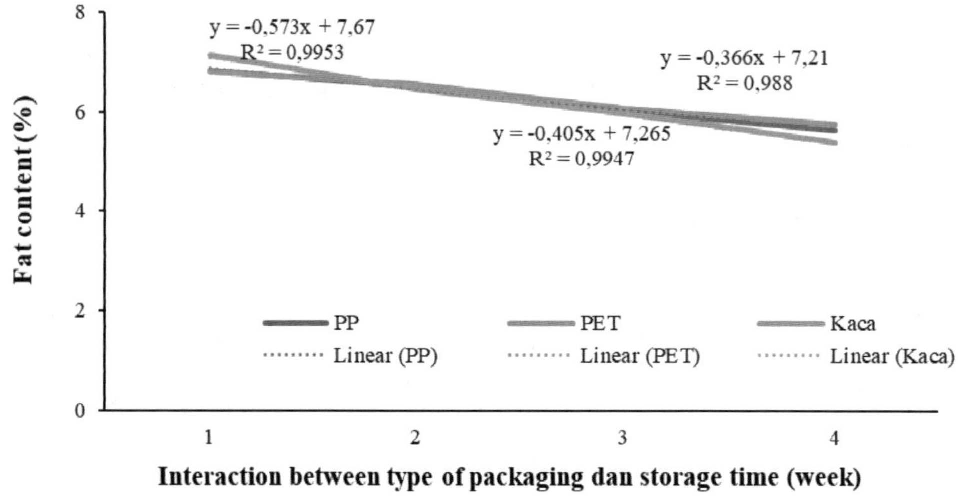


Figure 5. The fat content of cheese spread analogue during storage on several types of packaging

Cheese spread analogue packaged in glass has the highest reduction in fat content compared to other packaging types, which is evident from its linear regression equation (Fig. 5). This is likely because glass packaging is impermeable to air, with trapped air triggering more rapid fat hydrolysis. This results in the fat content of cheese spread analogue packaged in glass being lower than the same product stored in other types of packaging. This is in accordance with [15], who noted that probiotic products packaged in glass bottles have the highest increase in water content compared to other packages since this packaging is airtight and translucent. The air contained in a glass bottle will increase the water content of the product due to light transmitted from outside the packaging. This increased water content will accelerate the reaction of fat hydrolysis. The fat content of cheese spread analogue stored for 4 weeks is lower when compared to the fat content of cheese in general, where 100 grams of cheese contains 20.3% fat [16].

### *3.4. Free fatty acid content*

The highest FFA levels were observed in cheese spread analogue packaged in PET (3.43%), while the lowest was observed in spread cheese analogue packaged in glass (3.39%) (Fig. 6). This is because glass packaging has the ability to prevent the transmission of water vapour to the product better than the other two packaging materials. According to [17], glass packaging has a water vapour transmission rate of <0.01 $g/m^2/day$, which is lower than that of PP (<10 $g/m^2/day$) and PET (100 $g/m^2/day$) packaging. Water vapour absorbed into food products can cause fat hydrolysis, causing fat to break down into free fatty acids.

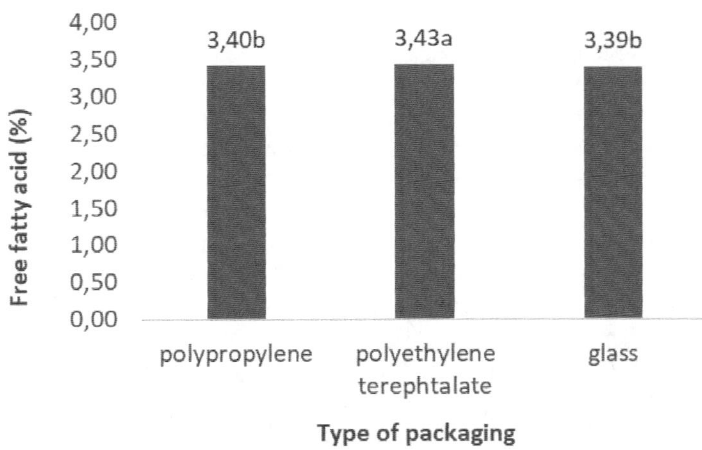


Figure 6. The free fatty acids of cheese analogue using different packaging types

Notably, cheese spread analogue stored at 10°C had higher FFA levels compared to the same product stored at -5 and 5°C. This is because the higher the storage temperature, the faster the chemical reaction occurs, which results in more rapid hydrolysis reactions. This is in accordance with [9], who noted that the higher the storage temperature, the greater the number of molecules having greater kinetic energy, thus making the particles involved in the reaction move faster. Higher temperatures also cause chemical damage reactions in food products to occur faster, while hydrolysis reactions can be accelerated with high temperatures as a catalyst [18].

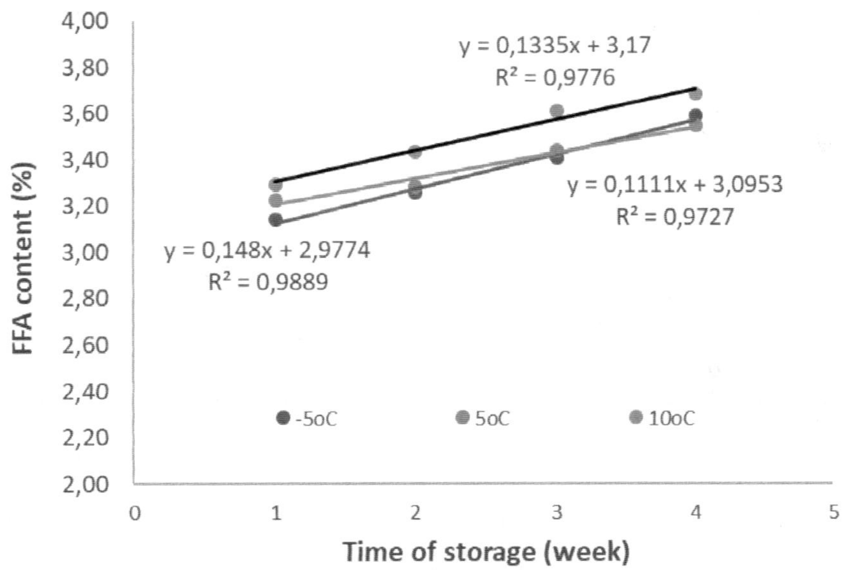


Figure 7. The free fatty acids of cheese analogue during storage at different temperatures

A previous study has noted that the maximum limit of free fatty acids in gruyere cheese products is 4.38% [19]. Based on these standards and the predictions using linear regression in Fig. 7, cheese spread analogue stored at -5, 5 and 10°C can be maintained for 8, 10 and 8 weeks, respectively.

### *3.5. Organoleptic characteristics*

The texture, appearance, colour and flavour of cheese spread analogue decreased during storage. Cheese analogue packed in PP and stored at 10°C (K1S3) exhibited the lowest texture value (4.2) at the end of storage (slightly smooth, lumps present, rather easy to spread). Cheese analogue packed in PET and stored at 5°C (K2S2) exhibited the highest texture value (5.6) at the end of storage (smooth, slightly lumpy, easily spread).

Changes in cheese spread analogue texture during storage are due to protein denaturation, which is marked by the appearance of lumps. This result is consistent with [20], who noted that denaturation can change protein properties. Changes due to denaturation are due to decreased enzyme or hormone activity, salt or acid solubility, crystal formation ability and clot formation stability.

The appearance of cheese analogue also decreases during the storage period. Cheese analogue in PP packaging stored at 10°C (K1S3) had the poorest appearance value (4.3) at the end of storage (slightly mouldy, less attractive, slightly shiny). Cheese analogue stored in glass packaging at 5°C (K3S2) had the best appearance value (5.4) at the end of storage (nearly mouldy, quite attractive, rather shiny).

The growth of fungi during storage is caused by the water content held by corn milk-based cheese analogue, which is relatively high and increases during storage [21]. Notably, high water content can support the growth of microorganism, while water activity in food affects the resistance of food to microbial attack.

The colour of cheese analogue decreased during the storage period. Cheese analogue stored in PP at 10°C (K1S3) had a lower colour value (4.5) compared to other treatments at the end of storage (yellow slightly speckled with green, less attractive, slightly dull). Meanwhile, cheese analogue in glass packaging stored at 5°C (K3S2) had a higher colour value (5.6) compared to other treatments at the end of storage (nearly mouldy, quite attractive, rather shiny).

The aroma of analogue cheese decreased during storage. Analogue cheese stored at 10°C in PET (K2S3) had a lower aroma value (3.9) than other treatments at the end of storage (very acidic, rancid). However, cheese analogue packaged in glass and stored at -5°C (K3S1) and 5°C (K3S2) had the same aroma value (5.4) at the end of storage, which is higher compared to other treatments (slightly acidic, slight cheese aroma).

The most effective packaging type for cheese analogue is glass packaging. At the end of a 4-week storage period, cheese analogue stored in glass exhibited better chemical and organoleptic characteristics when compared to other types of packaging, namely 69.60% moisture content, 6.91% dissolved protein content, 5.4% fat content and an FFA level of 3.65%. The organoleptic characteristics included a fine texture with small lumps, easy to spread, nearly appearing mouldy, bright, attractive, rather shiny, pale yellow colour, and has an acidic odour that is slightly rancid and does not smell like the distinctive aroma of cheese.

The optimum storage temperature for spread cheese analogue is 5°C, which has better chemical and organoleptic characteristics after a 4-week storage period compared to other storage temperatures (69.26% moisture content, 7.23% dissolved protein content, 5.68% fat content and 3.54% FFA level). Its organoleptic properties include a fine texture with small lumps, easy to spread, appearing almost non-mouldy, attractive and rather shiny pale-yellow colour, somewhat bright, and slightly acidic aroma with only a slight cheese aroma.

Analogue cheese remains suitable for consumption after up to 4 weeks of storage, with a water content of 68.71%, dissolved protein content of 7.62%, fat content of 6.50%, FFA content of 3.32% and pH of 4.8. Moreover, its texture is smooth, slightly lumpy and easily spread, with an appearance of being nearly mouldy, attractive and rather shiny, pale yellow, and bright with a slightly acidic aroma and only a slight cheese aroma.

## 4. Conclusion

Glass is the most effective type of packaging for the storage of corn milk-based cheese spread analogue, which has better chemical and organoleptic characteristics after a 4-week storage period compared to other types of packaging. The optimum storage temperature for corn milk-based cheese spread analogue is 5°C, which has better chemical and organoleptic characteristics which after a 4-week storage period when compared to other storage temperatures.

**Acknowledgement**
This research was supported by Jenderal Soedirman University through *Riset Unggulan Terapan* [267/UN23.14/PN/2019].

References

[1] Badem A and Ucar G 2016 *J. Food Dairy Technol.* **4** 44–48
[2] Aini N, Prihananto V, Sustriawan B, Romadhon D and Ramadhan R N 2019 *Int. J. Food Sci.* **2019** 1–8
[3] Aini N, Sustriawan B, Prihananto V and Heryanti T 2019 *IOP Conf. Ser. Earth Environ. Sci.* **255** 012016
[4] Aini N, Prihananto V, Sustriawan B, Astuti Y and Maulina MR 2018 *IOP Conf. Ser.Earth Environ. Sci.* **102** 012049.
[5] Colín-Chávez C, Soto-Valdez H and Peralta E 2014 *Food Packag. Shelf Life* **1** 170–8
[6] Aini N, Prihananto V, Wijonarko G, Astuti Y, Maulina M R, and Muthmainah M 2017 *Adv. Sci. Lett.* **23** 5796–5798
[7] Kowalik J, Lobacz A, Zulewska J and Dec B 2018 *Int. J. of Food Sci. and Tech.* **53** 1541-8
[8] Guinee T P, Feeney E P and Fox P F 2001 *Lait* **81** 475–85
[9] Bubelová Z, Tremlová B, Buňková L, Pospiech M, Vítová E and Buňka F 2015 *J. Food Sci. Technol.* **52** 4985–93
[10] Alves R M V, Van Dender A G F, Jaime S B M, Moreno I and Pereira BC 2007 *Int. Dairy J.*,. **17** 365–73
[11] Oliveira R B A 2016 *Trends in Food Sci. and Tech.* **57** 11–9
[12] Moser P, Telis V R N, de Andrade Neves N, García-Romero E, Gómez-Alonso S and Hermosín-Gutiérrez I 2017 *Food Chem.* **214** 308–18
[13] Ah J and Tagalpallewar G P 2017 *Journal of Food Science and Technology* **54** 3766–78
[14] Akhondzadeh A, Misaghi A and Khaschabi D 2007 *LWT - Food Sci. and Tech.* **40** 973–81
[15] da Cruz A G, Faria J de A and Van-Dender A G F 2007 *Food Res. Int.* **40** 951-56
[16] Dharaiya C N, Jana A H and Aparnathi K D 2019 *J. Food Sci. Technol.* **2019** 1-10
[17] Dixon J 2011 *Packaging materials: 9. Multilayer packaging for food and beverage*. Brussels: ILSI Europe,
[18] Ormanci H B and Colakoglu F A 2017 *J. Food Process. Preserv.* **41** 1–10
[19] Georgala A K, Kaminarides S E and Anifantakis E M 2006 *Aust. J. Dairy Technol.* **61** 26–31
[20] Schenkel P, Samudrala R and Hinrichs J 2013 *Int. J. Dairy Technol.* **66** 220–30
[21] Giri S K, Tripathi M K and Kotwaliwale N 2018 *Journal of Food Science and Technology.* **55** 1667-74.

# DNA Barcoding for mangrove identification

**R. Taufan Harisam[1,2]*, Asrul Sahri Siregar[1], Norman Arie Prayogo[1], Purnama Sukardi[1], Nguyen The Hung[3]**

[1]Fishery and Marine Faculty, Universitas Jenderal Soedirman, Jenderal Soedirman University, Purwokerto 53122, Indonesia
[2]Research and Community Service, Jenderal Soedirman University, Indonesia
[3]Thai Nguyen University of Agriculture and Forestry, Viet Nam

*Corresponding author : taufan.ltd@gmail.com

**Abstract.** Mangroves are distributed in the transition zone between sea and land, mostly in tropical and subtropical areas. They provide important ecosystem services and are therefore economically valuable. Mangrove species identification using traditional taxonomical methods is often burdened with taxonomic controversies. DNA barcoding provides a useful tool for species identification and phylogenetic reconstruction. rbcL and matK are short and unique DNA sequences, and also provide good identification for magrove. This critical review highlights the development of the use of molecular applications that is DNA Barcoding. We focus on observing the development of the use of DNA barcoding in the world, especially on mangroves. Our observations are limited to the use of rbcl and matk markers in some mangrove species in the world. In conjugation with newer and faster techniques such as high-throughout sequencing, rbcL and matK marker can serve as an effective modern tool in mangrove identification.

Keywords : rbcL, matK, DNA barcoding, mangrove

## 1. Introduction

Mangrove forests, in a broad sense, are a community of woody plants that grow in intertidal zones in the tropics and subtropics, which play an important role in maintaining the balance of coastal ecosystem zones (Dan et al, 2016). Mangrove forests are highly productive ecosystems with the same primary production level as tropical forests [1]; [25].

Species identification method living things have evolved from morphological identification to on molecular identification based on short pieces of DNA called "DNA barcode" [11]. DNA Barcoding is a very short standardized DNA sequence that can be used for identification to species level [8]. The concept of DNA barcoding has become one of the most important and significant scientific visions in the past decade. As an effective tool for species level identification, the concept of DNA barcoding has developed rapidly and gained popularity throughout the world. In 2009, Professor Paul Hebert and collaborators on duty at the University of Guelph, Canada, initially the use of DNA Barcoding. Around 2005, The concept of DNA barcoding was introduced into botanical research [3]; [5]. Further research is required to compare DNA barcode fragments to test their efficacy for species identification [14; 11].

Identification of mangroves based on morphological characteristics, such as leaf shape, flower shape, branching stems and root shape is very susceptible to errors [21;19]. There are many mangrove species morphologically that look almost the same, and the distribution area is in almost the same location. It is difficult to identify at the level of mangrove species that have similar morphological forms using external morphological forms of plants, such as *Sonneratia alba*, *Sonneratia caseolaris* (L.), *Sonneratia Hainanesis*, *Sonneratia paracaseolaris* or *Bruguiera sexangula* (Lour.) Poir and *Bruguiera sexangula* (Lour.), and fact it is difficult to understand the evolutionary relationship between mangrove species and traditional classification methods. The development of the world of biotechnology, especially in the molecular field is very rapid, one of which is DNA Barcoding which provides species-level identification results with a high level of accuracy at a relatively affordable price [27].

Technically, the use of DNA barcoding for mangrove identification is a comparison of short DNA sequences compared to sequences that already exist at Gene Bank. Some Gene which is commonly used to identify mangroves are Gene rbcL and matK. In this article, we highlight the development of DNA barcoding, especially the rbcL and matK markers for mangrove identification in the world.

## 2. The rbcl and matk markers: A marker for mangrove identification

Mangroves are unique plants and grow at tidal locations. There are 84 species of 24 genera and 16 families of mangrove plants in the world, including 70 species of 16 genera and 11 families of true plants and 14 species of 8 genera and 5 families of semi-mangrove plants [16]. Conventional mangrove identification often causes errors due to damage to plant parts and differences in perspective. One of the effective molecular markers is DNA barcode. In the last decade, DNA barcoding method has become very important, especially for mangrove identification. Many researchers have begun using DNA barcodes to explore mangroves in several countries. The Used of DNA Barcode markers for reasons of high efficiency and accuracy.

DNA Barcoding is designed to provide accurate identification at the species level through molecular markers based on short standard gene regions [7;18]. In 2009, the CBOL (Consortium for the Barcode of Life) Plant Working Group initially identified and recommended the use of the chloroplast-derived DNA barcode fragments rbcL and matK. The core barcode *matK* locates at the intron region in chloroplast lysine tRNA (*trnK*) gene, and is ~1,550 bp in length, encoding a mature enzyme that involves in type II intron splicing during RNA transcripts [18]. *matK* is a single-copy and one of the fastest evolving genes in protein encoding regions of the chloroplast genome [10]. Rbcl has supported it as a barcode fragment because of its universality, easy amplification and comparability [12]. However, the use of rbcL rarely finds variations at the species level, only at the species level above [13; 20;7;17]. The use of DNA barcoding in mangrove research has been carried out to determine the genetic characteristics of mangroves [18;9]. Further studies on the DNA barcoding especialy on rbcL and matK marker for the mangrove have mainly focused on DNA barcoding analysis and phylogenetic relation of several mangroves in China [8]. Loss of evolutionary unique species in the mangrove ecosystem has been reported and DNA barcoding provided phylogenetic information for developing unified mangrove management plan worldwide [5]. According to Suman [24], in their study rbcL and matK markers provided initial assessment data that would be useful for broader applications of DNA Barcoding in the ecological studies of Mangrove Plants.

Tabel 1. Some important publications on DNA barcoding for mangrove identification (Indexed scopus)

| Serial number | Topics | References |
|---|---|---|
| 1. | Mangroves in Guangdong Province, China | [8] |
| 2. | Mangrove plant species of Visakhapatnam Coast, Andhra, India | [24] |
| 3. | Genetic mutation in mangrove *Acanthus ilifolicius* Coastal Cilacap, Central Java, Indonesia. | [25] |
| 4. | Mangroves from Goa, West Coast, India | [18] |
| 5. | Phylogenetic, Sequence Analysis and Structural Studies of Maturase K Proteins from Mangroves | [26] |
| 6. | Phylogenetic relationships of the Rhizophoraceae in China | [22] |

## 3. Closing and future DNA barcoding research trends

DNA Barcode research for the mangrove field will continue. Molecular mangrove conservation is a challenge in the world of biological conservation. The influence of anthropogenic activity on mangrove ecosystems is increasing every year. There are can be effect for the extent of mangroves to narrow in the future. Rapid and precise identification of mangroves will continue because researchers need them for other fields such as forestry, coastal management, and exploration of wetland areas. Molecular research on mangrove associations still needs to be done. Important discoveries regarding DNA barcoding in sub-tropical and tropical countries are very important to be developed. The challenge of managing mangroves is not easy, especially molecular-based mangrove management to ensure its sustainability.

## 4. Conclution

Research publications on DNA barcoding in the mangrove field are still limited. rbcL and matK are effective markers for mangrove identification, but have not been proven to look for genetic variations at the species level. rbcL and matK markers provide preliminary assessment data that will be useful for broader applications of DNA barcodes in the ecological studies of Mangrove Plants.

## References

[1] Alongi, D. M. 2014. Carbon Cycling and Storage in Mangrove Forests. *Annu. Rev. Mar. Sci.* **6**, 195–219.

[2] CBOL Plant Working Group. A DNA barcode for land plants. Proc. Natl. Acad. Sci. USA 2009, 106, 12794–12797.

[3] Chase, M.W.; Salamin, N.; Wilkinson, M.; Dunwell, J.M.; Kesanakurthi, R.P.; Haider, N.; Savolainen, V. Land plants and DNA barcodes: Short-term and long-term goals. Philos. Trans. R. Soc. Lond. B. Biol. Sci. 2005, 360, 1889–1895.

[4] Dan, X.Q.; Liao, B.W.; Wu, Z.B.; Wu, H.J.; Bao, D.M. 2016. Resources, conservation status and main threats of mangrove wetlands in China. *Ecol. Environ. Sci.* **2016**, *25*, 1237–1243. (In Chinese).

[5] Daru B.H., Yessoufou K., Mankga L.T., Davies T.J. 2013. A global trend towards the loss of evolutionarily unique species in mangrove ecosystems. PLoS ONE. 2013;8(6):e66686.

[6] Fazekas A.J, Burgess KS, Kesanakurti PR, Graham SW, Newmaster SG, Husband BC, Percy DM, Hajibabaei M, Barrett SCH. 2008. Multiple multilocus DNA barcodes from the plastid genome discriminate plant species equally well. PLoS ONE 3:e2802.

[7] Fazekas, A. J. 2008. Multiple multilocus DNA barcodes from the plastid genome diseriminate plant species equally well. *PLoS ONE* **3**, e2802 (2008).

[8] Feng Wu , Mei Li , Baowen Liao, Xin Shi and Yong Xu. 2019. DNA Barcoding Analysis and Phylogenetic Relation of Mangroves in Guangdong Province, China. Forests 2019, 10, 56; doi:10.3390/f10010056

[9] Harisam, R.T., Haryono, F.E.D., Marhaeni, B., Prayogo, N.A., Soedibya, P.H.T. 2018. Genetic mutation in mangrove *Acanthus ilifolicius* base on DNA Barcode (rbcL and matK gen) in the different environment change in coastal Cilacap, Central Java, Indonesia. E3S Web of Conferences **47**, 05005.

[10] Hebert P.D., Cywinska A., Ball S.L. 2003. Biological identifications through DNA barcodes. Proc. Biol. Sci., R. Soc. ;270 (1512): 313–321.

[11] Hebert, P.D.; Cywinska, A.; Ball, S.L.; de Waard, J.R. 2003. Biological identifications through DNA barcodes. Proc. Biol. Sci. 2003, 270, 313–321.

[12] Hollingsworth, P.M. Refining the DNA barcode for land plants. 2011. Proc. Natl. Acad. Sci. USA 2011, 108, 19451–19452.

[13] Kress, W. J. & Erickson, D. L. A two-locus global DNA barcode for land plants: the coding *rbcL* gene complements the non-coding *trnH-psbA* spacer region. 2007. *PLoS ONE* **2**, e508.

[14] Kress, W.J.; Erickson, D.L.; Swenson, N.G.; Thompson, J.; Uriarte, M.; Zimmerman, J.K. Advances in the use of DNA barcodes to build a community phylogeny for tropical trees in a Puerto Rican forest dynamics plot. PLoS ONE 2010, 5, e15409.

[15] Kress, W.J.; Wurdack, K.J.; Zimmer, E.A.; Weigt, L.A.; Janzen, D.H. Use of DNA barcodes to identify flowering plants. Proc. Natl. Acad. Sci. USA 2005, 102, 8369–8374.

[16] Liao, B.W.; Zhang, Q.M. Area, distribution and species composition of mangroves in China. 2014. Wetl. Sci. 2014, 12, 435–440. (In Chinese).

[17]Newmaster, S. G., Fazekas, A. J. & Ragupathy, S. DNA barcoding in land Plants:evaluation of *rbcL* in a multigene tiered approach. *Can. J. Bot.* **84**, 335–341 (2006).

[18] Sadde A.A, Rahul Arvind Jamdade, Kundan Kumar. 2016. Assessment of mangroves from Goa, west coast India using DNA barcode. SpringerPlus 5:1554. DOI 10.1186/s40064-016-3191-4.

[19] Sahu S.K, Kathiresan K. 2012. Molecular Markers: an intricate tool for new insights in mangrove genetics. Int. J. Adv. Biotechnol. Res 3:847e863.

[20]Sass, C., Little, D. P., Stevenson, D. W. & Specht, C. D. DNA barcoding in the Cycadales: testing the potential of proposed barcoding markers for species identification of cycads. *PLoS ONE* **2**, e1154 (2007).

[21] Sheue C.R, Liu HY, Tsai CC, Rashid SMA,Yong JWH, Yang YP. 2009. On the morphology and molecular basis of segregation of *Ceriops zippeliana* and *C. decandra* (*Rhizophoraceae*) from Asia. Blumea 54: 220–227.

[22] Shi, Suhua., , Yelin Huang. 2002. Phylogenetic relationships of the Rhizophoraceae in China based on sequences of the chloroplast gene matK and the internal transcribed spacer regions of nuclear ribosomal DNA and combined data set. Biochemical Systematics and Ecology, 30(4), 309-319.

[23] Subrata Trivedi, Abdulhadi A. Aloufi, Abid A. Ansari, and Sankar K. Ghosh. 2016. Role of DNA barcoding in marine biodiversity assessment and conservation: An update. Saudi J Biol Sci. 2016 Mar; 23(2): 161–171. doi: 10.1016/j.sjbs.2015.01.001.

[24] Suman P, Raj TSD, Vani PR and Sreeramulu SH. 2018. DNA barcoding and biological assessment on a few mangrove plant species of Visakhapatnam Coast, Andhra Pradesh, India. Int J Pharm Sci Res 2018; 9(2): 795-18. doi: 10.13040/IJPSR.0975-8232.9(2).795-18.

[25] Taufan Harisam, Sugiyono, Rudhi Pribadi, Asrul Sari Siregar, Sri Marnani, Dewi Wisudyanti Budi Hastuti, Norman Arie Prayogo. 2018. Acute effects of crude oilfor three common mangrove seedling in segara anakan nature reserve (sanr) cilacap, Indonesia . E3S Web of Conferences 47, 04005 (2018).

[26] Thakar B., Sambhaji., J. Dhanavade, Maruti., D. Sonawane and Kailas. 2017. Phylogenetic, Sequence Analysis and Structural Studies of Maturase K Proteins from Mangroves. Current Chemical Biology, Volume 10, Number 2.

[27] Xu, S., He Z.,Zhang, Z., Guo, Z., Guo, W., Lyu, H., Li, L., Yang, M., Du, Z., Huang, Y.,Zhou, R., Zhong, C., Boufford, D.E., Lerdau, M., Wu, C.i., Duke, N.C., Shi, S. 2017. The origin, diversification and adaptation of a major mangrove clade (Rhizophoreae) revealed by whole-genome sequencing. National Science Review. https://doi.org/10.1093/nsr/nwx065.

[28] Yong Kang, Zhiyan Deng, Runguo Zang and Wenxing Long. 2017. DNA barcoding analysis and phylogenetic relationships of tree species in tropical cloud forests. Scientific Reports volume 7, Article number: 12564.

# Genetic Similarity of Local Mandarin Accessions (*Citrus reticulata*) Resulted from Endosperm Culture According to ISSR and Microsatelite Markers

Mia Kosmiatin[1], Chaireni Martasari[2], Rinanda Gandhi Ningrum Prasetia[3], Frizky Amelia Kurniawati[3], danPrita Sari Dewi[3]*

[1] Indonesian Center for Agricultural Biotechnology and Genetic Resources Research and Development (ICABIOGRAD) Bogor, Indonesia

[2] Indonesian Citrus and Subtropical Fruits Research Institute (ICSFRI) Malang, Indonesia

[3] Faculty of Agriculture, JenderalSoedirmanUniversity, Karangwangkal, Purwokerto Utara 53123, Indonesia

*Corresponding author: p_saridewi@yahoo.com

**Abstract**. Citrus is the third most important fruit commodity in Indonesia, especially seedless citrus. One technique to get seedless citrus is by endosperm culture possessing triploidy sets of the chromosome. Plants that have triploidy sets of the chromosome will usually become sterile or seedless plants. This research aimed to 1) compare ratio of the length and width of the leaves between three types of local mandarin (*Citrus reticulata*) as mother plants and mandarin accessions grown from endosperm cultures, 2) identify mother plant and mandarin accessions from endosperm cultures with Inter Simple Sequence Repeats (ISSR) markers and 3) identify mother plant and mandarin accessions from endosperm cultures with Simple Sequence Repeats (SSR) markers. The research was conducted at Indonesian Citrus and Subtropical Fruits Research Institute (ICSFRI) Batu started from October 2018 until January 2019. The plants used in the study were Garut mandarin and seven accessions grown from endosperm cultures. Identification of plants with morphological markers has several disadvantages due to environmental factors. Identification with molecular markers is needed to complete the morphological information obtained. SSR markers are more suitable to be used in this study compared to ISSR markers because the results are in accordance with the objectives of this study, which is not expected to have high genetic variation. Identification with SSR molecular markers is more suitable because it showed low genetic diversity or no genetic diversity between mother plants and plants endosperm cultures when compared to ISSR markers. The desired change in the plants produced by endosperm cultures is only in ploidy properties.

Keywords: Mandarin; endosperm culture; genetic uniformity; and molecular marker.

## 1. Introduction

Citrus was the third most important fruit commodity in Indonesia after Mango when viewed from crop area and production amount per year. Indonesia's citrus production in 2015 is estimated at 2.40 million tons and continues to increase by 2019 with a production estimate of 2.77 million tons. The average increase of citrus production over the next five years is estimated at 3.64% per year [1]. It was a challenge as well as opportunities for citrus farmers, citrus entrepreneurs, and governments in an effort to increase citrus production [2]. Genetic diversity played an important role in plant breeding because information and understanding of genetic diversity can help the success of crop breeding programmes. The potential nature of a variety should be known in advance so that the diversity of germplasm can be utilized in the variety Improvement Program [3].

Triploid citrus plants are plants whose chromosome number is a multiple of three of their basic chromosomes resulted in a sterile condition. Breeding techniques used to produce triploid citrus plants are crossing techniques between diploid citrus plants with tetraploid citrus plants and endosperm culture techniques. Citrus plants grown from endosperm culture produce triploid citrus plants even though there are plants with other ploidy levels due to several factors. This technique able to produce seedless fruit that can increase economic value without any other character changes such as leaf shape. Local mandarin used in this research is jerukkeprokGarut (*C. reticulata*) which is famous for its noble fruit qualities. Not all morphological characteristics can be used as a solid character to distinguish one group from another because there are traits of plants that are strongly influenced by environmental changes that include nutrition, temperature, humidity, and climate [4]. Molecular approach is needed to determine the diversity of tangerines and differentiate groups of oranges in genetic terms and can be seen through the arrangement of DNA bands that appear.

This research aimed to compare ratio of the length and width of the leaves between three types of mandarin (*C. reticulata*) as mother plants and local mandarin accessions (jerukkeprokGarut) grown from endosperm cultures, identify mother plant and the local mandarin accessions from endosperm cultures with Inter Simple Sequence Repeats (ISSR) markers, and identify mother plant and the local mandarin accessions from endosperm cultures with Simple Sequence Repeats (SSR) markers.

## 2. Materials and Methods

The research was conducted at Plant Breeding Laboratory Indonesian Citrus and Subtropical Fruits Research Institute (ICSFRI) Batu started from October 2018 until January 2019.

Plant materials were the mother plant of the local mandarin (jerukkeprokGarut), GT 12, GT 110, GT 2, GT 113, GT 224, GT 109, and GT 6. The materials used were ISSR

IOP Conf. Series: Earth and Environmental Science **406** (2019) 012019 doi:10.1088/1755-1315/406/1/012019

primers: ISSR 4 (HVH(TCC)5), and ISSR 5 ((TCC)5RY), SSR primers: TAA15 (Forward sequence: GAAAGGGTTACTTGACCAGGC and Reverse sequence: CTTCCCAGCTGCACAAGC), and TAA41 (Forward sequence: AGGTCTACATTGGCATTGTC and Reverse sequence: ACATGCAGTGCTATAATGAATG), distilled water, young leaves of the samples, concentrated HCl, concentrated NaOH, Dream taq (*Thermo Fisher Scientific*), loading dye, TBE 0.5X, agarose, Ethidium Bromide (EtBr), 70% alcohol, alcohol 96% , $ddH_2O$, TE 1X, chloroform isoamyl alcohol (Chisam), cold isopropanol, Sodium acetate, β-Merchaptoethanol, extraction buffer, NaCl, DNA marker (100 bp), CetylTrimethyl ammonium bromide (CTAB) solution, Tris-HCl buffer, Ethylene diamine solution tetraacetic acid (EDTA), and Polyvinyl Pyrulidone (PVP).

The tools used in identifying the genetic diversity of citrus are micro rulers, analytical scales, mortars, pastles, Eppendorf 2 μl tubes, Eppendorf 1.5 μl tubes, types (various sizes), micropipets, erlenmeyers, measuring cups, stirrers, hot plates, vortices, centrifuge, water bath, PCR plate, electrophoresis tank, UV Transilluminator, pH meter, autoclave, freezer, fridge, fume hood, gloves, masks, and sprayer.

The research process consisted of characterization of morphological marker, DNA isolation, quantitative and qualitative analysis of extracted DNA, PCR amplification, and data analysis. Total of seven plants grown from Garut mandarin endosperm culture and Garut mandarin mother plants were measured by the length and width of the leaves. The measurement results are calculated ratio. The ratio obtained can be analyzed using Statistical Package for the Social Sciences (SPSS) software. The results of the analysis of leaf length and width ratio are presented in the form of dendogram. DNA extraction and isolation use the CTAB method to obtain pure DNA. Quality test results using electrophoresis with 0.8% agarose and 0.5X TBE produce DNA with insignificant protein and RNA contamination, so that it can still be used in the PCR process. Quantity test results are performed by comparing DNA obtained with the ladder. The bands resulted from electrophoresis was scored and then analyzed using the Sequential Agglomerative Hierarchial and Nested-Unweighted Pair-Group Method with Arithmetic (SAHN-UPGMA) program in NTSYS software version 2.1. The analysis results were presented in the dendrogram.

## 3. Result and Discussion

### *3.1. Characterization Based on Morphological Markers*

Morphological observations were made by measuring the length and width of the leaves of each citrus accession.

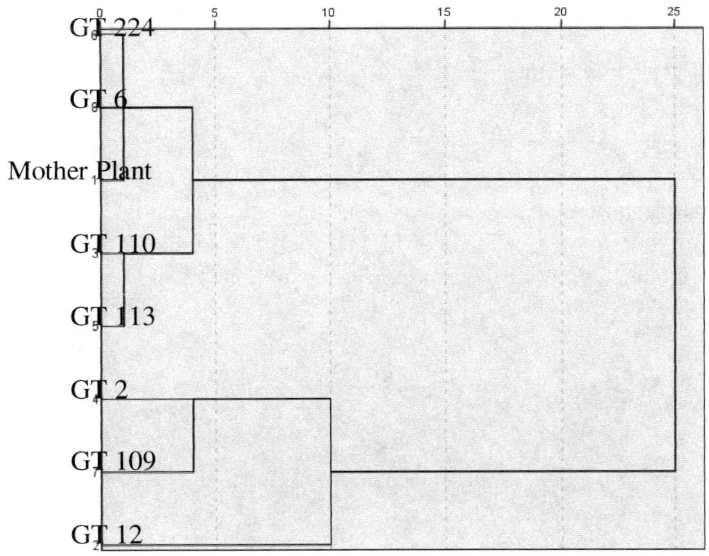


Figure 1. Morphological Dendogram of Garut Mandarins

The ratio of leaf length showed that the Garut mother plant is in one group with the accession of GT 224 and GT 6 because they have the same leaf length and width ratio of 2. This showed that the accession of GT 224 and GT 6 has a kinship with the Garut mother plant. Other accessions of Garut mandarins do not indicate any kinship with the mother plant of Garut mandarin. Bani et al. stated that morphological markers can be influenced by environmental factors such as the temperature and nutrition [5]. Therefore, identification of molecular characters was needed in testing plant diversity to complement morphological information.

### *3.2. Genetic Diversity Based on Microsatellite Markers*

Molecular markers were effective in genetic analysis and have been widely applied in plant breeding programs. The ISSR markers are PCR-based molecular markers, which amplify the area between two short nucleotide (microsatellite) replications [6]. The amount of DNA needed in the ISSR markers is 5-50 ng. The SSR or better known as microsatellite markers are molecular markers consisting of repetition units of 1-6 pairs of DNA bases with high variations. The use of SSR markers is relatively easy, very informative, locus specific, and able to read codominant traits, so that it can be applied for genetic diversity analysis [7].

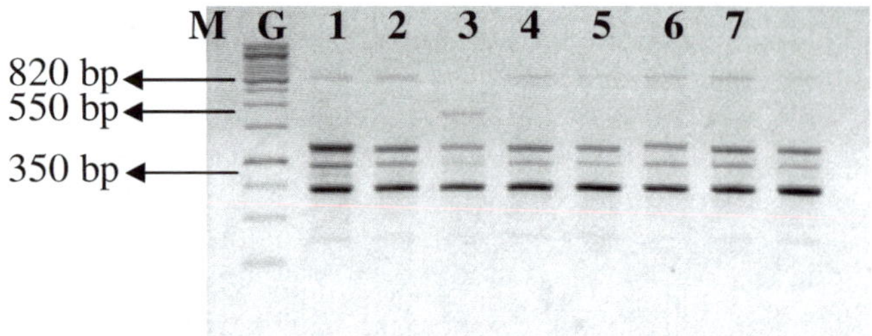


Figure 2.Amplification of the mother plant and the accessions grown from the endosperm cultures by using ISSR 4. M = ladder, G= Mother plant of Garut ,1 = the mother plant (jerukkeprokGarut), 1-7 = the accessions grown from the endosperm cultures, 1= GT 12, 2= GT 110,3= GT 2, 4= GT 113, 5= GT 224,6= GT 109, 7= GT 6.

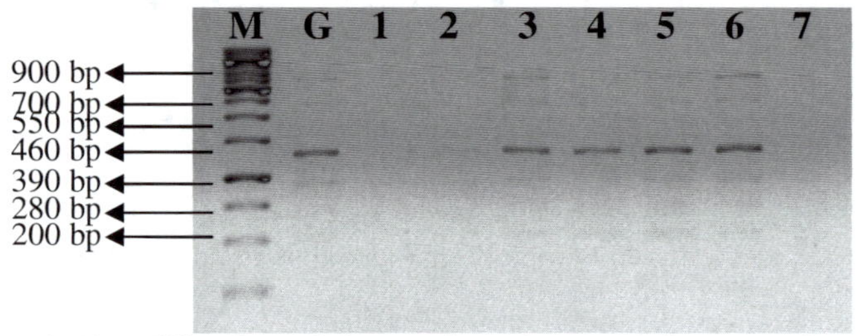


Figure 3. Amplification of the mother plant and the accessions grown from the endosperm cultures by using ISSR 5. M = ladder, G= Mother plant of Garut ,1 = the mother plant (jerukkeprokGarut), 1-7 = the accessions grown from the endosperm cultures, 1= GT 12, 2= GT 110,3= GT 2, 4= GT 113, 5= GT 224,6= GT 109, 7= GT 6.

Visualization of PCR results from Garut ISSR 4 produced 7 bands, 3 polymorphic bands and 4 monomorphic bands. The percentage of polymorphism that formed was 42.85%. Visualization of the results of the PCR Garut tangerine ISSR 5 produces 7 polymorphism bands so that the percentage of polymorphism from the Garut tangerine ISSR 5 formed is 100%. Based on these results it can be concluded that the ISSR markers produced high levels of polymorphism. This is consistent with the opinion of Yasin et al. which stated that ISSR markers produced higher levels of polymorphism and showed a higher level of genetic diversity [8].

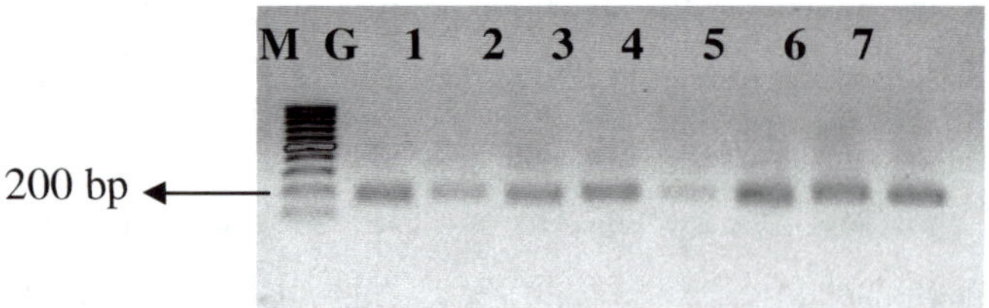


Figure 4.Amplification of the mother plant and the accessions grown from the endosperm cultures by using SSR TAA15. M = ladder, G= Mother plant of

Garut, 1 = the mother plant (jerukkeprokGarut), 1-7 = the accessions grown from the endosperm cultures, 1= GT 12, 2= GT 110,3= GT 2, 4= GT 113, 5= GT 224,6= GT 109, 7= GT 6.

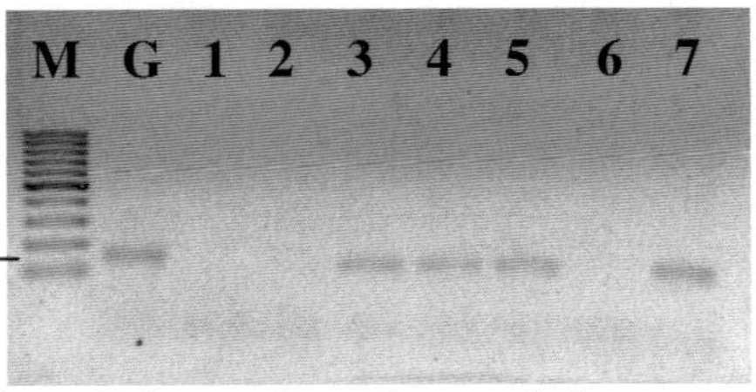


Figure 5. Amplification of the mother plant and the accessions grown from the endosperm cultures by using SSR TAA41. M = ladder, G= Mother plant of Garut , 1 = the mother plant (jerukkeprokGarut), 1-7 = the accessions grown from the endosperm cultures, 1= GT 12, 2= GT 110,3= GT 2, 4= GT 113, 5= GT 224,6= GT 109, 7= GT 6.

The results of PCR amplification and visualization of PCR results showed that the primary SSR TAA15 and TAA41 primers mostly produced monomorphic and uniform banding patterns. Suryanto states that samples that are not amplified with the primers used can be caused by several things such as primary mismatches, efficiency, and optimization of the PCR process [9].

The diversity between plant accessions can be seen from the number of bands and the thickness of DNA bands produced [8]. Genetic diversity or polymorphism has the meaning that there are two or more alleles at a locus in the population. Genetic diversity can be observed by observing through genetic characters with observed characteristics in the form of DNA, which is difficult to be influenced by the environment [10].

Molecular characterization results ISSR and SSR that have been presented show that molecular characterization with ISSR produces high genetic diversity. This can be seen based on the polymorphism band formed in the visualization of PCR results. Characterization with molecular SSR does not produce high genetic diversity. This is because the band formed in the visualization of PCR results is monomorphic.

Data analysis of PCR results visualization from ISSR and SSR primers was performed using the NTSYS 2.1 application. Tape scoring is done based on the formed band, the visible ribbon is given a value of 1 while the non-visible ribbon is given a value of 0.

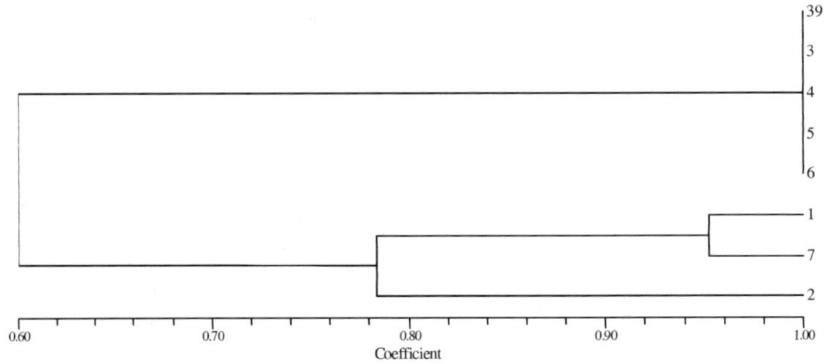


Figure 6.Dendogram of Garut mandarins with primer ISSR. 39= Mother Plant of Garut, 1= GT 12, 2= GT 110,3= GT 2, 4= GT 113, 5= GT 224,6= GT 109, 7= GT 6.

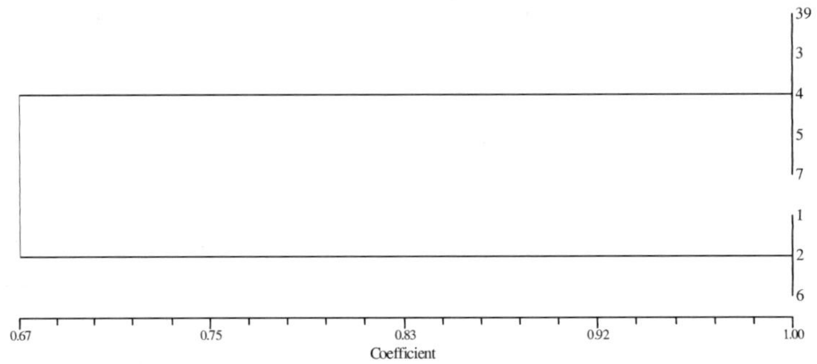


Figure 7.Dendogram of Garut mandarins with primer SSR. 39= Mother Plant Garut, 1= GT 12, 2= GT 110,3= GT 2, 4= GT 113, 5= GT 224, 6= GT 109, 7= GT 6.

Dendogram analysis on Garut mandarin with ISSR primers at a coefficient of 0.80 resulted in three groups. The first group is the mother plant of Garut, GT 2, GT 113, GT 224, and GT 109. The second group is the accession of GT 12 and GT 6. The third group is the accession of GT 110. Dendogram analysis results on Garut mandarins with SSR primers on the coefficient 0.83 there are two groups. The first group is the mother plant of Garut, GT 2, GT 113, GT 224, and GT 6. The second group is the accession of GT 12, GT 110 and GT 109.

Most of the mandarins produced by endosperm are in the same group as their parent plants. This shows that the mandarin plants resulting from endosperm culture produce low genetic diversity or do not produce genetic diversity. Desired changes in plants resulting from endosperm culture are only in the ploidy nature. Endosperm mandarins are expected to produce fruit without seeds (seedless). Kosmiatin and Ali explain the use of triploid plants to be important because they do not involve genetic modification, so they are more easily accepted by the market. The endosperm culture technique is the most efficient technique for getting seedless citrus. The regeneration system has been mastered and segregation is low so that the chances of getting the same seedless oranges with the parents are higher [11]. Based on the dendogram and the previous explanation, SSR markers are more appropriate to be used in this study because it is seen from the visualization of the results of the PCR and the dendogram results of data analysis show a low level of genetic diversity. ISSR markers are included in dominant inheritance, while SSR markers are included in codominant inheritance. Purnomo and Rejekistate that codominant markers data are generally more precise than dominant one [12].

## 4. Conclusion

a. Identification of plants with morphological markers has several weaknesses due to environmental factors. Identification with molecular markers is needed to complete the morphological information obtained and more accurate the results.
b. Identification with molecular markers SSR is more suitable because it shows a low genetic diversity or no genetic diversity between the mother plants and plants resulting from endosperm culture when compared with the ISSR markers.

**Acknowledgement**

We are grateful and also thank to Director of Indonesian Citrus and Subtropical Fruits Research Institute (ICSFRI) Malang, East Java.This study was supported by Ministry of Agriculture of Indonesia.

**References**

[1]Pusat Data danSistemInformasiPertanian. 2015. *Outlook KomoditasPertanianSubsektorHortikulturaJeruk*. Jakarta. http://epublikasi.setjen.pertanian.go.id/epublikasi/outlook/2015/Hortikultura/Outlook%20Jeruk%202015/files/assets/common/downloads/Outlook%20Jeruk%202015.pdf

[2] Soelarso, R. B. 1996. *BudidayaJerukBebasPenyakit*. Kanisius, Yogyakarta.

[3]Nurdianawati, Safarinda, N.Wicaksana, danAnas. 2016. Analisiskesesuaianmarka SSR (*Simple Squence Repeats*) untukidentifikasikeragamangenetikpadakacangbambaraasalJawa Barat. *JurnalAgrikultura*. 27(2): 120-123. http://jurnal.unpad.ac.id/agrikultura/article/view/10528

[4]Hardianto, E. Mujiarto, dan E. S. Sulasmi. 2007. Kekerabatan genetik beberapa spesies jeruk berdasarkan taksonometri. *Jurnal Hortikultura*. 17 (3): 203-216.http://ejurnal.litbang.pertanian.go.id/index.php/jhort/article/view/1006

[5]Bani, P. W, B. S.Daryono, danPurnomo. 2017. Penandamolekuler*Inter Simle Sequence Repeat*untukmenentukanketahanantanamanjagungterhadappenyakitbulai. *JurnalFitopatologi Indonesia*. 13(4): 127-135. http://journal.ipb.ac.id/index.php/jfiti/article/view/18660/13170

[6]Yulita, K. S, F. Ahmad, dan D.Martanti. 2014. Analisiskeragamangenetikkentanghitam(*Plectranthusrotundifolius* (Poiret) Sprengel) berdasarkanmarka RAPD dan ISSR. *BeritaBiologi*. 13(2): 127-135. https://pdfs.semanticscholar.org/8850/442b0f4eeb5e42ca198f84bb9eea60c0490c.pdf

[7]Hafizah, R. A., R.Adawiyah, R. M.Harahap, S. Hannum, dan P. JSantoso. 2018. Aplikasimarka SSR padakeragamangenetik durian (*Duriozibethinus*Murr) di Kabupaten Deli Serdang, Sumatera Utara. *Journal of Biology*. 11(1): 49-56. http://journal.uinjkt.ac.id/index.php/kauniyah/article/view/5668

[8]Yasin, M., D.Septiadi, N.Kendarini, dan D.Agisimanto. 2017. Keragamangenetikhasilaplikasikolkhisinpadatanamanjeruksiam cv. Pontianak (*Citrus nobilis*) secaramorfologidanmolekuler. *J. ProduksiTanaman*. 5(11): 1835-1844. http://protan.studentjournal.ub.ac.id/index.php/protan/article/view/577

[9]Suryanto, D. 2003. *MelihatKeanekaragamanOrganismeMelaluiBeberapaTeknikGenetikaMolekular. Skripsi*.Universitas Sumatera Utara.

[10]Langga, I. F., M. Restu., danT. Kuswinanti. 2012. Optimalisasisuhudan lama inkubasidalamekstraksi DNA tanamanbitti (*Vitexcofassus*Reinw) sertaanalisiskeragamangenetikdenganteknik RAPD-PCR. *JurnalSainsdanTeknologi.* 12(3): 265–276. http://pasca.unhas.ac.id/jurnal/files/0b939e0181da0eccc16cd29267c1545a.pdf

[11]Kosmiatin, M. dan A.Husni. 2018. Perakitanvarietasjeruktanpabijimelaluikonvensionaldannonkonvensional. *JurnalPenelitiandanPengembanganPertanian.* 37(2): 91-100. http://ejurnal.litbang.pertanian.go.id/index.php/jppp/article/view/9673

[12]Purnomo, E.,dan R. S.Ferniah. 2018. Polimorfismecabairawitdancabaigendotdenganpenanda RAPD (*Random Amplified Polymorphic DNA*). *BerkalaBioteknologi.*1(1): 1-5. https://ejournal2.undip.ac.id/index.php/bb/article/view/2176

# Composite Image with a Geographic Information System Approach

**Arief Kelik Nugroho[1)], Ipung Permadi[2)]**

[1,2]Department of Informatics, Faculty of Engineering,
Universitas Jenderal Soedirman, Indonesia

Email: [1]arief.nugroho@unsoed.ac.id ,[2] ipung.permadi@unsoed.ac.id

**Abstract.** Geographic information systems are used for spatial data processing. Data processing results from the system can be used as a tool to determine the condition of an area. By using several components one of which is coloring in geographical conditions in the region. The process of combining several colors to form a composite image that can show the clarity of the object. The results obtained from this study reveal the clarity of objects in the Banyumas Regency region by using levels in the band composition in spatial imagery

**Keywords:** *GIS,level,band,image,composite.*

## 1. Introduction

Geographical information systems can be utilized by system users to deal with geographic or geographic issues. Geography has geographic relations, including the surface of the earth's landscape. This section explores relations and characteristics, with computer support used to analyze the image so that the specificity of the image can be sharpened, so users can find out information from the image being processed. The information obtained results from the creation, handling, and storage so that it can be used. Composite image is a combination of multiple channel images. Compilation of composite images is intended to obtain a better visual picture. So that the object observation and sample selection can be improved. RGB is a technique or a way to display a composite image, in one image display there are three single images. RGB or Red Green Blue, or Red Green Blue is used as a technique for displaying images because the three colors are the three basic colors. And the combination of these three colors produces other colors. The difference in color in the composition of this composite image makes it easy to be able to see the appearance of different objects and can distinguish various objects clearly. Because each channel in the image has a different wavelength and shows a different color .

The multispectral system produces coverage of regional images recorded on various spectral channels (bands). With this system, an area recorded on n channels will produce n images, each of which has a different spectral variation. Each channel is actually sensitive to certain features, according to the basis for choosing the width and interval spectrum as in the spectral reflection curve from laboratory observations. Sensitivity to certain features (eg variations in vegetation density) logically also contains insensitivity to other features (eg differences in soil color). However, if channels with wide wavelength intervals are used, the overall appearance is not well differentiated. Thus, to be able to distinguish various types of objects properly, several channels are used together. Information on an 8-bit basis is stored in bytes. A byte is a unit of information consisting of 8 bits.

## 2. Methods

To get the image from the color composition, the following method is used :
Apply image compression, ie by compressing the pixel values of the original range (for example 0-15, 0-32, etc.) to 0-5, on all spectral paths involved then determine the channels that are colored red (and then all values the pixels are called new-red $NP_s$ , which ones are green ( new-green $NP_s$ ), and which ones are blue ( new-blue $NP_s$). Apply pixel values to composite images (NP comp ) based on new-red NP, new-green NP, new-blue NP. Present a composite image to the monitor screen with predefined color palette guidelines for composite pixel values[1].

## 3. Basic Theory

Information on an 8-bit basis is stored in bytes. A byte is a unit of information consisting of 8 bits. For an 8-bit system (= 1 byte) , each data (pixel) is stored in separate bytes. In other words, every 1 pixel will be saved as 1 byte. A value of 1 kilobyte (1KB) is equal to 210 = 1024. If an image consists of 500 columns and 1200 pixel rows, 600,000 bytes are needed. Image-storage systems with row-columns (raster / tesrelation systems) are stored with clear and consistent addresses according to position in rows and columns[1].

### *3.1. Sequential Band (BSQ)*

Image with the sequential band format (BSQ) generated from each channel is saved as a separate file or file. The data storage sequence is carried out from the first line of channel 1, second line, third line, last line . Data is stored as channel 1 file and the process is carried out from the first row to the second channel until the last row.

| | | | | |
|---|---|---|---|---|
| 10 | 11 | 20 | 40 | 1 |
| 12 | 2 | 5 | 7 | 10 |
| 2 | 3 | 2 | 2 | 2 |
| 10 | 14 | 71 | 99 | 81 |
| 0 | 5 | 42 | 36 | 15 |

**Figure 1**. Channel 1

| | | | | |
|---|---|---|---|---|
| 2 | 14 | 32 | 30 | 1 |
| 18 | 4 | 2 | 8 | 14 |
| 3 | 4 | 4 | 3 | 3 |
| 20 | 22 | 33 | 44 | 55 |
| 51 | 50 | 45 | 31 | 14 |

**Figure 2**. Channel 2

| | | | | |
|---|---|---|---|---|
| 70 | 30 | 30 | 31 | 80 |
| 15 | 64 | 65 | 32 | 81 |
| 75 | 70 | 70 | 70 | 32 |
| 23 | 20 | 11 | 10 | 5 |
| 0 | 0 | 0 | 4 | 42 |

**Figure 3**. Channel 3

**Table 1**. BSQ Storage Format

| **BSQ format** | **Pixel value** |
|---|---|
| Channel 1 | 10 11 20 40 1 12 2 5 7 10 2 3 2 2 2 ... 15 |
| Channel 2 | **2** 14 32 30 1 18 4 2 8 14 3 4 4 3 3 ... 14 |
| Channel 3 | **70** 30 30 31 31 80 15 64 65 32 81 75 70 70 70 .. 42 |

*3.2. Band Interleaved By Line (BIL)*

The storage process is carried out from the first row of channel 1 followed by row 1 to channel 2, first line to channel n followed by line 2 to channel 1, line 2 to channel 2, line 2 to channel n and so on until image *n*. The BIL format for all image data in *n* will be saved as one file.

**Table 2.** BIL Storage Format

| BIL Format | Pixel value |
|---|---|
| Channels 1,2,3; Row 1 | 10 11 20 40 1 2 14 32 30 1 70 30 30 31 31 80 |
| etc. | |

*3.3. Band Interlayed By Pixel (BIP)*

Storage is carried out from pixel 1 line to 1 channel, pixel 1 line 1 channel 2, ... pixel 1 last line channel 1, last pixel last line channel 2 and so on.

*3.4. Run-Length Encoding (RLE) and Block Encoding*

Storage is done by re-expressing the number of consecutive pixels with the same value as one value pair. Block encoding is a compression method that resembles RLE but its application is dimensional (not just along lines).

*3.5. Landsat Band 8*

Image as raw data is processed separately in each band, image processing requires conversion and merging of bands into image data sets so that an image is composed of a combination of several bands which is then called a composite image.

**Table 3.** Landsat Composite

| Closing Type | Combination of Spectral Channels |
|---|---|
| The waters | Bands 1, 5 & 8 / Bands 2, 3 & 4 |

IOP Conf. Series: Earth and Environmental Science **406** (2019) 012020 doi:10.1088/1755-1315/406/1/012020

| | |
|---|---|
| Settlement | Bands 2.5 & 8 |
| Agriculture | Bands 2, 3 & 4 |
| Forest | Bands 2, 5 & 8 |
| Salt | Bands 2, 3 & 4 |
| Dry Vegetation | Bands 2.5 & 8 |
| Irrigated Vegetation | Bands 2, 5 & 8 |

## 4. Result And Discussion

### *4.1. Image pre-processing*

Pre-processing is done to prepare the image data as initial processing before processing more images. This procedure aims to correct the image data that is experiencing distortion or error into its original form. There is work being done at the stage of pre-processing of the image are: correction of geometric and correction radiometric. Landsat 8 image data generally has been corrected geometrically can be seen from its parallelogram shape in accordance with the direction of motion recording vehicle. While the radiometric correction needs to consider the objectives, the segmentation and majority method actually requires the original image so as to facilitate the segmentation of recorded objects. In pre-processing often encounters an image mosaic stage because the study area is on two different recording sheets, as the Banyumas district is in path 120 raw 65 and path 121 raw 65. This condition requires that the interpreter combines the two sheets after being united will be cut again according to the study area.

Figure 4 is two different landsat images, while the study area is in both images so it needs to be mosaics.

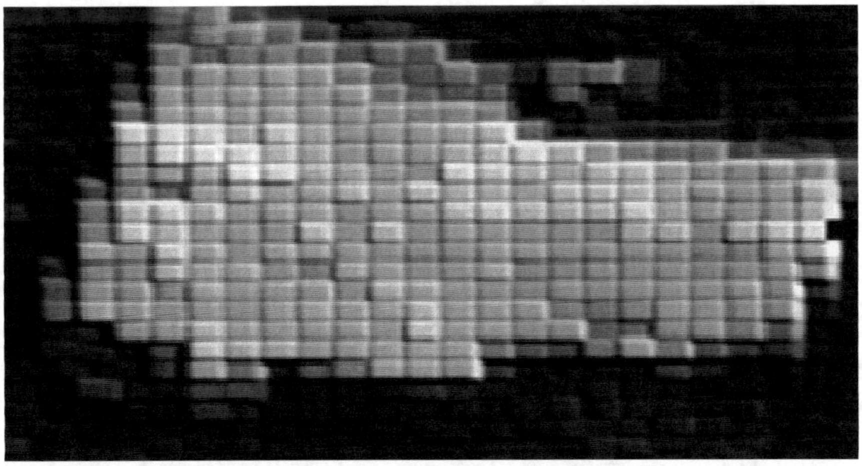

**Figure 4**. Path 120 raw 65 and path 121 raw 65 (Landsat image 8)

Figure 5 below is the result of a mosaic of two images.

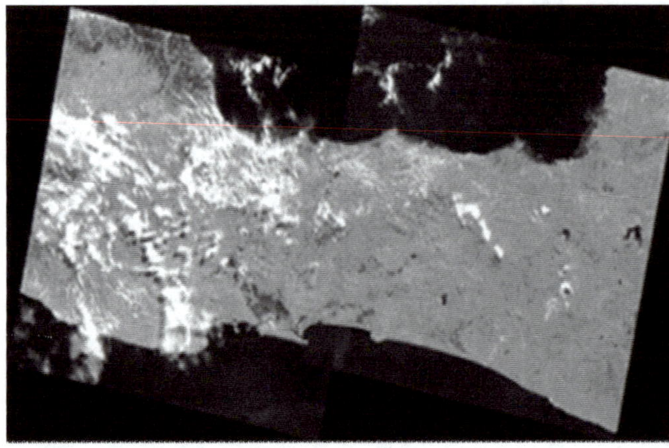

**Figure 5**. Mosaic Landsat Image 8

Limitations on the ability of the hardware to work will greatly affect the success of the work process, one of which is caused by very large data memory with limited hardware capacity.
Limitation of image coverage in accordance with the study area will reduce the workload, in other words this procedure is an attempt to select data for further processing in research analysis and ignore data not included in the study area.

### 4.2. *Image Composite*

The composite image used is the color composition of the band that is able to highlight the geomorphological theme . Noting the image used is Landsat 8 image, with 11 bands owned making it easier to do color composites. Identification for the theme emphasizes how to distinguish basic objects such as water, soil and vegetation. These three basic objects each have sensitivity in each band so it is necessary to avoid bands 1, 2 and 3 because the band will be able to obscure the basic information on Landsat 8 images that are classified as medium scale. Bands 1, 2 and 3 each have their expertise on water objects, settlements and vegetation so that the results of visualization have not been able to characterize the object specifically.
Figure 6 below shows the composite landscape view 467 used.

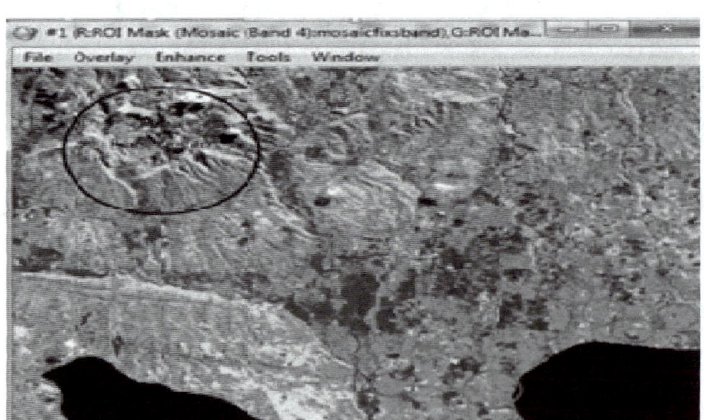


**Figure 6**. Composite 467 Landsat image 8

Composite recommendations for identification on Landsat 8 imagery so that it gives rise to the uniqueness of the color of objects and the distinct difference is composite 467. This composite presents three main objects expressly, vegetation is represented in green and its gradation shows the condition of vegetation where dark green indicates dense or healthy vegetation and light green indicates needle-leafed vegetation or grassland.

**Figure 7**. Image of a part of Banyumas ( bright green agriculture )

**Figure 8.** Image of a portion of the Banyumas area ( Vegetation: bright red)

Band ratio is used to produce certain effects in relation to the spectral aspect of the vegetation, the reduction of the shadowing effect, and the lithology. Through this material, a new image with pixel values is generated which is the quotient of the pixel value of channel A and channel B.

An image can emphasize the aspect of vegetation density, especially for the infrared channel close to the red channel. Besides that, in general, penis material can suppress the effects of shadows, for example on vegetated slopes.

**Figure 9.** Image of a portion of the Banyumas area ( Vegetation: Bright green)

In addition to image sharpening, there are still other transformations that are often used to produce new information. This transformation can be grouped into two, namely: a) Transformations that can sharpen certain information, but at the same time eliminate or suppress other information; and b) Transformations that 'summarize' information by reducing the dimensionality of data. Unlike the case with various sharpening algorithms, this particular transformation operates more on the spectral dominant. Another feature is that in many cases, this transformation involves several spectral channels at once.

**Figure 10**. Image of a portion of the Banyumas region (humidity)

Light blue indicates rice fields whereas settlements and land objects are visually dominant in red color and gradation. The following is an example of the result of calculating the number of pixel counts based on channel 1 in the table 1 below.
Value of new pixels resulting from contrast stretching.

**Table 4**. Pixel values from channel 1

| Pixel value | | | | | | | | | |
|---|---|---|---|---|---|---|---|---|---|
| 2 | 26 | 0 | 2 | 0 | 2 | 5 | 9 | 12 | 0 |
| 36 | 33 | 2 | 0 | 2 | 2 | 9 | 7 | 0 | 0 |

IOP Conf. Series: Earth and Environmental Science **406** (2019) 012020 doi:10.1088/1755-1315/406/1/012020

| | | | | | | | | | |
|---|---|---|---|---|---|---|---|---|---|
| 29 | 36 | 2 | 0 | 2 | 9 | 12 | 12 | 2 | 12 |
| 31 | 36 | 5 | 0 | 0 | 19 | 17 | 21 | 0 | 9 |
| 31 | 29 | 5 | 5 | 0 | 0 | 19 | 19 | 2 | 14 |
| 5 | 15 | 7 | 5 | 2 | 0 | 0 | 19 | 12 | 12 |
| 7 | 36 | 5 | 17 | 5 | 2 | 0 | 2 | 9 | 14 |
| 17 | 17 | 17 | 17 | 23 | 23 | 2 | 0 | 12 | 12 |
| 21 | 14 | 19 | 21 | 23 | 26 | 36 | 33 | 0 | 2 |
| 12 | 9 | 9 | 26 | 23 | 26 | 33 | 31 | 2 | 2 |

The result of the composite in a channel 2 represented in table 5 .

**Table 5.** Pixel values from channel 2

| Pixel value | | | | | | | | | |
|---|---|---|---|---|---|---|---|---|---|
| 2 | 34 | 0 | 0 | 0 | 0 | 5 | 5 | 5 | 0 |
| 36 | 31 | 2 | 0 | 0 | 2 | 5 | 2 | 0 | 2 |
| 36 | 36 | 2 | 0 | 0 | 5 | 7 | 5 | 0 | 7 |
| 34 | 34 | 0 | 0 | 2 | 14 | 14 | 17 | 0 | 5 |
| 31 | 34 | 2 | 0 | 0 | 0 | 14 | 2 | 0 | 10 |
| 2 | 31 | 2 | 2 | 2 | 0 | 0 | 14 | 7 | 5 |
| 0 | 2 | 0 | 2 | 0 | 2 | 2 | 5 | 5 | 7 |
| 14 | 17 | 12 | 17 | 26 | 29 | 2 | 2 | 7 | 5 |
| 17 | 7 | 12 | 19 | 29 | 31 | 29 | 34 | 0 | 2 |
| 7 | 5 | 5 | 29 | 29 | 34 | 29 | 34 | 0 | 2 |

The result of the composite in a channel 3 represented in table 6 .

**Table 6**. Pixel values from channel 3

| Pixel value | | | | | | | | | |
|---|---|---|---|---|---|---|---|---|---|
| 0 | 33 | 0 | 0 | 0 | 0 | 31 | 29 | 31 | 0 |
| 33 | 31 | 0 | 0 | 0 | 0 | 24 | 2 | 0 | 0 |
| 31 | 33 | 0 | 0 | 0 | 33 | 31 | 31 | 0 | 24 |
| 27 | 31 | 0 | 0 | 0 | 22 | 20 | 18 | 0 | 33 |
| 24 | 29 | 0 | 2 | 2 | 2 | 18 | 0 | 0 | 22 |
| 2 | 29 | 0 | 2 | 0 | 2 | 0 | 15 | 24 | 33 |
| 0 | 0 | 0 | 2 | 0 | 0 | 0 | 33 | 33 | 31 |
| 20 | 18 | 18 | 15 | 31 | 33 | 2 | 0 | 29 | 33 |
| 18 | 24 | 15 | 24 | 27 | 31 | 29 | 27 | 0 | 0 |
| 27 | 33 | 33 | 31 | 33 | 31 | 29 | 27 | 0 | 2 |

## 5. Conclusion

The image as raw data is processed separately in each band, by using the band's color composition it can highlight the contents of the image, (for example land) and provide basic object information on the image. In utilizing spatial imagery as a data source to be displayed on a geographic information system, it is very influenced by the complexity of the objects stored in the image. The results obtained from this study reveal the clarity of objects in the Banyumas Regency region by using levels in the band composition in spatial imagery.

## 6. References

[1]. Danoedoro, P.(2002).Pedoman Praktikum Pemrosesan Citra Digital. Yogyakarta: Fakultas Geografi Universitas Gadjah Mada.
[2]. Tsou, M H. Smith, J.(2011).Free and Open Source Software for GIS Education. Department of Geography:San Diego State University.
[3]. Palma, F. Amato, F. Nolè, G. Martellozzo, F. Murgante, BA. (2016).SMAP Supervised Classification of Landsat Images for Urban Sprawl Evaluation. *ISPRS Int. J. Geo-Inf.* ()
[4]. http://landsatapp.s3-website-us-west-2.amazonaws.com (accessed on 1 November 2018)
[5]. Correia,R. Duarte,L.Teodoro,AT.Monteiro,A.Processing Image to Geographical Information Systems (PI2GIS)—A Learning Tool for QGIS. *Educ. Sci*
[6]. Bunch,M T. Kumaran, T V.Using Geographic Information Systems (GIS) For Spatial Planning and Environmental Management in India: Critical Considerations. *International Journal of Applied Science and Technology,*2(2),40-54.
[7]. *Warren, S. (2004).The Utopian Potential of GIS.Cartographica, 39(1):5-16.*
[8]. Pantazis, G.(2017).Multi Parametric Evaluation of Back Propagation Artificial Neural Network in Determination of Geoid Undulations. *International Journal of Applied Science and Technology*,7(2),38-45.

IOP Conf. Series: Earth and Environmental Science **406** (2019) 012021 doi:10.1088/1755-1315/406/1/012021

# Optimum frequency identification of anomalous geomagnetic signal related to earthquake precursor in Sumatra Island

**I Sokacana[1], M S Rosid[1,*], and S Ahadi[2]**

[1]Departement of Physics, FMIPA Universitas Indonesia, Depok 16424, Indonesia
[2]Laboratory of Potential Geophysics and Time, Meteorology Climatology and Geophysics Agency, Jakarta, Indonesia

*Email: syamsu.rosid@ui.ac.id

**Abstract.** Many scientific studies has detected ULF (ultra-low frequency) emission before an earthquake. An earlier study of Loma Prieta 7.1 M earthquake in 1989 shows ULF anomaly ranged between 0.01-10 Hz before an earthquake. A similar study of Kushiro earthquake 6.4 Mw shows that the ULF anomaly can be detected at a frequency range of 0.022-0.1 Hz. Furthermore, a study conducted in Sumatra Island from 2007-2012 shows the ULF anomaly at 0.01-0.06 Hz. Newer study on the 2009 Padang earthquake and Mentawai 2010 earthquake shows ULF emission between 0.012-0.022 Hz. Based on the aforementioned researches, a study is conducted on frequency range of 0.01-0.09 Hz to identify the optimum ULF frequency. Study conducted in this case uses magnetic diurnal variation data from Sumatra region from 2017 to 2018, especially earthquakes recorded at Gunung Sitoli (GSI), and Sicincin (SCI) stations. Previous studies above provide an overview of ULF ranges associated with earthquakes. Based on all above, an analysis is conducted to find the precursors using data of past earthquake events by utilizing Fast Fourier Transform to transform time-domain data to frequency domain, bandpass filters to eliminate noise, power ratio of *Z/H* to determine the time window, and Single Station Transfer Function to determine the location window. Lastly, moving average is used to assist determination of time window. Magnetic anomalies between 0.01-0.03 Hz, with amplitude exceeding monthly average, were observed before the earthquakes. Analysis in this frequency range shows occurence of earthquake 5-6 days after geomagnetic anomalies were detected. This study shows that location window prediction is possible with 100% accuracy. Therefore, 0.01-0.03 is currently the optimum frequency for precursor analysis.

## 1. Introduction

At present, earthquake precursors have been better studied. An earthquake can be predicted within the range of approximately 30 days from the detected anomaly, and allows prediction of the location of the epicenter of the earthquake to the prediction of magnitude. A study of collaborative monitoring results of non-seismic earthquake precursors by BMKG and Kyushu University using the power ratio polarization method of *Z/H* components at frequencies of 0.012 Hz and 0.022 Hz geomagnet [1].

However, in the end, the damage potential due to earthquakes remains inevitable, this event can be "predicted" to reduce the impact of damage that occur and save as many human lives as part of natural disaster mitigation.

## 2. Materials and Methods

### *2.1 Transforming magnetic diurnal variaton into frequency domain*

There are 5 (five) interesting Sumatra's earthquake magnetic anomaly data in which are in the range of 3.904°N to 6.319°S and 94.087°W to 101.250°E. The earthquakes occurred on 12 July 2016, 21 November 2017, 25 May 2017, 13 June 2018 and 25 June 2018. The data is a daily magnetic variation per second that is transformed using Fast Fourier Transform (FFT) into the frequency domain. However, only 3 (three) of the earthquakes of November 21st, 2017, June 13th, 2018, and June 25th, 2018 will be analyzed and interpreted in this study as the others two are showing too much noise disturbance.

**Table 1.** Lists of detail earthquake parameters in this study.

| Mw | Year | Month | Day | HH | MM | Lon. | Lat. | Depth | Location |
|---|---|---|---|---|---|---|---|---|---|
| | | | | WIB (GMT+7) | | | | | |
| 5,3 | 2017 | 11 | 21 | 06 | 41 | 96,93 E | 1,37 N | 15 Km | 52 Km Northwest Nias, North Sumatra |
| 5.9 | 2018 | 6 | 13 | 06 | 08 | 98,61 E | 2,06 S | 41 Km | 105 km Southwest KEP-Mentawai. West Sumatra |
| 4,9 | 2018 | 6 | 25 | 13 | 22 | 100,08 E | 2,86 S | 24 Km | 107 km Southeast KEP-Mentawai, West Sumatra |

### *2.2 Filtering frequency spectrum*

Signals resulted from FFT are sorted by creating a frequency bandwidth filter in the range of 0.01-0.03 Hz, 0.03-0.05 Hz and so on up to 0.3 Hz (with range of 0.02 Hz). To get the optimum frequency, the frequency of the ULF signal is then displayed in a graph to be compared in the cross corelation process. The signal that is the focus of study is the signal of the $Z$ component of the magnetic field, since the $H$ component is more sensitive than the $Z$ component to the external field [1].

### *2.3 Determination of onset time*

Onset time is the time which a geomagnetic anomaly is detected for the first time. This parameter is then used as the point of reference for incoming earthquakes. Using polarization ratio of $Z/H$ it is possible to determine the onset time as introduced by Ahadi, *et al.* [1]. The value of $Z/H$ ratio can also be used for predicting magnitude of upcoming earthquake.

### *2.4 Determining source location of the anomaly*

The direction of the magnetic anomaly can be determined by using Single Station Transfer Function (SSTF) [2]. The function is essential to determine source direction of the detected anomaly from reference station, and then the result is presented in an azimuthal fashion (see an example in figure 1).

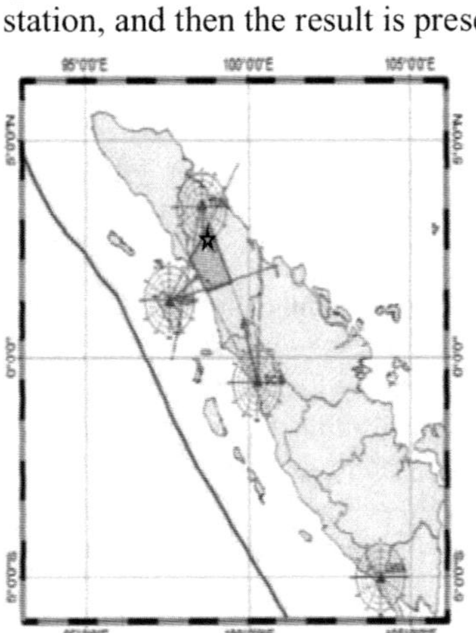

**Figure 1.** Example result of SSTF from multiple stations in Sumatra (BMKG, 2018) [3].

### *2.5 Magnitude calculation*

There are three elements needed to calculate an equation that can be used to determine the magnitude of an earthquake from magnetic data. Those are value of $Z/H$ polarization ratio ($A$), magnitude of earthquake ($M$), and hypocenter distance of past earthquake from reference station ($D$). The parameters then combined on a fitted model plot (shown in figure 2) to produce an equation as follows,

$$M = 0.088 \times A + 0.002 \times D + 4.334 \quad (1)$$

The equation was obtained from 30 earthquakes in the Sumatra region with standard deviation of ± 0.2 and correllation coefficient of $r^2 = 0.724$.

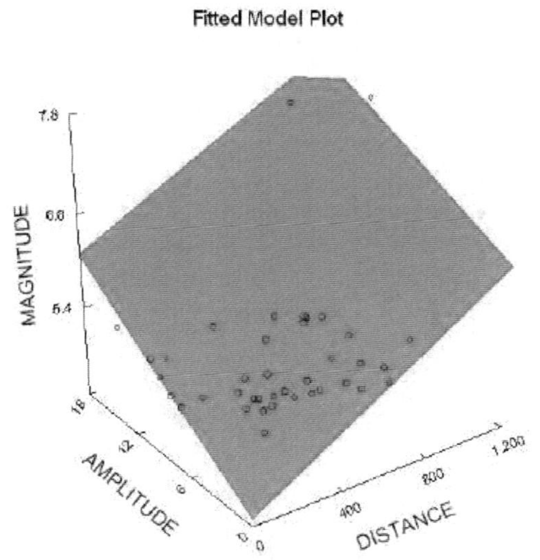


**Figure 2.** Fitted plot model of magnitude calculation from magnetic data [4].

*2.6 Comparing and corroborating results with secondary data*

The results of earthquake monitoring are matched with each other to see whether they support one another or not, as a control between one another. If a discrepancy between stations occurs, it must be repeated the process and evaluated.

At the initial stage, it is necessary to match the prediction data of the precursors with the earthquake event data that has occurred. Matched variables are: (1) calculation of earthquake magnitude prediction with earthquake event magnitude values (2) epicenter prediction through the SSTF method [2] with earthquake event locations, (3) time of earthquake event that occurred after the anomaly was still in a period of fewer than 30 days [1], and (4) Disturbance Storm Time index in that time span.

The next step is to compare the *Z*/*H* plot at a particular frequency and related observation stations, for example 0.01-0.03 Hz at the TUN station compared to the MLB station at the same frequency and time. In other words, onset time is corrected by magnitude prediction, location prediction and Disturbance Storm Time index from another reference station.

## 3. Results and Discussion

*3.1 Earthquake of November 21st, 2017 5.3 Mw*

The earthquake on November 21st, 2017 occurred with magnitude of 5.3 Mw. The closest station to the epicenter are Gunungsitoli and Padangpanjang station. However, in this event, no magnetic anomalies are occurring and recorded at the same time (see figure 3). Therefore, it is impossible to calculate the magnitude potency and predicting earthquake foreshock zone since the onset time was not determined on higher frequency range, the value of *Z*/*H* polarization ratio was too high and will give incorrect value for equation (1).

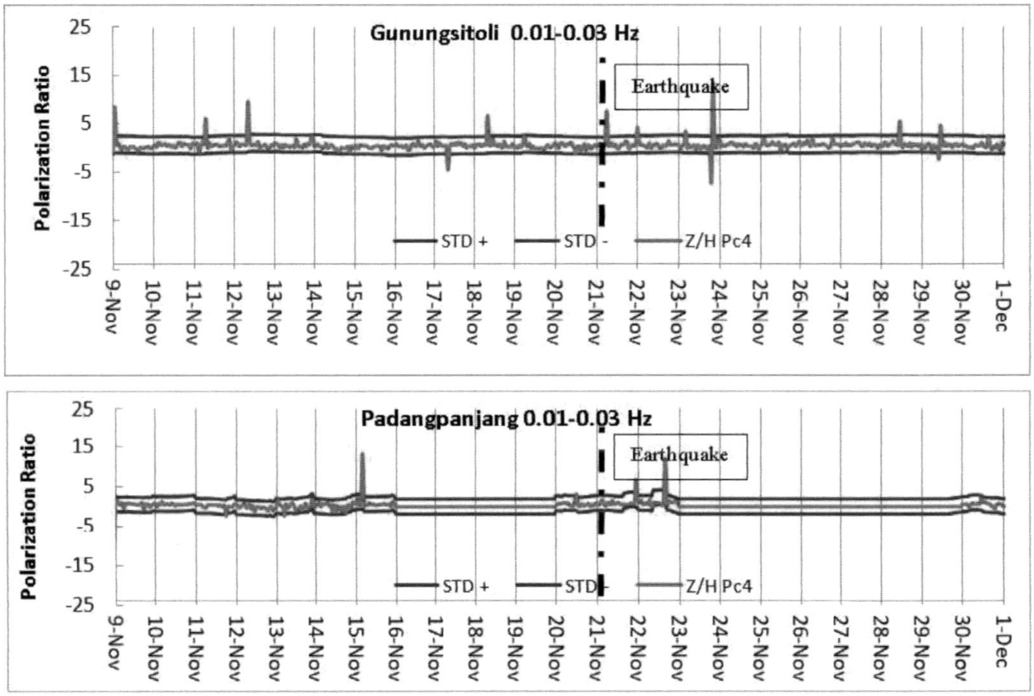


**Figure 3.** *Z/H* polarization ratio for November 21st, 2017 earthquake.

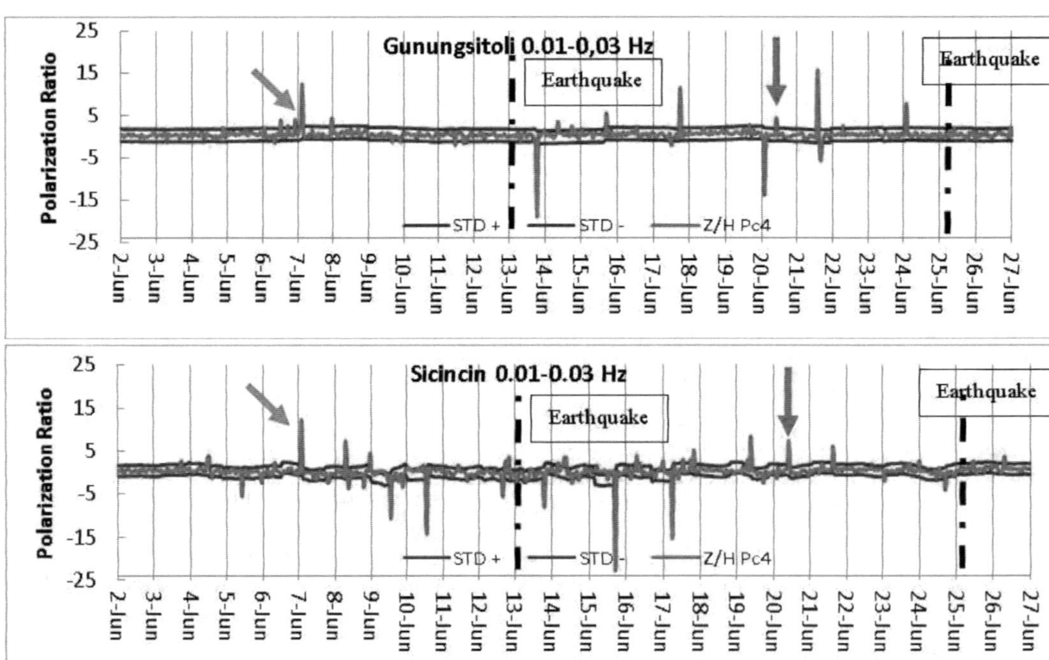


**Figure 4.** *Z/H* polarization ratio of June 13th, and June 25th, 2018 earthquake. Onset time of June 13th marked by green arrow. Onset time of June 25th marked by red arrow.

*3.2 Earthquake of June 13th, 2018, 5.9 Mw*

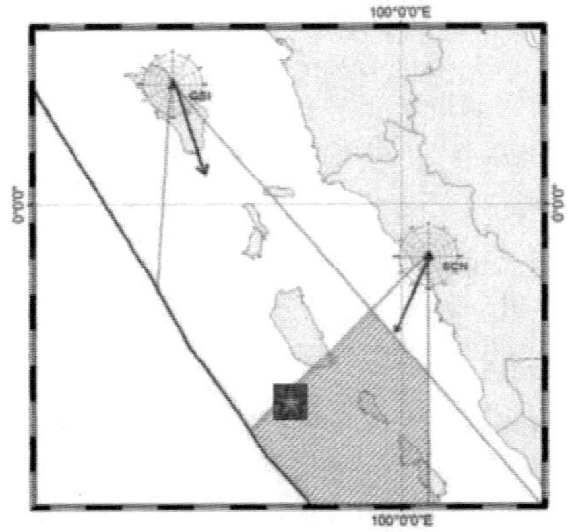


**Figure 5.** Epicenter of June 13th, 2018 earthquake in predicted zone.

The results of determining the onset time through *Z/H* polarization in the 0.01-0.03 spectrum show that there are indeed anomalies with that occur at the same time, the earthquake occurred 6 days after anomaly (of 7 June – 13 June) as shown in figure 4. At the Sicincin station, anomalies are also found at the same time as the Gunung Sitoli station. Further analysis shows that magnitude prediction of 5.9 Mw earthquake prediction deviates by 0.1 to 0.6. However, the location prediction is inside the boundary area of earthquake preparation zone, therefore the location prediction is successful.

Based on equation (1), the parameters of the magnitude calculations are as follows,

- Sicincin 0.01-0.03 Hz: A = 12 nT, D = 251 Km, and M = 5.8 Mw;
- Gunung Sitoli 0.01-0.03 Hz: A = 2 nT, D = 400 Km, and M = 5.3 Mw.

*3.3 Earthquake of June 25th, 2018, 5.0 Mw*

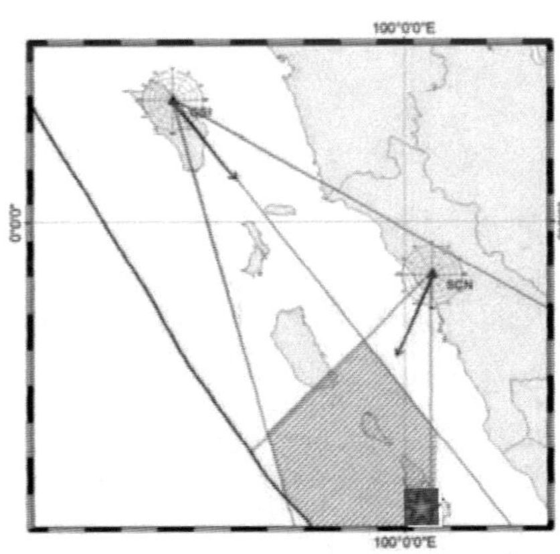


**Figure 6.** Epicenter of June 25$^{th}$, 2018 earthquake in predicted zone.

On 5.0 Mw earthquake, data from both stations presented 0.4 deviation in magnitude prediction. However, the location prediction is correct and the earthquake occurred 5 days after anomaly (June 20th – June 25th). Based on equation (1), the parameters of the magnitude calculation are as follows,

- Gunung Sitoli 0.01-0.03 Hz: A = 4 nT, D = 450, and M = 5.4;
- Sicincin 0.01-0.03 Hz: A = 7 nT, D = 250, and M = 5.4.

Two of the three earthquakes shows good results. The two earthquakes are on 13th and 25th June 2018. The predicted magnitude of the earthquake missed the range, even though the location prediction is correct. Whereas on the earthquake November 21st, 2017 predictions of magnitude and exact location are unable to be determined. So if at least two parameters (2/3 parameters) are correct or close, the prediction is considered a success. Also, the WDC index shows no significant disturbance of external magnetic field or storm, as seen in figure 7. Meanwhile, results from higher frequency ranges are showing very high *Z*/*H* polarization ratio values and show significant amount of noise. So it is no longer relevant for earthquake prediction analysis. The noises from higher frequency range are expected to exist since it is part of the earth's natural geomagnetic activites and variation amplitudes can be quantified in tenths of nT for the diurnal variation and hundreds, or exceptionally, thousands of nT in the case of strong magnetic storms [4][5]. At higher frequencies there are various other phenomena, for example geomagnetic pulsations, take place, this could explain why the 0.01-0.03 Hz frequency range is the best frequency range in this study.

Magnitude prediction and calculation in this study is still imperfect. The magnitude is not always exactly the same, because there are outliers from the magnitude data used in the fitting, this is why deviation is expected. Onset time prediction is also imperfect, sometimes there are no earthquake after detected anomaly. However prediction of foreshock location is expected to be accurate because it is directly extracted from the MAGDAS sensors using SSTF. Lastly, for future study it is necessary to have more MAGDAS stations that is closer to each other with maximum distance of 400 kilometers for the best data quality.

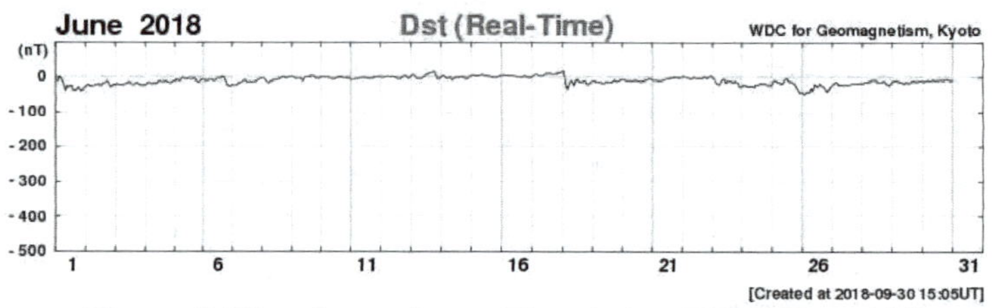


**Figure 7.** Disturbance Storm Time index (WDC Kyoto)[6].

## 4. Conclusion

In this study, the ULF signal most associated with earthquake precursors is in the frequency range of 0.01 - 0.03 Hz. The frequency of 0.01-0.03 Hz gives the most optimal amplitude response than the other frequencies in this study, anomalies detected on higher frequency range are more difficult to be analyzed for magnitude prediction, epicenter prediction, and onset time determining because of noise disturbance.

## Acknowledgement

This study is supported by Indonesian Meteorology, Climate and Geophysics Agency (BMKG) and was made possible by past relations with MAGDAS PROJECT, Ukyoto, and WDC of Geomagnetism Kyoto.

## References

[1] Ahadi, S., Ibrahim G., Puspito N.T., Saroso S., (2014), Determinations Onset time of Earthquake Precursor by Analyzing ULF-EM Emission Signal in Sumatra Region, *Proceeding Conference on Applied Electromagnetic Technology (AEMT),* Lombok, 11 - 15 April.

[2] Hattori, K., (2004), ULF Geomagnetic Changes Associated with Large Earthquake, *Journal Terrestrial, Atmospheric and Oceanic Sciences (TAO)*, **Vol. 15**, No.3, 329-360.

[3] BMKG, (2018), Precursor Review: Sumatra Region 2018, BMKG division of Potential Geophysics and Time.

[4] Syirojudin M., Masturyono, Murjaya, J., Ahadi, S., Yoshikawa, A. (2017), Improving the Analysis Method of ULF Geomagnetic Data for Earthquake Precursor Monitoring in the Sumatera Region. CTBC: Science and Technology Conference.

[5] Kivelson, M. and Russel, C.T. (1995), *Introduction to Space Physics*. Cambridge University Press, pp. 568.

[6] Merrill, R.T., McElhinny and McFadden (1998), *The Magnetic Field of the Earth*. Academic Press, pp. 549.

[7] Geomagnetic Equatorial DST Index, WDC for Geomagnetism, Kyoto DST Index service, <http://wdc.kugi.kyoto-u.ac.jp/dstdir/>.

# Identification of the El Niño Effect on Lake Toba's Water Level Variation

Hendri Irwandi[1,2], M. Syamsu Rosid[1,*], and Terry Mart[1]

[1]Physics Department, FMIPA Universitas Indonesia, Kampus Depok 16424, Indonesia
[2]Deli Serdang Climatology Station, Climatology, Meteorology and Geophysical Agency (BMKG)
Jln. Meteorologi Raya No. 17 Sampali Medan, North Sumatra, Indonesia

[*]Email: syamsu.rosid@ui.ac.id

**Abstract.** Lake Toba is located in the province of North Sumatra, Indonesia which is the largest lake in southeast Asia and is the largest volcanic-type lake in the world. The lake's countenance level is decreased and this can have a negative impact on utilization of power plants and agricultural irrigation. This study will see whether the water level is decreased due to the influence of El Niño activity. The methods used are statistical and descriptive analysis. The data used is the water level of Lake Toba in period of 1957-2016, precipitation data uses NOAA Climate Prediction Centre (CPC) data with a resolution of 0.50° × 0.50° from January 1957 to December 2013 (56 years) to analyse rainfall anomalies. El Niño's activities have an effect on Lake Toba water level, where the moderate El Niño > 12 months duration resulted in a reduction of 1.5-2 m, while the strong El Niño with a duration of > 10 months resulted in a decline of 2-2.5 m.

**Keywords**: Lake Toba, El Niño, CPC, Water Level.

## 1. Introduction

Lake Toba is located in North Sumatra Province, Indonesia. The lake is the largest volcanic lake in the world [1, 2] formed since 1.3 million years ago [2]. The ancient eruption of Mount Toba was stated as the eruption of the largest volcano in the world [3, 4]. Lake Toba's surface area is 1,124 km$^2$, while it's land area is 2,486 km$^2$ (see Figure 1). The surface of the lake is at an altitude of 903 m above sea level with a length of approximately 50 km and a width of about 27 km, with a flat depth the mean is 228 m [5, 6]. Lake Toba functioned as a water transportation infrastructure, community water source [7] and the main one as a hydroelectric power station to supply electrical energy needs [8].

Nowadays, Lake Toba faces an environmental crisis. There are some problems of widespread deforestation, drought, decreased water levels, water quality, and loss of biodiversity [9]. The decline in the level of water level of Lake Toba occurred quite significantly, especially in 1984 to 1987 [10] and in 1997 El Nino activity resulted in a long dry season in Indonesia which resulted in a decrease in rainfall in the Lake Toba watershed, resulting in the decline of Lake Toba water level [7, 11, 12]. Indications of the effect on decreasing lake water level, in general, are due to the role of climate variability, one of which is El Nino [13].

IOP Conf. Series: Earth and Environmental Science **406** (2019) 012022 doi:10.1088/1755-1315/406/1/012022

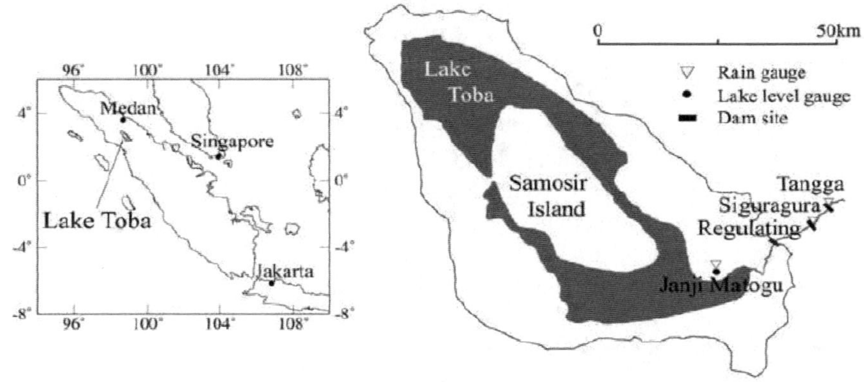


Figure 1. Lake Toba catchment area [14].

The atmospheric component associated with El Niño is called "Southern Oscillation." Scientists often refer to the phenomenon where the atmosphere and oceans collaborate with ENSO stands for El Niño - Southern Oscillation. Under El Niño conditions, sea surface temperatures in the East Equatorial Pacific are hotter than normal conditions. Whereas during La Niña conditions, sea surface temperatures in the East Equatorial Pacific are lower than normal conditions as shown in Figure 2 [15]. The Oceanic Niño Index (ONI) is obtained by calculating the average anomaly of sea surface temperature in the Niño region of 3.4 (190° East – 240° East and 5° N – 5° S) based on the El Niño definition of Figure 3 [15]. ENSO is a combined phenomenon of atmospheric ocean interactions that causes annual climate variations in the world [16]. The ENSO event itself was caused by the process of interaction of the atmospheric ocean in the Pacific region which not only changed the direction of atmospheric circulation but also the pattern of currents and climate in the Indonesian maritime continent [17].

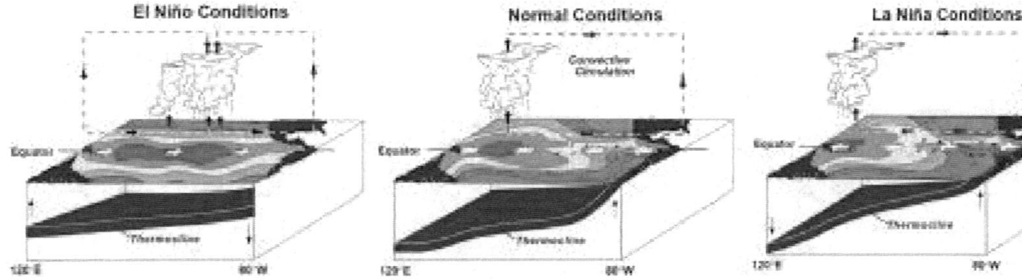


Figure 2. ENSO conditions [18].

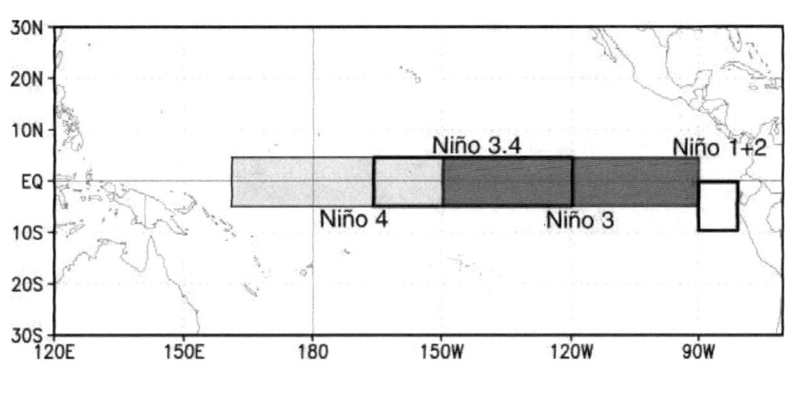

Figure 3. Niño Region [19].

The intensity of ENSO is divided into three categories, namely: El Nino Weak with an index range of 0.5–1 °C, El Nino Moderate with an index range of 1–2 °C, and El Nino is strong with an index range of > 2 °C. Weak La Nina with index range (-0.5) - (- 1) °C, Moderate La Nina with index range (-1) - (- 2) °C, and La Nina Strong with index range < -2) °C [20].

Indonesia received negative influences from ENSO from April until its peak in August and September. The effect will disappear in December. The influence of ENSO occurred in almost all regions of Indonesia except for the northern part of Sumatra and parts of Kalimantan [21]. The strong El Niño has a different effect in each region in Sumatra, the El Niño influence is stronger in the southern region of Sumatra than in the northern part of Sumatra [22]. The active impact of El Niño affects the decrease in rainfall in the DJF, JJA and SON periods in the East Coast, East Slope and Mountains regions with an average percentage decrease in annual rainfall of up to 7% [23].

This study aims to identify and explain the effect of El Nino on rainfall in the Lake Toba catchment area which is indicated to cause a decrease in rainfall and an impact on the decline in Lake Toba's water level.

## 2. Data and Methods

This study uses Oceanic Niño Index (ONI) data to determine the time and duration of El Niño events. ONI was obtained from the SST (Sea Surface Temperature) anomaly results in the Niño region 3.4 (190° - 240° E and 5° N - 5° S). El Niño is identified with an index value > 0.5 which occurs for 5 consecutive months. To see the activity, the duration of ONI with an SPL anomaly >1° C will be classified into 4 classifications including: duration < 3 months, 3-6 months, 7-12 months and >12 months. Monthly rainfall data uses NOAA Climate Prediction Center (CPC) data with a resolution of 0.50° × 0.50° from January 1957 to December 2013 (56 years) to analyze rainfall anomalies in the regions of Indonesia (6° N-11°S and 95°-141°E), North Sumatra ( 0.8° S-4.66° N and 96.66°-100.8° E) and Lake Toba catchment areas (2.18°-2.95° N and 98.43°-99.28°E).

Lake Toba water lavel data for the 1957-1980 period were obtained from the Lake Toba and Sungai Asahan Hydrology books [7], while the 1981-2016 period was obtained from PT Inalum. The data variation of Lake Toba water level are analyzed whether there is a significant relationship to rainfall anomaly data in the Lake Toba catchment area. Monthly climate data is used to show the local characteristics of Lake Toba watershed, while observational rainfall data around Lake Toba watershed is used to make corrections rather than NOAA CPC data.

## 3. Results and Discussion

### *3.1. Oceanic Niño Index Analysis*

ONI analysis results show that in the period 1957-2016 there were 15 El Nino events where generally in moderate category with a percentage of 60%, the strong category 27% and a weak 13%. The moderate El Nino is happened in 1957/58, 1963/64, 1965/66, 1968/69, 1987/88, 1991/92, 1994/95, and 2002/03. The strong El Nino is happened in 1972/73, 1982/83, 1997/98, and 2015/16, while the weak El Nino is happened in 1977/78 and 2004/05.

Duration of ONI which has anomaly temperature > 1 °C was done to show the strength of El Nino activity. The duration < 3 months occurred in 1977/78, 1993/94, and 2006/07, the duration 3-6 months occurred in 1963/64, 1968/69, 1994/95, 2002/03, 2009/10, the duration of 7-12 months occurred in 1965/66, 1972/73, 1982/83, and 1997/98, while the duration for > 12 months occurred in 1987/88 and 2015/16 (see Figure 4).

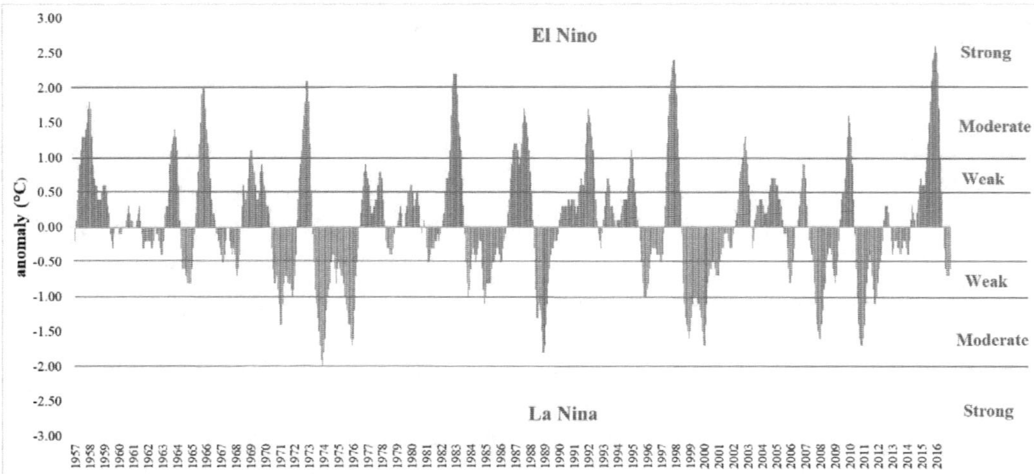


Figure 4. Oceanic Niño Index (ONI) during 1957-2016 period.

### *3.2. Rainfall Anomaly of ENSO period*

The results of rainfall anomaly analysis in Indonesia based on NOAA CPC data shows that during the active El Nino period, a decrease in monthly rainfall ranged from 0-100 mm occurred in 1957, 1963, 1965, 1972, 1982, 1987, 1991-1994, 1997, and 2002. Significant decreases in rainfall occurred in 1972, 1983 and 1997. In this period an increase in monthly rainfall ranged from 0-150 mm, increased rainfall occurred in 1973, 1983, 1988, 1995, 1998-1999, 2007 and 2010, a significant increase occurred in 1998-1999 and 2010 as shown in Figure 5.

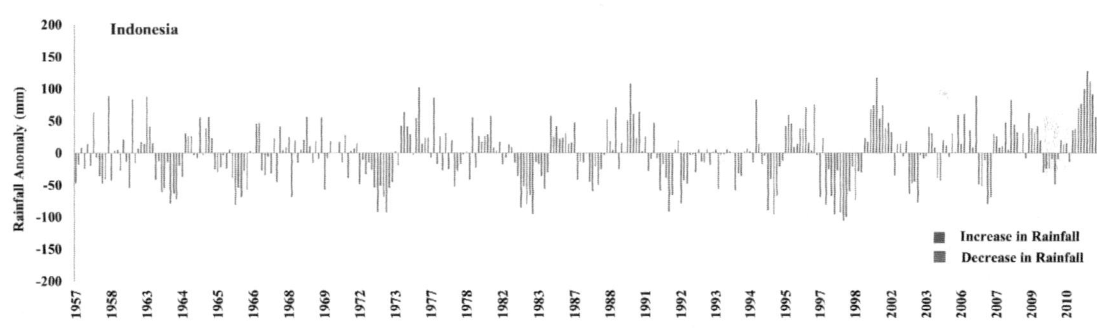


Figure 5. Rainfall anomaly in Indonesia during 1957-2013 period.

The analysis of rainfall anomaly in the North Sumatra region based on NOAA CPC data shows that in the active El Nino period, a decrease in monthly rainfall ranged from 0-100 mm occurred in 1958, 1963, 1982, 1992, 1994, 1997 and 1998. The decreasing rainfall significantly occurred in 1963 and 1997. Generally, in this period there was an increase in monthly rainfall in North Sumatra. Increasing in monthly rainfall ranged from 0-150 mm are occurred in 1957, 1965, 1966, 1983, 1987, 1988, 1999, 2003, 2007 and 2010. A significant increase occurred in 1999 and 2007 as shown in Figure 6.

IOP Conf. Series: Earth and Environmental Science **406** (2019) 012022 doi:10.1088/1755-1315/406/1/012022

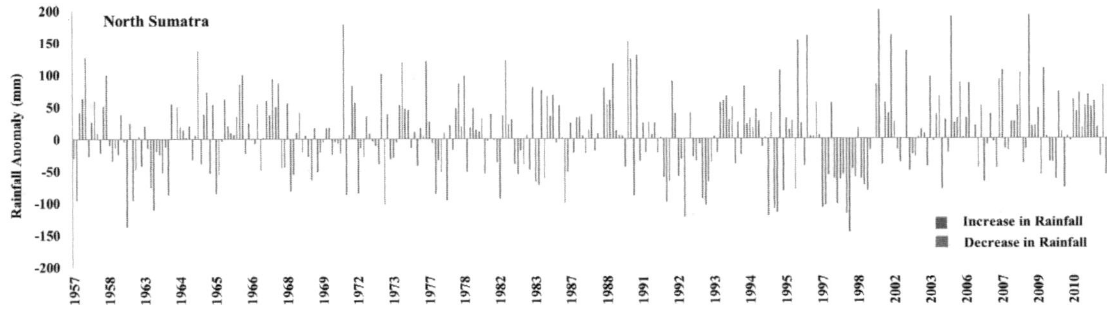


Figure 6. Rainfall Anomaly in North Sumatra during period of 1957-2013.

The analysis of rainfall anomaly in the Lake Toba catchment area based on NOAA CPC data shows in the active El Nino period, decreasing in monthly rainfall ranged from 0-150 mm occurred in 1958, 1963, 1968, 1977, 1991, 1992, 1995, 1997/98 and 2015. Significant decrease in rainfall occurred in 1997 and 2015. Here, the NOAA CPC data has been corrected by manual rainfall observation data. In this period the area has also an increase in monthly rainfall ranged from 0-150 mm, occurred in 1957, 1965/66, 1972/73, 1978, 1987/88, 1993/94 and 2016 (see the Figure 7).

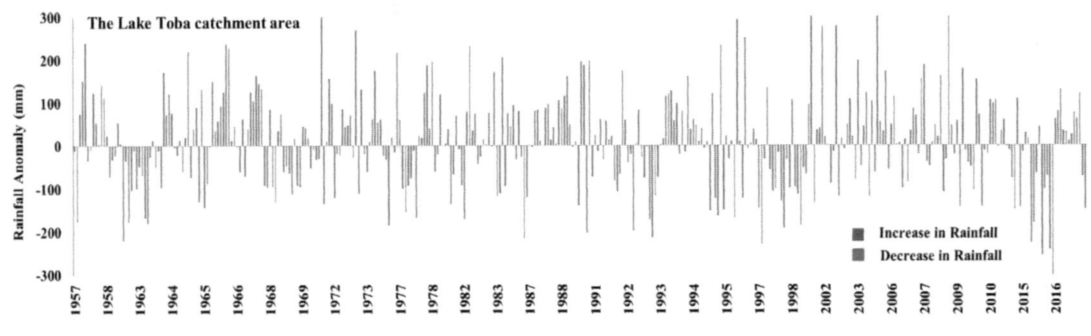


Figure 7. Rainfall anomaly in the Lake Toba catchment area during period of 1957-2016.

*3.3. Climate conditions of Lake Toba catchment area*

Climate conditions of Lake Toba catchment area represented by the Parapat Geophysics Station. The area is classified as a wet tropical climate with temperatures ranging of 18-28 °C. The average minimum temperature is 18.4 °C, the average maximum temperature is 27.1 °C, and the average temperature is 21.5 °C. The average annual high rainfall is 175-225 mm. Lake Toba catchment area is equatorial rain type with peak rain (rainy season) occurring in April and November, while the rain valley (dry season) occurs in February and June. The data are shown in Figure 8.

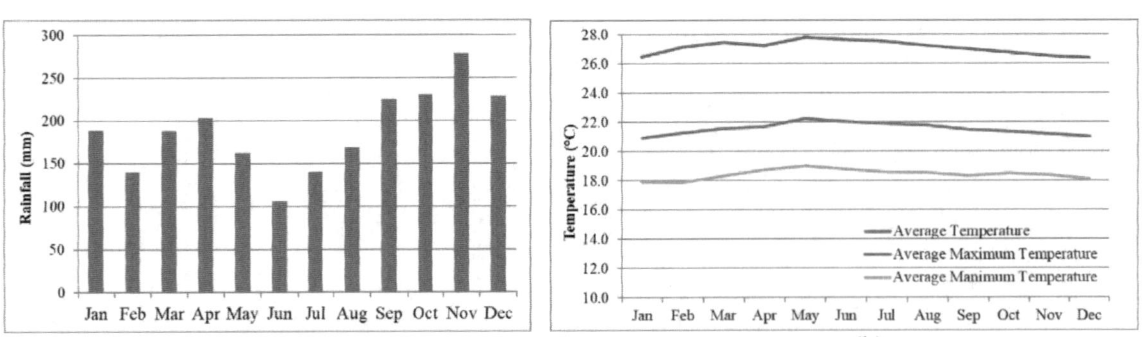


(a) (b)

Figure 8. (a) Air Temperature, (b) Normal rainfall catchment area of Lake Toba.

### *3.4. Analysis of trends in water levels and rainfall in Lake Toba catchment area*

The analysis of Lake Toba's water level showed a declining trend with a downward trend of 2.4 cm per year. In the period of 1957-1984 the average height of the water level was around 905.01 m, whereas in the period 1984-2016 the average height of the Lake Toba water level decreased by around 904.01 m. The water level decline occurred in the period of 1986/87, 1990/91, 1997/98, 2001/02, 2006/07, 2011 and 2015/16. Significant decreases occurred in 1986-1988 and 1997/98 with decreases of 2-2.5 m. Annual rainfall in the Lake Toba catchment area shows the same pattern as the reduction in rainfall in the period 1986, 1997/98 and 2015. There is a good correlation between rainfall and Lake Toba's water level as shown clearly in Figure 9.

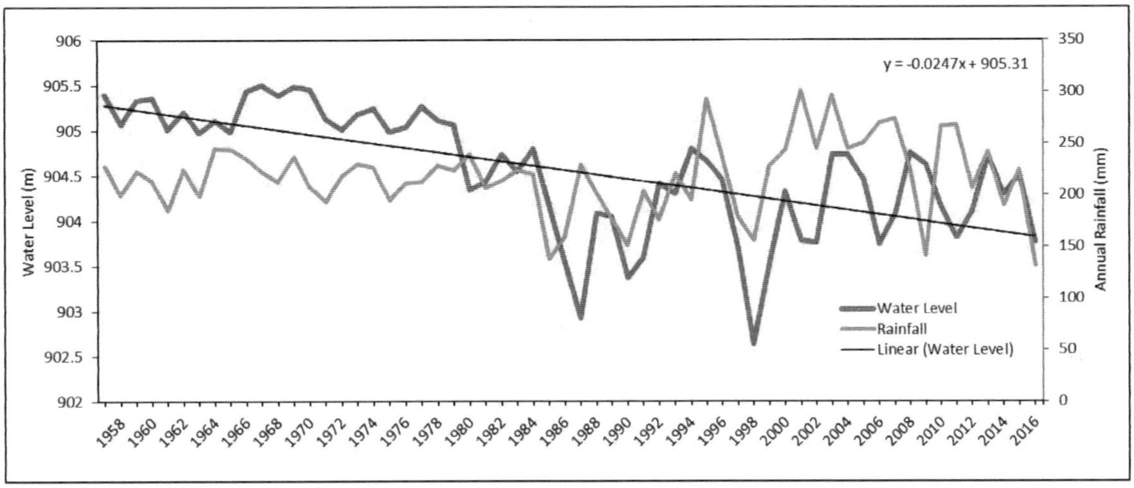


Figure 9. There is similar pattern between water level and rainfall trend in the catchment area of Lake Toba. The Toba water level tend to decrease significantly during this period.

## 4. Conclusions

- The downward trend in Lake Toba's water level ranges from 2.4 cm/yr in the period of 1957-2016
- In the period 1957-1983 there was no indication of the influence of El Nino, whereas in the period 1984-2016, El Nino activity was indicated to have an effect on the decline in Lake Toba's Water Level
- El Nino affects the decrease in rainfall in the Lake Toba catchment
- El Nino is indicated to have an influence on Lake Toba's Water Level in 1987-1988, 1997-1998 and 2015-2016
- Strong El Nino with a duration of > 12 months has an impact on the decrease in Lake Toba Water Level by 1.5-2 m
- El Nino Very strong with a duration of > 10 months has an impact on the decrease in Lake Toba's water level by 2-2.5 m

## References

[1] Soeprobowati, T. Retnaningsih., Suedy, S.W. A. 2017. Diatoms Water Quality of Toba Lake and Its Management. *Proceedings of the 16th World Lake Conference*, p. 549.
[2] Chesner, C.A. 2012. The Toba caldera complex. *Journal Quaternary International*. **258**: p. 5-18.
[3] Williams, M. 2012. The~ 73 ka Toba super-eruption and its impact: history of a debate. *Journal Quaternary International*, 2012. **258**: p. 19-29.

[4] B. Wagner, M. J. Leng,T. Wilke, A. Böhm, K. Panagiotopoulos, H. Vogel, J. Lacey, G. Zanchetta, and R. Sulpizio. 2013. Potential impact of the 74 ka Toba eruption on the Balkan region, SE Europe. *Journal Climate of the Past Discussions*. **9**: p. 3307-3319.
[5] Lukman. and Ridwansyah. 2010. Kajian kondisi morfometri dan beberapa parameter stratifikasi perairan Danau Toba. *Limnotek*. **17**(2): p. 158-170.
[6] Nontji., A. 2016. Danau-danau Alami Nusantara.
[7] Loebis., J. 1999. Hidrologi Danau Toba dan Sungai Asahan. *Puri Fadjar Mandiri*. Jakarta.
[8] Sihotanga, H., J.M. Yanuar., Purwanto., Widiatmaka., Basuni, Sambas. 2012. Model Konservasi Sumberdaya Air Danau Toba. *Jurnal Pengelolaan Sumberdaya Alam dan Lingkungan*. **2** No. 2: p. 65-72.
[9] Saragih., B. and S. Sunito. 2001. Lake Toba Need for an integrated management system. *Lakes & Reservoirs: Research and Management*.
[10] Acreman, M.C., J.R. Meigh., and K.J. Sene. 1993. Modelling the decline in water level of Lake Toba, Indonesia. *Adwmces in Water Resources*. p. 207-222.
[11] Oeliem., H. 2000. Kajian Penurunan Muka Air Danau Toba. North Sumatra University.
[12] Tanakamaru, H., T. Kato., and K. Takara. 2004. Water Balance Analysis And Water Level Simulation of Lake Toba Indonesia. *Proceeding of the 2nd Asia pacific Association of Hydrology and Water Resources Conference*. **Vol.12** p. 107-115
[13] Okonkwo, C., B. Demoz, and S. Gebremariam. 2014. Characteristics of Lake Chad level variability and links to ENSO, precipitation, and river discharge. *ScientificWorld Journal*. p. 145893.
[14] Sianturi, T. 2004. Degradasi Danau Toba. *Jurnal Penelitian Bidang Ilmu Pertanian*. **Volume 2 No. 1**.
[15] Trenberth, K.E. 1997. The definition of el nino. 1997. *Bulletin of the American Meteorological Society*. **78**(12): p. 2771-2778.
[16] McBride, J.L., M.R. Haylock, and N. Nicholls. 2003. Relationships between the Maritime Continent heat source and the El Niño–Southern Oscillation phenomenon. *American Meteorological Society*. **16**(17): p. 2905-2914.
[17] Aldrian, E. and Susanto, R. Dwi.,2003. Identification of three dominant rainfall regions within Indonesia and their relationship to sea surface temperature. *Royal Meteorological Society*. **23**(12): p. 1435-1452.
[18] McPhaden, M.J., 2014. Evolution of the 2002/03 El Niño. *Bulletin of the American Meteorological Society*. **85**(5): p. 677-696.
[19] Kozar, M.E., Mann,M.E. , Camargo, Suzana. J., Kossin, James. P., and Evans, Jenni.L. 2012. Stratified statistical models of North Atlantic basin-wide and regional tropical cyclone counts. *Journal of Geophysical Research: Atmospheres*. **117**(D18).
[20] Null, J. 2018. El Niño and La Niña Years and Intensities. *Golden Gate Weather Services*.
[21] Aldrian, E. 2002. Spatial patterns of ENSO impact on Indonesian rainfall. *Jurnal Sains & Teknologi Modifikasi Cuaca*, Vol. **3**: p. 5-15.
[22] Khoir, Aulia.N., D, Yuni., N, A. Rozanah., and Safril, Agus. 2018. The influence of strong El Niño to seasonal variability in Sumatera. *International Symposium on Earth Hazard and Disaster Mitigation* (ISEDM).
[23] Irwandi, H., Pusparini, N., Ariantono, J.Y., Kurniawan, R., Tari, C.A., Sudrajat, A.2018. The Influence of ENSO to the Rainfall Variability in North Sumatra Province. *IOP Conference Series: Materials Science and Engineering*. IOP Publishing.
[24] Schneider,U., Becker, A., Finger, P., Christoffer, A. M., Markus Ziese, M., Rudolf, B. 2014 GPCC's new land surface precipitation climatology based on quality-controlled in situ data and its role in quantifying the global water cycle. *Journal Theoretical Applied Climatology*. **115**(1-2): p. 15-40.

# PRODUCTION RISK OF PEANUT FARMING IN DRY LANDSECOND PATTERN AFTER CORN WITH BIOCHARON BLITAR DISTRICT, INDONESIA

**Asnah[1], Masyhuri[2] and Jangkung Handoyo Mulyo[2,3]**
[1] Tribhuwana Tunggadewi University, Jl. Telaga Warna, Tlogomas Malang
[2] Faculty of Agriculture, University of Gadjah Mada, Jl. Flora, Bulaksumur, Yogyakarta
[3]Centerfor Population and Policy Studies, Jl. Tevesia, Bulaksumur, Yogyakarta
*E-mail : asnah.unitri@gmail.com , mmasyhuri@hotmail.com , JHandoyoM@gmail.com

**Abstract.** The reduction in land due to the transfer of functions from agriculture to non-agriculture, requires stakeholders to intensify the management of dry land and infertile land which was not originally a priority. Therefore, every opportunity for adaptive technology development needs to be utilized as well as possible to encourage increased productivity. This study aims to determine the impact of the use of biochar used in the first cropping maize farming on production and risk of production of the second cropping peanut farming in dry land. The study uses a basic survey method, by determining the location by purposife. The number of samples was 150 people determined by the proportional random sampling method. The data analysis method uses the production function model and the production risk function using the Cobb-Douglass ordinary least square (OLS) method. The results found that the use of biochar in the first cropping maize farming had an impact on the production and risk of the second cropping pattern of peanut production. As a comparison, the factors of production in peanut farming which previously had the effect of reducing production; in the second cropping pattern after maize had an effect on increasing production and reducing production risks include seeds, phosphor fertilizer, potassium fertilizer, pesticide, and labor in the family. Only workers outside the family have the effect of reducing production and increasing risk.

Keywords: Biochar; Peanut Farming; Second Pattern

## INTRODUCTION

Agricultural politics is a combination of government attitudes and actions taken with the aim of influencing the course of the agricultural sector to be more productive, efficient, advanced and increase farmers' income and welfare more evenly [1]. Integration between agricultural development and agricultural politics as a policy instrument is needed, therefore, good, correct and sustainable farm management must be an important part of agricultural development activities. Agricultural development is an integral part of regional development and national development. The harmony between these developments should maintain, where regional development as part of national development must align with achieving increased agricultural productivity, farmer welfare, suitability of agricultural land, market suitability and sustainability of production systems and environmental sustainability.

Along with the development and rate of industrial improvement, at present most of the land in Java has experienced degradation, which has an impact on decreasing soil fertility. Severely degraded land is the cause of the low productivity of plants. Referring to [2], the characteristics of degraded land base on variations in parent material. They are including : percentage of dust fraction on soil texture, type of land use / management, land slope and conservation; as well as variable chemical content, among others : soil acidity (pH), organic C, total N, availability of P, Ca, Mg, K, Na, cation exchange capacity (CEC), saturation level, Al and H. exchange. Dry land and degraded land are the same land have low fertility.

Differences in the characteristics of farming on fertile land and on dry land are clearly seen among others in the aspects of irrigation, land productivity and the extent and potential in Indonesia (Table 1). Areas with dry land have a larger amount than arable land, and as such must always look for opportunities to develop adaptive technology in their respective locations, so that they will be able to encourage increased land and plant productivity. The aim is to improve the welfare of farmers and prevent land use change, as well as avoid land abandonment,because of the transfer of the farmers' profession and migration to look for other better sources of income outside the region.

Dry land has great potential to develop, therefore efforts and strategies that can encourage increased productivity are needed. Research by [3], found that a strategy that needed to develop for peanut production in the dry land of East Sumba, East Nusa Tenggara, was to increase yields through intensive use of production factors and expansion of planting areas. In addition, intensive farming management strategies are also needed, by applying new superior varieties and planting technology, increasing business scale and cropping index.

Research conducted by the Research Institute for Bean and Tuber Crops Research Institute, [4] on acid dry land in Central Lampung, found that the assembling of cultivation technology has been able to increase peanut production by 18% to 83%. A variety of strategies in the management of dry land need to be done, especially those that are adaptive in order to be applied by farmers properly and successfully. The application of intercropping technology of groundnuts and shallots by [5], in Parigi Moutong, Central Sulawesi, found that variety had an important role in the success of farming in dry land, where the variety of Tuban and Bison peanuts, as well as Palasa local shallot varieties can produce the highest benefit with a B / C value > 1.

The use of adaptive technology in the development of peanuts on dry land, will greatly help farmers increase productivity which is expected, can take place in the long - run. The application of cultivation technology, which includes the regulation of cropping patterns, planting systems, intensive use of production factors, the use of superior seeds or a combination of these technologies, are expected to spur increased productivity and income of farmers. Therefore, this study was conducted, with the aim to determine the impact of the use of biochar as a production factor applied to maize on the first cropping pattern, on the production and production risk of peanut farming in the second cropping pattern.

IOP Conf. Series: Earth and Environmental Science **406** (2019) 012023 doi:10.1088/1755-1315/406/1/012023

## RESEARCH METHODOLOGY

This research was conducted using a basic survey method [6], with the help of a research instrument in the form of a questionnaire as a guideline for data collection. Research locations in Kalitengah Village, Panggungrejo District, Blitar Regency, were selected using a purposive method, in accordance with [7]; and [8]; [9] and [10]. The number of samples was 150 people taken by proportional random sampling from 652 population members of the combined farmer groups.

The method of data analysis uses multiple regression analysis of the Cobb-Douglass OLS model. The mathematical formulation in the research data analysis method, follows [11]:

a. Production Function:

Ln Pro = $\beta_o + \beta_1$Ln Llhn + $\beta_2$ Ln Bnh + $\beta_3$ Ln PpkN + $\beta_4$Ln PpkP + $\beta_5$ Ln PpkK + $\beta_6$ Ln Pp Kd + $\beta_7$ Ln Pest + $\beta_8$ Ln TK DK +$\beta_9$ Ln TK LK + e .......... (1)

Information :

Formula (1) is used for the analysis of early season peanut farming (as a comparison) and the second cropping pattern (which is planted after corn with biochar).

| | |
|---|---|
| Ln Pro | = peanut production (kg) |
| Ln Llhn | = land area (ha) |
| Ln Bnh | = seed (kg) |
| Ln Ppk N | = Nitrogen fertilizer (kg) |
| Ln Ppk P | = Phosphor fertilizer (kg) |
| Ln Ppk K | = Kaliumfertilizer (kg) |
| Ln PpkKd | = the amount of manure (kg) |
| Ln Pest | = the amount of pesticide (ltr) |
| Ln TKDK | = the number of workers in the family (HKO) |
| Ln TKLK | = the number of workers outside the family (HKO) |
| $\beta_o$ | = *intercept* |
| $\beta_1$-$\beta_9$ | = estimator parameter coefficients |
| e | = *error term*(random / residual error) |

**b. Production Risk Function**

$$\varepsilon^2 = \theta_o + \theta_1 Ln\,\Pr o + \theta_2 LnLhn + \theta_3 LnBnh + \theta_4 LnPpkN + \theta_5 LnPpkP + \theta_6 LnPpkK + \theta_7 LnPpkKd + \theta_8 LnPest + \theta_9 LnTkdk + \theta_{10} LnTklk + \varepsilon \quad \cdots \quad (2)$$

Information:

$e^2$ = Residual squares of the production function.

## RESULT AND DISCUSSION

Theoretically, the use of production factors expect to have a significant effect in increasing production and reducing production risk. However, in the empirical conditions, this study produced findings that were not entirely the same as the expected value as in the theory. The complete analysis result presents in (Table 2) and (Table 3).

The production risk function is obtained by analyzing the residual production function, which is squared and becomes the dependent variable, then regresses to all the independent variables by adding the production variable as the independent variable. The symbol or notation used in the analysis of the production function in the discussion of the results of this study is f, while the production risk function is given h for the symbol.

The coefficient of determination on the results of the analysis of the production function of the first season peanut farming (as a comparison) and the second cropping pattern after corn farming with biochar, respectively 0.9821 and 0.9465. The magnitude of the coefficient of determination shows, the independent variables included in the model, namely: land area, seeds, nitrogen fertilizer, phosphorus fertilizer, potassium fertilizer, manure, pesticides, labor in the family and labor outside the family are able to explain the dependent variable first season peanut production and second cropping patterns respectively at 98.21 percent and 94.65 percent. While the remaining 1.79 percent and 5.35 percent, are explained by other factors outside the model.

In the production risk function, the coefficient of determination was 0.7287 in the first season and 0.9457 in the second cropping pattern after corn. Thus, the free variable of production, land area, seeds, nitrogen fertilizer, phosphorus fertilizer, potassium fertilizer, manure, pesticides, family labour and outside family labour, are able to explain the dependent variable production risk of 72.87 respectively percent in the first season and 94.57 percent in the second cropping pattern after corn. While the remaining 7.13 percent and 5.43 percent are explaining by other factors outside the model. Simultaneously, the selected independent variables included in the model both in the analysis of the production function and production risk function of peanut farming, based on the f test (simultaneous test) have a significant effect on the variable depending on $\alpha = 1\%$ or a 99 percent confidence level.

Production factors that have a significant effect on the production of peanut farming in the first season are land area, seeds, phosphorus fertilizer, family labor and outside family labor. While the factors of production of nitrogen fertilizer, potassium fertilizer, manure and pesticides have no significant effect on production. Furthermore, in the second cropping pattern of peanut farming after maize, the factors of production that had a significant effect on production included land area, pesticides, family labor, and potassium fertilizer and outside family labor. While the factors of production that haveno significant effect were seed, nitrogen fertilizer, phosphorus fertilizer and manure.

Based on the results of individual analysis of the independent variables, production has a significant effect on production risk, on the first season of peanut farming and the second cropping pattern after corn with $\alpha = 1\%$ or a confidence level of 99 percent. A sign of positive expectation, which means that the risk of production in peanut farming increases with increasing production. The expected condition is that production increases but risk decreases, however in this research, result production risk increases with increasing production. Such conditions can occur due to the operation of demand and supply mechanisms or market mechanisms. As peanut production increases, so does the number of peanuts offered in the market. Assuming ceteris paribus, if the demand for peanuts in the market is in a fixed condition, an increase in the number of peanuts offered because of increased production will drive down the selling price of peanuts in the market. This means increasing the risk of production in the farming of dry land peanuts that must be borne by farmers. [12] states that an increase in amount of goods offered to the market due to increased production, while demand remains, it would encourage a surplus, where the amount of goods offered is more than the amount of goods demanded so that the price of goods will fall. The decline in the selling price of output causes farmer income to decrease, thereby affecting the ability of farmers to finance further of the farming which will also decline.

The results of the analysis of the production function of the first season peanut farming and the second cropping pattern have a coefficient sign on the production function is positive and the risk function of production is negative or in accordance with the signs of hope and supporting theory. Thus, the variable land area increases production while reducing production risk. In the first season peanut farming, the coefficient of land area in the production function is 1.1055 and in the second cropping pattern is 0.7861, both of which have a significant effect on production with $\alpha = 1\%$ or a confidence level of 99 percent. Each increase in land area of 1 percent will increase production of 1.1055 percent in the first season and 0.7861 percent in the second cropping pattern.

In the production risk function, increasing the area of land will reduce the risk of production in peanut farming in dry land, respectively by 20.1086 percent in the first season with $\alpha = 1\%$ and 0.0695 percent in the second cropping pattern, even though in the cropping pattern second has no significant effect. The results of the study of [13], [14] , are consistent with the results of the analysis on the risk of peanut farming production in the first season, and the second cropping pattern after maize; where land area has a positive effect on production and negatively affects production risk. In other words, land area has an effect on increasing production and reducing production risk. The results of the analysis on previous cropping patterns, increasing the area of land on corn farming with biochar showed significant results can increase production and production risk, whereas in the first season peanut farming, increasing land area significantly increases production while reducing production risk, and this is in accordance with a sign of expectation.

Seed as an independent variable has a significant effect on production and production risk in the first season of peanut farming, each with a significance level of 90 percent ($\alpha = 10\%$) and 99 percent ($\alpha = 1\%$). Each coefficient has a negative sign on the production function and positive on the production risk function. That is, increasing seeds by 1 percent will reduce production by 0.0640 percent and increase production risk by 1.9698 percent. At the farm level, the amount of peanut seeds per hectare is close to the recommended dosage. However, the results of this study found that adding seeds actually reduces production and increases production risk. Previous research conducted by [15] found that seed production factors were risk reducing or reduced risk, whereas in this study the results were risk increasing or increased production risk. It can be understood that differences in location and farm management can be a differentiating factor even though the variables used are the same.

The nitrogen fertilizer variable in the first season of peanut farming and the second cropping pattern after corn with biochar had no significant effect on both the production function and the production risk function. While the variable phosphorus fertilizer used by farmers in the first season of peanut farming has a significant effect on production and production risk at a 99% confidence level ($\alpha = 1\%$). The coefficients respectively - 0.1461 in the production function and 3.6895 in the production risk function, which can be interpreted as the addition of phosphorus fertilizer 1 percent, will reduce the production of peanut first season by 0.1461 percent and increase production risk by 3.6895 percent. In this case, although the peanut is cultivated on the same land as corn in the previous cropping pattern, the impact of the use of phosphor fertilizer on plant growth and yield is different. The cause of the difference in response is the difference in crop needs for macro nutrients derived from phosphorus.

Adding nutrients to the soil through fertilization activities is carried out based on soil and plant, which needs for nutrients. Nutrient needs in soil and plants can be fulfilled from nature or through fertilization during cultivation. If the availability of nutrients is sufficient, then the addition is not necessary. [16] , explained in a study conducted that the soil supplies 13 important nutrients, out of 16 essential nutrients needed for plant growth, especially food crops, however the presence of these nutrients does not always explain that the characteristics of the soil / land it is fertile. This indicates, land fertility is site specific or crop specific. Thus, the difference in the response of phosphorus fertilization to the two commodities cultivated on the same land, but with different seasons, can occur as the results of this study. In the second cropping pattern, the use of phosphorus fertilizer even though it has a coefficient mark as expected but does not significantly influence both production and production risk. The use of phosphorus fertilizer as a macro nutrient element, gave impacts both the soil and plants. The research of [17] in Research Institute for Agricultural Research and Development, discussed the important role of phosphorus / pospat fertilizer. It was found, the role and function of these fertilizers in the process of respiration and photosynthesis, nucleic acid preparation, formation of plant seeds and fruit producers, is stimulating root development, so plants will be more resistant to drought, and speed up the harvest so that it can avoid the risk of delay in harvest time.

Potassium fertilizer variable in the analysis of the production function of peanut farming in the second cropping pattern after corn with biochar, has a significant effect on production at a 95% confidence level ($\alpha = 5\%$). The regression coefficient of -0.0154 shows and increase of 1 percent potassium fertilizer will reduce peanut production by 0.0154 percent. On the other hand, the analysis of the risk function of the variable potassium fertilizer did not significantly influence peanut farming in both the first season and the second cropping pattern after corn.

In the empirical condition, potassium fertilizer is not widely used by farmers, (the dose of its use is still far below the recommended dose). Habits of farmers at the study site do not use potassium fertilizer in farming, even if using only a little. This is understandable given the relatively high price of non-subsidized potassium fertilizer so that not all farmers consider it important to use it. The research location is an area with rocky dry land conditions and topography varies from flat to hilly, and contains minerals, so that the most likely land already has potassium content naturally. Thus, the addition of doses to a certain extent will reduce production. The activity can be carried out by farmers is to add soil enhancers to increase potassium absorption. Thus, the use of biochar needs to do consistently. The average use of biochar in the previous cropping corn farming pattern is still below the recommended dosage of agronomic research results, where the average farmer uses only 1.2 tons per hectare, while the recommended dosage is 30-50 tons per hectare.

The use of variable labor within and outside the family in peanut farming has a coefficient sign in accordance with the sign of expectation. In the first season, the coefficient on the production function is -0.0633 for labor in the family, which means an increase of 1 percent of labor in the family will reduce production by 0.0633 percent. On the other hand, the coefficient for labor outside the family is 0.0484, which means an additional 1 percent of labor outside the family will increase production by 0.0484 percent, both of which have a significant influence on the level of confidence of 95 percent ($\alpha = 5\%$). In the second cropping pattern the coefficients in the production function are 0.0881 for the use of labor in the family and -0.0559 for workers outside the family, both of which are significant at 90 percent confidence level ($\alpha = 10\%$). The coefficient resulting from the analysis shows that the addition of 1 percent of labor in the family will increase production by 0.0881 percent, and the addition of 1 percent of labor outside the family will reduce production by 0.0559 percent. In the analysis of the risk function of the first season's production, the use of labor in the family has a coefficient of 1.2146 and labor outside the family is -0.9923, significant at 99 percent confidence level (α = 1%). Whereas in the second cropping pattern only the use of labor outside the family has a significant effect on production risk at a 95% confidence level (α = 5%), with a coefficient of 0.0271. Thus, the addition of 1 percent of labor in the family will increase production risk by 1.2146 percent, but the use of labor outside the family reduces the risk of production by 0.9923 percent, in the first season. Whereas in the second cropping pattern, the addition of 1 percent of the use of labor outside the family which had a significant effect in reducing production also had a significant effect on increasing production risk by 0.0271 percent. The results of this study are consistent with the results of other studies conducted by [13] , that labor increases risks both in organic and conventional farming.

At the farmer level, the use of labor in peanut farming also indicates to be excessive. Thus, what needs to be increased is the allocation of labor usage, and not the addition of the physical number of workers. The existence of farmer groups where farmers are bound to be members in it has many benefits of cooperation in various activities. Therefore, in every start farming habits in cooperation also carry over to the use of labor in farming. The release of the head of the Central Statistics Agency [18] in Tempo 5 May 2017, which was also strengthened by data from the Indonesian Central Statistics Agency (2017), stated that the number of people working in the agricultural sector was still quite large (39.68 million), albeit with lower wage rates compared to other sectors. The number of labor absorbed in the agricultural sector during the period of February 2016 to February 2017 increased by 0.12 percent. Thus, these conditions have contributed to the results of this study. The results of other studies that are consistent with this study and found that labor variables have a significant effect on production include conducted by [19]; [14]; [20]; and [21].

The use of pesticide input variable only significantly influences the production function and risk function of the second cropping pattern of peanut farming, with a 95 percent confidence level (α = 5%), a regression coefficient of 0.0679, which means that adding 1 percent of pesticides will increase production 0, 0679 percent. While the production risk function is significant at a 90 percent confidence level (α = 10%), the regression coefficient of -0.1189 indicates that the addition of a 1 percent pesticide that can increase production will also reduce production risk by 0.1189 percent. This research is in accordance with the expected sign, where the addition of pesticide inputs can increase production while reducing risk.

Empirically, farmers use pesticides in amounts that are classified as low as or, even lower than the recommended dosage. Farmers are wise enough to use pesticides so that they have a good impact on farming management, this is because in addition to reducing production costs can also avoid the adverse effects of immunity on plant-disturbing organisms, predator deaths and environmental damage as a result of excessive use of pesticides. Thus, the use of pesticides in dryland peanut farming can still be increased and can increase production while reducing risk. [22] , that pesticide is able to control plant-disturbing organisms (OPT) and effectively save crop losses conducted the study in accordance with the results of this study. On the other hand, [23] suggests that the use of pesticides in the field needs to follow the correct procedures to ensure safety and increase the effectiveness of use.

The variable manure only has a significant effect on the risk function of peanut farming production second cropping pattern after corn with biochar, at a 90% confidence level (α = 10%). The regression coefficient of -11.1189 means that the addition of manure 1 percent will reduce the risk of peanut production by 0.1189 percent. Although no significant effect, the use of manure has the potential to increase the production; it is seen from the regression coefficient that has a positive sign of hope. The need for manure in one hectare of land according to the guidelines for cultivating peanuts in agrokomplek [24], ranges from five to 10 tons per hectare. The use of techniques was given at the same time as planting. Empirically, the use of manure can still increase; bygiven the dose at the farm level is still very low. [13] in their study found that the use of animal manure inputs increases the risk of organic farming, but reduces the risk of conventional farming.

Biochar was using by farmers in the previous cropping corn farming. The use of biochar has a significant effect on increasing corn production, and has the potential to reduce production risk. This shows, in the regression coefficient, which has a negative expectation sign. Although empirically the use of biochar is still far below the standards and recommendations of the results of previous agronomic research. However, the willingness and willingness of farmers to produce and use biochar in managed farms is an important positive thing to appreciate. The benefits gained in the short term can motivate farmers to obtain greater benefits in the end. '

Biochar used by farmers in the previous cropping corn farming also had a good effect on the production of second cropping peanuts after corn, where there was an increase in production of 11.90 percent from 898.97 kg / hectare the previous season, to 1,005.95 kg / hectare in the second cropping pattern after corn farming with biochar. Several researchers in previous studies, both physically and biologically, have conveyed the positive impact of using biochar. The impact of the use of biochar on improving soil quality, including soil acidity (pH), cation exchange capacity (CEC) of soil physical properties and field capacity by providing biochar of more than 50 tons per hectare was found by [25]. Whereas the increase in production due to the impact of the use of biochar was reported by [26] and [27]. In his study found the benefits of biochar in improving the physical properties of the soil. Furthermore, [28] also found the use of charcoal as ameliorant to improve the physical and chemical properties of tropical soils, whereas in another study [29], reported the impact of biochar on improving soil biological properties.

## CONCLUSION

The use of biochar as a soil amendment in the first cropping maize farming has a positive impact on the farming of the second cropping peanut after maize. The positive impacts in question include increasing peanut production 11.90 percent, increasing the quality of nutrient uptake so that the variable use of fertilizer production factors, which initially had a significant effect on reducing production, in the first season of peanut farming, changed to increase production, on peanut farming the second cropping pattern after corn farming with biochar. Manure which initially had no significant effect on production risk, turned into a significant effect reducing production risk in the second cropping pattern after corn farming with biochar. The pesticides, which were initially having a significant effect on increasing production risk, have changed to significantly reduce production risk. There is a substitution of land production factor with seed that leads to intensification, where in the first season increases production, but in the second cropping pattern after maize farming with biochar turns into a significant effect in reducing production. However, there is a change in production risk from the initial significant effect of increasing risk, to reduce risk, although not significant. On the other hand, the seed, which initially had a significant effect on reducing production, changed to significantly increase the production, so that the substitution of adding seeds did not need to be followed by additional land, due to the functioning of biochar as an amendment, resulting in intensification of the use of seed inputs. Likewise, the use of labor in families substituted with workers outside the family, so there is no need to increase the physical number of workers, but it is necessary to increase the allocation of use.

## REFERENCES

[1] Hanafi Rita, *Pengantar Ekonomi Pertanian.* Yogyakarta: Andi, 2010.

[2] B. S. dan O. H. Sitorus, S.R., "Kriteria dan Klasifikasi Tingkat Degradasi Lahan di Lahan Kering A Preliminary Criteia and Classificntion of Land Degradation Level on Dryland Study : Dryland in Bogor Regency ) Degradasi lahan adalah proses penurunan produktivitas Proses degradasi iahan," *J. Tanah dan Iklim*, vol. 34, 2011.

[3] F. Rozi and I. Sutrisno, "Peluang Pengembangan Kacang Tanah Di Lahan Kering Nusa Tenggara Timur," *Bul. Palawija*, vol. 14, no. 2, pp. 71–77, 2016.

[4] BALITKABI, "Hasil Kacang Tanah di Lahan Kering Masam Meningkat Hingga 80 Persen. Retieved from https://balitkabi.litbang.pertanian.go.id/berita/hasil-kacang-tanah-di- lahan-kering-masam-meningkat-hingga-80-persen/," p. 1, 2009.

[5] A. I. dan S. Syafruddin, "Adaptasi bawang merah dan kacang tanah pada lahan kering di parigi moutong, sulawesi tengah," *J. Pengkaj. dan Pengemb. Teknol. Pertan.*, vol. 21, no. 1, pp. 25–35, 2018.

[6] Sugiyono, *Metode Penelitian Bisnis.* Alfabeta, 2008.

[7] Masri Singarimbun dan Sofian Effedie, *Metode Penelitian Survei.* Jakarta: LP3ES, 1989.

[8] S. Arikunto, *Prosedur Penelitian, Suatu Pendekatan Praktik.* Jakarta: Rineka Cipta, 2010.

[9] M. Nazir, *Metode Penelitian*. Jakarta: Ghalia Indonesia, 2011.

[10] B. Bungin, *Metodologi Penelitian Kuantitatif*, Edisi Kedu. Jakarta: Kencana Prenada Media Group, 2005.

[11] R. E. Just and R. D. Pope, "Production Function Estimation and Related Risk Considerations," *Am. J. Agric. Econ.*, vol. 61, no. 2, pp. 276–284, 1979.

[12] N. Gregory Mankiw, "CONSUMER DURABLES MD THE REAL INTEREST RATE," vol. 1148, no. 1148, p. 240, 1983.

[13] M. D. C. and A. O. L. Gardebroek, Cornelis, "Analysing Production Technology and Risk in Organic and Conventional Dutch Arable Farming Using Panel Data," *J. Agric. Econ.*, vol. 61, no. 1, pp. 60–75, 2010.

[14] J. dan S. Rinaldi, "Analisis Risiko Produksi dan Faktor yang Mempengaruhinya Pada Usahatani Kakao di Bali," in *Prosiding Seminar Nasional : Kedaulatan Pangan dan Pertanian*, 2014, pp. 637–642.

[15] K. E. Sularso, "Usahatani kentang dengan teknik konservasi teras bangku di dataran tinggi dieng kabupaten wonosobo jawa tengah (," *J. Pembang. Pedesaan*, vol. 10, no. 2, pp. 115–127, 2010.

[16] A. F. Handayanto, Eko., Nurul Muddarisna, *Pengelolaan Kesuburan Tanah.* Malang: Universitas Brawijaya Press, 2017.

[17] Normahani, "Mengenal Pupuk Fosfat dan Fungsinya bagi Tanaman," *Balai Penelit. Pertan. Lahan Rawa*, no. May 2015, pp. 2015–2017, 2015.

[18] Suharyanto, "Sektor Pertanian Serap Banyak Tenaga Kerja.," *Tempo, bisnis*, p. 2017, 2017.

[19] Junaedi, "Efisiensi Produksi, Perilaku Petani dan Daya Saing Usahatani Kapas Rakyat di Sulawesi Selatan," Universitas Gadjah Mada, 2013.

[20] S. H. S. dan A. P. Kusnadi, Nunung, Netti Tinaprilla, "Analisis Efisiensi Usahatani Padi di Beberapa Sentra Produksi Padi di Indonesia," *J. Agro Ekon.*, vol. 29, no. 1, pp. 25–48, 2011.

[21] S. H. dan S. Kurniawan, A.Y., "Analisis Daya Saing Usahatani Jagung pada Lahan Kering di Kabupaten Tanah Laut," *J. Forum Pascasarj.*, vol. 31, no. 2, pp. 93–103, 2015.

[22] Sudarmo, *Pestisida*, vol. 43, no. 8. 1991.

[23] R. C. Loehr, "Pollution Control for Agriculture," in *ebook elsevier*, Second Edi., Elsevier, 1984, p. 467.

[24] Tim, "Pedoman Budidaya Kacang Tanah (Arachis hypogaea L)," *Agrokomplekskita*, 2017.

[25] K. Y. Chan, L. Van Zwieten, I. Meszaros, A. Downie, and S. Joseph, "Agronomic values of greenwaste biochar as a soil amendment," *Aust. J. Soil Res.*, vol. 45, no. 8, pp. 629–634, 2007.

[26] M. Yamato, M., Okimori, Y., Wibowo, I.F., Anshori, S. And Ogawa, "Effects of The Application of Charred Bark of Acacia Mangium on The Yield of Maize, Cowpie and Peanut and Soil Chemical Properties in South Sumatera Indonesia," *J. Soil Sci. Plant Nutr.*, vol. 52, pp. 489–495, 2006.

[27] K. Y. Chan, L. Van Zwieten, I. Meszaros, A. Downie, and S. Joseph, "Using poultry litter biochars as soil amendments," *Aust. J. Soil Res.*, vol. 46, no. 5, pp. 437–444, 2008.

[28] B. Glaser, J. Lehmann, and W. Zech, "Ameliorating physical and chemical properties of highly weathered soils in the tropics with charcoal - A review," *Biol. Fertil. Soils*, vol. 35, no. 4, pp. 219–230, 2002.

[29] M. A. Rondon, J. Lehmann, J. Ramírez, and M. Hurtado, "Biological nitrogen fixation by common beans (Phaseolus vulgaris L.) increases with bio-char additions," *Biol. Fertil. Soils*, vol. 43, no. 6, pp. 699–708, 2007.

[30] Haryono, "Tema : ' Intensifikasi Pengelolaan Lahan Suboptimal dalam Rangka Mendukung Kemandirian Pangan Nasional ' Palembang , 20-21 September 2013," in *Intensifikasi Pengelolaan Lahan Suboptimal dalam Rangka Mendukung Kemandirian Pangan Nasional*, 2013, no. September, pp. 20–21.

[31] R. Efendi, "Mempertahankan dan meningkatkan produktivitas lahan kering dan produksi jagung dengan sistem penyiapan lahan konservasi," pp. 978–979, 2009.

[32] Anonymous, "Kabupaten Blitar Dalam Angka," Kabupaten Blitar, 2013.

[33] E. R. Walpole, *Introduction to statistics.* 1968.

IOP Conf. Series: Earth and Environmental Science **406** (2019) 012023 doi:10.1088/1755-1315/406/1/012023

Table1. Difference between Fertilization and Dryland Farming

| Characteristic | Fertile Land | Dry Land |
|---|---|---|
| Irrigation | Technical irrigation, semi - technical, simple | Rainfed |
| Fertility | Generally fertile, not easily degraded/damaged | Generally, less fertile and easily degraded |
| Productivity | Moderate to highand generally can be maintained. | In general, low to moderate and easily reduced |
| Production | In general, moderate to high, can be maintained and improved. | In general, low to moderate, and easy to decline and difficult to increase. |
| Potentials (Large) | Moderate to large but continues to decline | Large and many are not yet optimally managed, and continue to decline |

Source: [30], [31] and [32].

IOP Conf. Series: Earth and Environmental Science **406** (2019) 012023 doi:10.1088/1755-1315/406/1/012023

Table 2. Production Function and Risk Function of Farming in the First Season Dry Peanut Farming.

| Variable | | Coefficient | | T-Ratio | |
|---|---|---|---|---|---|
| Production (f) | Risk Function (h) | Production Function (f) | Risk Function (h) | Production Function (f) | Risk Function (h) |
| C | C | 7.6779*** | -141.3030*** | 25.2465 | -11.3715 |
| | Production | | 16.5770*** | | 11.6654 |
| Land area | Land area | 1.1055*** | -20.1086*** | 15.6910 | -10.4078 |
| Seed | Seed | -0.0640* | 1.9698*** | -1.3969 | 2.6078 |
| Nitrogen fertilizer | Nitrogen fertilizer | 0.0103 | -0.0869 | 0.2497 | -0.1779 |
| Phosphor fertilizer | Phosphor fertilizer | -0.1461*** | 3.6895*** | -3.7583 | 5.9251 |
| Kalium fertilizer | Kalium fertilizer | -0.0031 | 0.0361 | -0.6560 | 0.6440 |
| Manure | Manure | 0.0004 | 0.0215 | 0.1327 | 0.6412 |
| Pesticide | Pesticide | -0.0036 | 0.1895 | -0.3085 | 0.8104 |
| Family labour | Family labour | -0.0633** | 1.2146*** | -1.9140 | 2.8774 |
| Outside family labour | Outside family labour | 0.0484** | -0.9923*** | 2.2859 | -3.8185 |
| R-squared | | 0.9821 | 0.7287 | | |
| Adjusted R-squared | | 0.9809 | 0.6827 | | |
| F-statistic | | 854.8655 | 15.8485 | | |
| Prob(F-statistic) | | 0,0000 | 0,0000 | | |

Source: Primary Data Processed (2016)
Information :
***) Significant at the level $\alpha = 1\%$ t-table $\alpha = 1\% = 2{,}576$
**) Significant at the level $\alpha = 5\%$ t-table $\alpha = 5\% = 1{,}645$
*) Significant at the level $\alpha = 10\%$ t-table $\alpha = 10\% = 1{,}282$ [33].

Table 3. Production Function and Risk Function of Peanut Farmingon Dry Land Second Cropping Pattern after Corn

| Variable | | Coefficient | | T-Ratio | |
|---|---|---|---|---|---|
| Production function (f) | Risk function (h) | Production function (f) | Risk function(h) | Production function (f) | Risk function (h) |
| C | C | 5.8167*** | 8.0469*** | 8.6234 | 7.1287 |
| | Production | | 1.1885*** | | 4.8203 |
| Land area | Land area | 0.7861*** | -0.0695 | 5.4150 | -0.3863 |
| Seed | Seed | 0.1137 | 0.0247 | 1.1100 | 0.2071 |
| Nitrogen fertilizer | Nitrogen fertilizer | 0.0156 | -0.1379 | 0.2395 | -1.1347 |
| Posphor fertilizer | Prosphorfertilizer | 0.0576 | -0.0038 | 0.8298 | -0.3209 |
| Kalium fertilizer | Kalium fertilizer | -0.0154** | 0.0009 | -1.8866 | 0.1405 |
| Manure | Manure | 0.0050 | -0.0774*** | 0.9908 | -2.5399 |
| Pesticide | Pesticide | 0.0679** | -0.1189* | 1.9343 | -1.3656 |
| Famiy labur | Family labour | 0.0881* | 0.0327 | 1.4550 | 0.5860 |
| Outside family labour | Outside family labour | -0.0559* | 0.0271** | -1.5065 | 2.1792 |
| R-squared | | 0.9465 | 0.9457 | | |
| Adjusted R-squared | | 0.9430 | 0.9369 | | |
| F-statistic | | 275.3257 | 107.9095 | | |
| Prob(F-statistic) | | 0.0000 | 0.0000 | | |

Source: Primary Data Processed (2016)

Information:

***) Significant at the level of $\alpha = 1\%$ t-table $\alpha = 1\% = 2{,}576$

**) Significant at the level of $\alpha = 5\%$ t-table $\alpha = 5\% = 1{,}645$

*) Significant at the level of $\alpha = 10\%$ t-table $\alpha = 10\% = 1{,}282$ [33].

# Relationship of Sedimentation Rate to The Structure of Macrozoobenthos Community On Transitional in Ciletuh Bay, Sukabumi District, West Java

**Yuniarti[12], Yudi Nurul Ihsan[1], Syawaludin A. Harahap[1], and Delilla Suhanda[3]**

[1]Departement of Maritime Affairs, Faculty of Fisheries and Marine Sciences Padjadjaran University. Bandung-Sumedang Main Road, Km.21 Jatinangor, West Java, Indonesia.

[2]Centre for Consevation and Management of Maritime Areas Study, Faculty of Fisheries and Marine Science. Bandung-Sumedang Main Road, Km.21 Jatinangor, West Java, Indonesia.

[3]Marine Science Major, Faculty of Fisheries and Marine Sciences Padjadjaran University. Bandung-Sumedang Main Road, Km.21 Jatinangor, West Java, Indonesia.

Corespondent Author: yuniarti@unpad.ac.id

**Abstract.** Ciletuh is a coastal area located in Sukabumi Regency, West Java, which is currently included in the Geopark region in West Java. Ciletuh is a name of a bay that faces directly to the Indian Ocean, hence the similarity of its ocean characteristic. At present Ciletuh Bay has undergone massive sedimentation due to the large amount of sediment entering the surrounding bay coming from Ciwaru and Cimarinjung River and the existence of many floating net cages around the Ciletuh Bay, all of which empties into Ciletuh Bay. High sedimentation rate will disrupt environmental conditions especially organism found in sediments such as benthos. This research was designed using descriptive analytic methods in the form of field observations to determine the research station, sediment sampling and supporting parameters carried out at the research station point. The study was conducted at 5 sampling stations based on different characteristics. The purpose of this research is to see the relationship between sedimentation rate and structure of the macrozoobenthos community. Result presented that the species dominating the bay areas are *Nassarius* sp., *Cerithium* sp. *Clypeamorus* sp. *Rhinodavis* sp. *Certhiidae* , *Coralliophila* sp., *Turricula* sp., *Olivia caerullea*., T*hais* sp., *Phos roseatus*, *Hemifusus* sp. *Diplomariza* sp, *Neritadae*, *Donax* sp., *Pyramidella* sp., *Terebia* sp. Sedimentation rate calculated as 0,11 gr/cm$^2$/day; 0,13 gr/cm$^2$/day; 0,07 gr/cm$^2$/day; and 0,3 gr/cm$^2$/day. Based on sediment grain size, types of sediment founded on each station are (1) gravely mud, (2) sandy mud, (3) sandy mud with slight gravels, (4) mud with slight gravels, and; (5) sandy silt.

***Keywords:*** *Ciletuh Bay, Community Structure of Macrozoobenthos, Sedimentation Rate.*

## 1. Introduction

Ciletuh Bay is an area that has a unique diversity of geological rocks that makes this region one of the geopark areas in the Sukabumi region, West Java (Hardiyono et al 2015). Other than its amusing geopark area to be studied, the waters in Ciletuh Bay is also an intriguing area to study its oceanography due to the condition of the waters which are influenced by the characteristics of the Indian Ocean and notable rivers that flows into its water body, created a mixture of fresh water and sea water (Roswanty and Purnomo, 2014). The waters in Ciletuh Bay are often used by the surrounding community for fishing activities such as shrimp, grouper fish and lobsters farming that affect Ciletuh Bay's water quality and disrupt the life of macrozoobenthos that live in these waters. The level of sedimentation in Ciletuh Bay is considered high due to high human activities that bring negative impacts such as fishing activity when the water is receding to be hampered, the imbalance of life of aquatic organisms, the potential for accumulation of organic matter carried from rivers, and can cause overflow of water masses river so that the land area becomes flooded (Pamuji et al 2015).

The purpose of this study was to determine the value of the sedimentation rate and texture of the sedimentation and the relationship of the sedimentation rate to the structure of the macrozoobenthos community in the waters of Ciletuh Bay, Sukabumi Regency, West Java.

## 2. Materials and methods

This research was conducted in October 2018 in the waters of Ciletuh Bay using 5 sampling locations where the location of Station 1 is on Kunti Island; Station 2 of the Shrimp Pond; station 3 Mandra Island; station 4 Muara Sungai Ciwaru and station 5 Muara Sungai Cimarinjung and the processing of sediment samples was conducted at the Marine

IOP Conf. Series: Earth and Environmental Science **406** (2019) 012024 doi:10.1088/1755-1315/406/1/012024

Conservation Laboratory and macrozoobenthos analysis at the Ecology Laboratory of Padjajaran University. A picture of the location and the sampling point of the study can be seen in Figure 1.

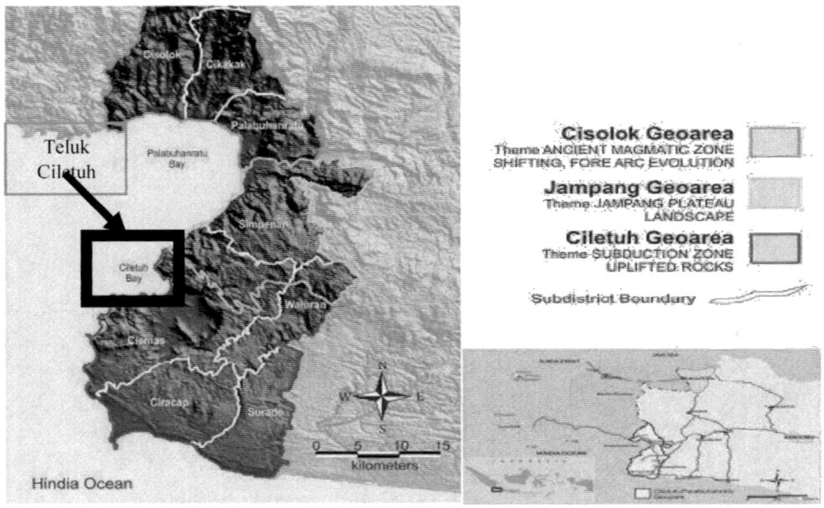


Figure 1. Research Location

The research method used was a survey method with random sampling or purposive sampling at five stations including data on water quality (temperature, brightness, salinity, DO, pH and flow velocity), sediment samples for sedimentation rate analysis with sediment trap tools, macrozoobenthos samples by using ekman grab and filter. Retrieval of water quality temperature, DO, pH and salinity was repeated 3 times then for the collection of current data (in 1 minute) and brightness, the device was removed in the water column, calculated the length of the rope being drawn and repeated 3 times. Sediment sampling for the analysis of sedimentation rates is taken using a sediment trap embedded in the bottom of the water on each station, left for 10 days then the results of the sediment trapped inside were removed and put into the ziplock to be weighed in mass. In sampling of macrozoobenthos, sediment will be taken using an ekman grab tool then the sediment was inserted into the filter to divide macrozoobenthos from the sediment then stored the samples in ziplock and gave it a 4% formalin solution as preservative.

Sediment sample analysis is done by separating sediment grains based on grain size using a shieve shaker tool with texture classification using KUMMOD SEL software based on the Wentworth scale, calculations on sedimentation rates using the formula from APHA 1971:

$$\text{Sedimentation rate} = \frac{10000}{\pi r2} (A\text{-}B) \ (\text{grams/m}^2\text{/week})$$

Notes:

A = weight of aluminum foil + sediment after heating (grams)

B = initial weight of aluminum foil (grams)

r = radius of *sediment trap* (cm)

**Individual Abundance**

$$KI\ (ind/m^3) = a\ ni$$

Notes:

a : Conversion number

ni : Amount of individuals in species i

**Diversity Index**

$$H' = -\sum pi \ln pi$$

Notes:

H' = Species diversity index

S = The number that makes up the community

Pi = Ratio of the number of individuals in species i (ni) to the number of individuals in the community (N)

**Uniformity Index**

$$e = H'/Hmax$$

Notes:

e = Uniformity Index

H' = Diversity Index

Hmax = ln S

**Domination Index**

$$C = \sum (ni/N)^2$$

Notes:

C = Domination Index

N = Number of individuals in all species

ni = Amount of individuals in species i

Data analysis that obtained then analyzed using regression-correlation method.

## 3. Results and Discussion

### 3.1. Water Quality Parameters

Physical and chemical parameters of the waters will influence the presence of macrozooenthos both directly and indirectly. Some parameters such as temperature, salinity, pH, dissolved oxygen, water transparency, and depth will support the fertility of the waters. Water Quality Parameters is presented in Table 1.

| Station | latitude | longtitude | Salinity | Temperature | pH | DO | Brightness | Depth |
|---|---|---|---|---|---|---|---|---|
| 1 | -7.186861° | 106.431861° | 35 | 29,5 | 7,97 | 34 | 200 | 200 |
| 2 | -7.189389° | 106.443500° | 34 | 30 | 7,85 | 7,2 | 35 | 80 |
| 3 | -7.182806° | 106.448667° | 34 | 28 | 7,74 | 6,4 | 80 | 90 |
| 4 | -7.182845° | 106.457651° | 32 | 28 | 7,3 | 5,9 | 70 | 210 |
| 5 | -7.171157° | 106.464503° | 30 | 27 | 7,55 | 5,9 | 50 | 180 |

Temperature values in the measurement range between 27-30°C. Based on the temperature value, the waters of the Ciletuh bay are included in the appropriate category to support the survival of macrozoobenthos. Ronsenberg (1987) in Nybakken (1992), states that temperature influences the movement, metabolism rate and macrozoobenthos mortality. In addition, temperatures that can endanger the life of macrozoobenthos range between 35-40°C (Welch, 1980).

Salinity value in Ciletuh Bay ranges from 30-35°/₀₀. Salinity affects the life of macrozoobenthos including the rate of growth and survival of aquatic biota (Yeanny, 2007). This salinity level is still in the normal limit of coastal water and mixed water.

pH value of waters is one of the important parameters in determining water quality (Prescott et al., 2004). The pH of Ciletuh Bay waters ranges from 7.3 to 7.97, the pH value is included in the category of not acidic or non-basic or can be said to be normal. Aquatic organisms have different abilities in tolerating aquatic pH. Mortality can be caused by a lower pH than it is caused by a high pH (Nuriyawan et al., 2016)

Concentration of dissolved oxygen ranged from 5.9-34 mg/L. The highest concentration was found at station 1, which reached 34 mg/L. This is due to the presence of sunlight that enters the body of the waters thus helping photosynthesis in supplying oxygen to the waters (Salmin, 2005). According to Araoye (2009), the diffusion of oxygen into water bodies naturally slows down except in conditions of strong turbulence, therefore some of the important sources of oxygen are through the process of photosynthesis by aquatic organisms and plants.

The results of the measurement of water transparency range between 35 - 200 cm. the turbidity level of a waters is influenced by suspended and colloidal materials contained in water, for example sediment particles, organic matter, plankton or microorganisms (Mason, 1981).

### 3.2. Grain Size

Results of the calculation of the percentage and type of sediment texture at each station in Ciletuh Bay can be seen in Table 2.

Table 2. Percentage and Texture Type of Sediments at Each Station in Ciletuh Bay

| Station | Sediment Fraction (%) | | | Sediment Type |
|---|---|---|---|---|
| | Gravel | Sand | Silt | |
| 1 | 5,4 | 46,1 | 48,4 | Gravely Mud |
| 2 | 0 | 35 | 65 | Sandy mud |
| 3 | 0 | 11,3 | 88,7 | Sandy mud, slight gravels |
| 4 | 1,7 | 5 | 93,3 | Mud, slight gravels |
| 5 | 0 | 14,3 | 85,7 | Sandy Silt |

Based on the results of sediment analysis in Ciletuh Bay, Sukabumi Regency, West Java, gravel fraction values of 1.7 - 5.4% were obtained. The percentage of sand fraction ranges from 5 - 46.1%. And the percentage of silt ranges between 18.4 - 93.3%. Based on these data it can be seen that the percentage of silt is the largest sedimentary fraction compared to the gravel and sand sediment fraction, so that after analysis using the

IOP Conf. Series: Earth and Environmental Science **406** (2019) 012024 doi:10.1088/1755-1315/406/1/012024

KUMMOD (Soil Texture Segmentation) software (Mahbub, 2006) it can be seen that the type of sediment texture in Ciletuh Bay is silt.

Sediment with silt fraction dominates at each stations, this can be caused by erosion activity in the highlands so that carrying soil particles carried by the river flow towards the estuary. Another factor that can be the cause of more clay fractions is due to the current speed at each station classified in the medium category so that the particles easily settle in the waters. If the current is weak, the particles that settle are dust and silt particles (Nybakken, 1992).

The amount of sediment mass that is transported through a unit of area in units of time is called the sediment rate. Based on the calculation of sediment rate, the data presented in Table 3 is obtained.

Table 3. Data from Sedimentation Rate Analysis Results

| Date of Sediment Trap installation | Date of Sediment Trap Retrieval | Station/Plot | Sedimentation Rate Accumulates ($gr/cm^2$ per day) |
|---|---|---|---|
| 7/10/2018 | 21/10/2018 | Station 1 | 0,11 |
| 7/10/2018 | 21/10/2018 | Station 2 | 0,13 |
| 21/10/2018 | 2/11/2018 | Station 3 | 0,07 |
| 21/10/2019 | 2/12/2018 | Station 4 | 0 |
| 21/10/2018 | 2/11/2018 | Station 5 | 0,3 |

Results of sedimentation rates measurements during the study ranged from 0.3 to 0.11 $gr/cm^2$ per day. Based on these data the sedimentation rate is influenced by the speed of the current. According to Augustine et al., (2013) slower flow velocity will cause particles that are not washed up to settle and form the basic elements of water, while fast currents will wash away dissolved particles. In addition, the high rate of sediment in Ciletuh Bay is influenced by sediment deposition in the bottom waters and the mixing of fresh water and sea water so that it turns into a flooded area or shallow water. The highest sedimentation rate is at station 2 and the lowest is at station 5. Station 4 does not have the sedimentation rate because there are obstacles when installing and taking sediment traps, this is because at station 4 it has a strong current so the sediment trap is not strong enough to be installed at the station.

## Makrozoobenthos

Based on the calculation results of the macrozoobenthos community structure, the data obtained is presented in Table 4.

Table 4. Individual Abundance Calculation Results Data

| Seasonal | Index | | | |
|---|---|---|---|---|
| Change | KI (ind/m$^3$) | H' | e | C |
| Station 1 | 250 | 0,68 | 0,98 | 0,5 |
| Station 2 | 3750 | 1,69 | 0,7 | 0,289 |
| Station 3 | 625 | 1,32 | 0,95 | 0,28 |
| Station 4 | 250 | 0,68 | 0,98 | 0,5 |
| Station 5 | 375 | 1,09 | 0,99 | 0,33 |

From the results of the research that has been done, the value of individual abundance index (KI), diversity (H'), uniformity (e) and dominance (C) that obtained each have different results according to each station. In station 1, each one of the macrozoobentos was identified with the genus *Nassarius sp.* and *Cerithium sp.* In station 2 type of macrozoobenthos that were found *are lipeamorus sp., Rhinodavis sp., Certhiidae sp., Coralliophila sp., Turricula sp., Cerithium sp., Olivia sp., Nassarius sp., Phos roseatus*, and *Hemifusus sp.* At station 3 there were obtained macrozoobenthos with genus of *Diplomariza sp., Nassarius sp., Neritadae sp.,* and *donax sp.* Station 4 obtains macrozoobenthos with the types of *Nassarius sp.,* and *Cerithium sp.* At the last station, station 5, macrozoobenthos with the types of *Pyramidella sp., Nassarius sp.* and *Terebia sp.* were obtained.

## Relationship of Sedimentation Rate to the Structure of Macrozoobenthos Community

The results of the linear regression calculation between the sedimentation rate of abundance, diversity, uniformity, and the dominance of macrozoobenthos in Ciletuh Bay Sukabumi Regency are presented in Figure 2.

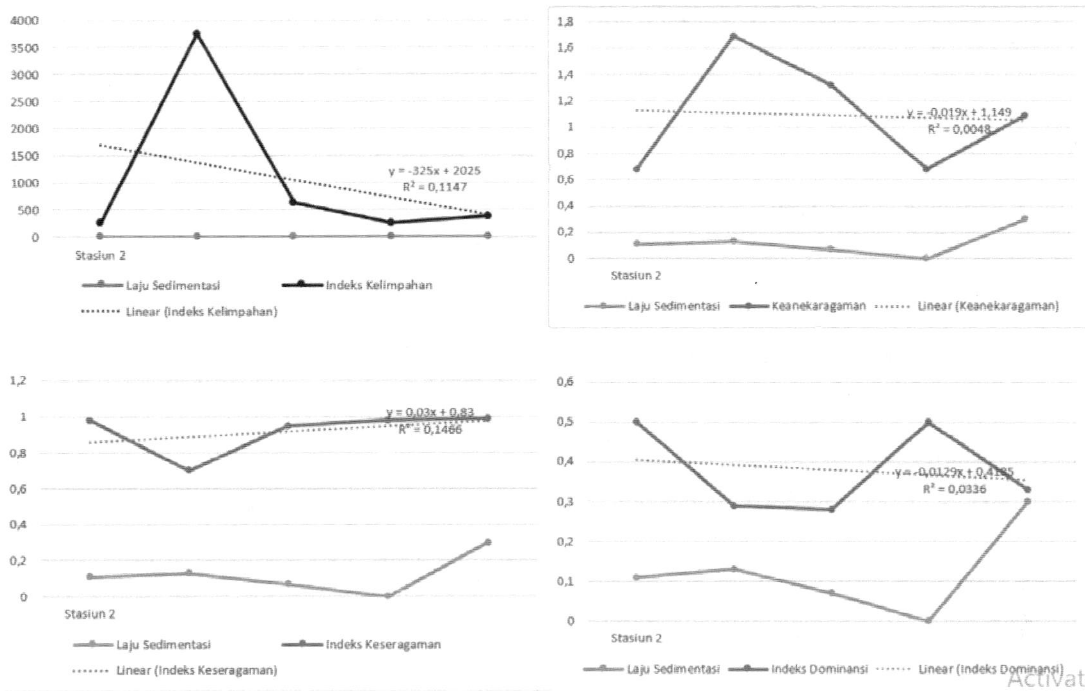


Figure 2. Sedimentation Rate Regression with Index of Abundance, Diversity, Uniformity and Dominance of Macrozoobenthos

Correlation rate of sedimentation to the abundance index of macrozoobenthos individuals is 0.1147, the correlation value of sedimentation rate to macrozoobenthos diversity is 0.0048, the correlation value of sedimentation rate with macrozoobenthos uniformity is 0.1466, the correlation value of sedimentation rate to macrozoobenthos dominance is 0.0336 then the graph shows the relationship between the sediment rate and the index of abundance, diversity, uniformity and dominance of macrozoobenthos. The higher the rate of sedimentation, the higher the abundance, diversity, uniformity and dominance of macrozoobenthos. This is because sediments that are carried by currents and settle at the bottom of the cation generally carry low nutrient content and are needed by macrozoobenthos for life.

## Acknowledgement

This Research was totally supported by Universitas Padjadjaran, Indonesia by HIU Project (Hibah Internal Unpad). We thank to Delilla Suhanda, Riska Zandhi, Fani Wulan Sari,

Hazman Hiwari, and Ridlo Setiadeswan for sampling in Ciletuh Bay. Many thanks to reviewers for their comments that greatly improved the quality of the manuscript.

## Conclusion

1. The highest sedimentation rate is found at station 2 with sediment rate of 0.13 $gr/cm^3/day$ with sandy mud sediment fraction and the lowest is at station 5 with sediment rate of 0.3 $gr/cm^3/day$ with sandy silt fraction.
2. The highest abundance and diversity were 3750 $ind/m^3$ and 1.69 found in station 2 with the dominant macrozoobenthos type being *Nassarius sp.* with a low level of uniformity.
3. *Nassarius* sp. Is the type of macrozoobenthos that is most often found at each station.
4. A very strong correlation was obtained between the sedimentation rate and the uniformity and abundance index of macrozoobenthos.

## Citation and References

[1] Araoye, P.A. 2009. The Seasonal Variation of pH and Dissolved Oxygen (DO2) Concentration in Asa Lake Ilorin, Nigeria. International Journal of Physical Science 4(5): 271-274.

[2] Hardiyono, A., Syarfi, I., Rosana, M. F., Yuningsih, E. Y., Herry, & Andriany, S. S. (2015). Geotourism Potential in the Ciletuh Bay Area, Sukabumi, West Java. Bulletin of Scientific Contribution , 13(2):119-127.

[3] Mason, C.F. 1981. Biology of Freshwater Pollution. Longman. New York. 250p.

[4] Nybakken, J.W. 1992. Marine Biology An Ecological Approach. Translated by Eidman and Bengen. PT. Gramedia Jakarta

[5] Pamuji, A., Muskananfola, M. R., & A'in, C. (2015). Effect of Sedimentation on Macrozoobenthos Abundance at Betahwalang River Estuary, Demak Regency.. Indonesian Journal of Fisheries Sciences and Technology (IJFST), 10(2): 129-135.

[6] Roswanty, M. R., & Purnomo, P. W. (2014). Sedimentation Level in Wudung River Estuary, Wedung District, Demak. Maquares, 3(2):129-137.

[7] Salmin. 2005. Dissolved Oxygen (DO) and Biological Oxygen Demand (BOD) as Indicators to determine Aquatic Quality. Oseana Journal. Volume 30 (3): 21-26

IOP Conf. Series: Earth and Environmental Science **406** (2019) 012024 doi:10.1088/1755-1315/406/1/012024

[8] Welch, E.B. 198.. Ecological Effect of Waste Water. Cambride University Press. Cambride. 337p

[9] Yeanny, M.S. 2007. Macrozoobenthos Diversity in Belawan River Estuary. Department of Biology. Faculty of Mathematics and Natural Sciences, North Sumatra University. Meda. 2(2): 37-4

# Spatial distribution of soil water content in potato horizontal-ridge profile under various ridge dimensions

**Krissandi Wijaya, Purwoko Hari Kuncoro, Ardiansyah, Poppy Arsil, and Hilda Aryani**

Study Program of Agricultural Engineering, Department of Agricultural Technology, Faculty of Agriculture, Jenderal Soedirman University
Jl. dr. Soeparno Karangwangkal Purwokerto 53123

E-mail: kwijaya77@gmail.com

**Abstract.** Vertical (slopping)-ridge is generally applied to cultivate potato crop in tropical highland area with intensive use of non-organic fertilizer, by which the soil erosion as well as environmental degradation might be significantly accelerated. On the other hand, horizontal (contour)-ridge has been very effective to reduce the soil erosion in potato cropping field, but yet slightly ineffective to support the optimal crop production, due to the waterlogging in the ridge profile. Dimension of the horizontal-ridge is expected to affect water distribution as well as the waterlogged condition, on which a specific study need to be focused, in order to develop an appropriate drainage system on the ridge. This study was aimed to characterize the spatial distribution of soil water in the horizontal-ridge profile under various ridge dimensions. Totally 9-potato-plots of 300 x 300 $cm^2$ with 5% slope were prepared in Serang village, Purbalingga with various dimensions and replications of the horizontal ridges: 30 x 30 x 30 $cm^3$ (G30), 30 x 40 x 30 $cm^3$ (G40), and 30 x 50 x 30 $cm^3$ (G50). Of each plot, the dielectrically (volumetric)-water content of the horizontal-ridge soil within the grid of 80 x 80 $cm^2$ at the depth of 5, 10, and 20 cm were measured every months using EC5 moisture sensors and EM50 data logger, and then averaged. Based on the given grid, the data were plotted in contour pattern and were then analysed using semivariogram (Gstat and GNUplot) to characterize the spatial distribution and variability, respectively. The results showed that the spatial distribution of dielectrically (volumetric)-water contents tended to be inter-correlated within each ridge dimension, and the data were increased with soil depth and ridge dimension increment. Accordingly, the data were spatially-correlated with *sill* (*C, semivariance*) ranged from 0.00084 – 0.00429 $cm^3$ $cm^{-3}$ and *range of influence* (*a*) ranged from 0.624 – 2.809 m, in which the G50 had most representative and stable trend of the spatial variability (*C* and *a* values) among others.

**Keywords:** Horizontal ridge; potato crop; ridge dimension; soil water content; spatial variability

IOP Conf. Series: Earth and Environmental Science **406** (2019) 012025 doi:10.1088/1755-1315/406/1/012025

## 1. Introduction

Potato is a horticultural product having potential market sell. The product is now becoming the second highest export demand of 5.5 million ton per year or equivalent to USD of 4.23 million [1]. In other hand, the domestic production is limited still, thus some efforts is required to increase the production sustainably, for instance, by applying the appropriate cultivation methods.

In tropical area, potato is generally cultivated in highland areas under slopping-land and -ridge to meet suitable soil physical as well as aeration condition for the good production [2]. The method, however, causes severe soil erosion that accelerate land degradation either on site or in surrounding environments [3,17,23]. Therefore, introducing an alternative method is important to minimize the degradation and to increase the production simultaneously.

A horizontal (contour)-ridge has been introduced and implemented during a last decades. For instance, [3] reported that applying the horizontal ridge in highland potato cultivation might reduce the run-off and soil loss (soil erosion) by 17-34% and 67-73%, respectively, as compared to the vertical-ridge, depend on slopes and cultivation seasons. However, the production tended to reduce by 12-23% in rainy season [2,3]. This was expected due to waterlogging within the ridge profile [4].

Based on the problems above, understanding the spatial distribution of soil water within the horizontal-ridge profile is essential [3,4], as the initial step to determine an appropriate drainage technique as well as to reduce the waterlogged in the ridge-profile. Furthermore, the land management practices including tillage, mulching, and ridging or mounding tends to affect water content and redistribution in soil [5,6,7,18,21]. Unfortunately, the research on this issue especially the relationship between ridging practice and soil-physical properties distribution has been rarely documented.

This study was aimed to characterize the spatial distribution of soil-water in the horizontal-ridge profile under various ridge dimensions. The study might clarify the spatial distribution pattern of soil water in the ridge profile, on which the drainage technique can be then appropriately implemented.

## 2. Materials and method

### 2.1. Experimental site

The research was conducted on a potato cropping plot located at Serang village, Purbalingga (7°14'31'' S, 109°16'50''E) with a typical soil of Andisol (Table 1). As shown in Figure 1, the plot was divided into 9 sub-plots (3 m x 3 m large) having 4 horizontal-ridges each (as replication, 75 cm ridge interval) under various ridge dimensions (widths): 30 x 30 x 30 cm (G30), 30 x 40 x 30 cm (G40), and 30 x 50 x 30 cm (G50) and 5% field slope. On these sub-plots, organic fertilizers: chicken manure and Petroganik (a local commercial product: 12.5% C-organic, 10-25 C/N ratio) were applied with the rate of 10 and 20 ton $ha^{-1}$, respectively in order to meet equivalent rate of NPK with those of the inorganic fertilizer usually used. At the horizontal-ridges of each sub-plot, potato crop was then sowed with cropping interval of 40 cm. Furthermore, plastic sheet (80 cm in height) was vertically installed along each sub-plot edges, and sediment collector was put at each sub-plot lower side.

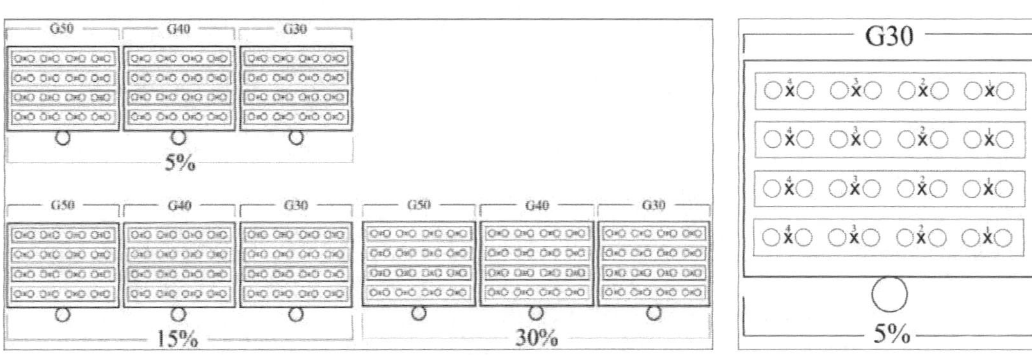


**Figure 1**. Experimental plots and design/treatments (left side), and samples points (right side).

**Table 1**. Physical and chemical properties of Serang's Andisol soil

| **Parameter** | **Dimension** | **Value** |
|---|---|---|
| Texture | | *Loam* |
| Sand | g g$^{-1}$ | 37.44 |
| Silt | g g$^{-1}$ | 48.18 |
| Clay | g g$^{-1}$ | 14.38 |
| Filed capacity | % | 44.07 |
| Permanent wilting point | % | 19.83 |
| C-organic | % | 5.60 |
| pH | | 4.96 |
| Total-N (Available-N) | % (ppm) | 0.52 (57.16) |
| Total-P (Available-P) | % (ppm) | 0.49 (0.61) |

(*Source*: [4])

### 2.2. Sample collection and measurement

Throughout a cultivation period (3 months), the (dielectrically)-volumetric water content ($\theta$) of soil at 5, 10, and 20 cm depth were measured on the horizontal ridges of each sub-plot (nearby potato crops, following the grid/rectangle of 75 cm x 75 cm) every month using EC5 moisture sensor, of which the data were recorded by EM50 data logger (Decagon Inc.) (Figure 1). The disturbed soil samples were also taken from the similar locations at the same depth and time as the undisturbed soil samples collection. The samples were analyzed to determine basic physical properties of the soil (Tabel 1).

### 2.3. Data analysis

Of each sub-plot, the (dielectrically)-volumetric water content ($\theta$) of soil within 3 months at 3 depths sampling were calculated based on [8]'s method and then averaged to meet more representative sub-plot data. Following the given grid/rectangle form, the data were then plotted in contour pattern to clarify the spatial distribution.

The spatial variability of the dielectrically (volumetric) water contents were analysed using semivariogram ($\gamma(h)$) (Spherical model, Gstat and GNUplot) as shown in Equation 1 [9] and 2 [10]. Furthermore, the missing data were equipped using Kriging interpolation (in Surfer 16).

$$\gamma(h) = \frac{1}{2N(h)} \sum_{i=1}^{N(h)} [z(x_i) - z(x_i + h)]^2 \quad (1)$$

where, $N(h)$ is the number of sample pairs separated by a distance $h$, $z(x_i)$ is the measured sample value at point $x_i$, and $z(x_i + h)$ is the measured sample value at a distance $h$ from $x_i$.

$$\gamma(h) = \begin{cases} C_o + C\left[\frac{3h}{2a} - \frac{h^3}{2a^3}\right]; & 0 \le h \le a \\ C_o + C; & h \ge a \end{cases} \quad (2)$$

where, $h$ is the distance of sample pairs, $C_o$ is the nugget effect, $C_o + C$ is sill, and $a$ is the range of influence (for spherical model, effective distance ($a'$) = 2/3$a$).

## 3. Results and discussion

### 3.1. Spatial distribution of the volumetric water content of soil

The spatial distribution of the volumetric water content ($\theta$) of soil within the horizontal ridges of 30 x 30 x 30 cm dimension (G30) at the depth of 5, 10, and 20 cm is shown in Figure 2. As can be seen in the figure, the $\theta$ increased with depth increment. For all the depths, the highest $\theta$ values (0.330 – 0.420 $cm^3\ cm^{-3}$) were spatially distributed along the western and southern side toward the center of the sub-plots, while the lowest $\theta$ values (0.215 – 0.285 $cm^3\ cm^{-3}$) were spatially distributed around the eastern – northern corner of the sub-plots. Amongst, the $\theta$ values were spatially correlated one to another.

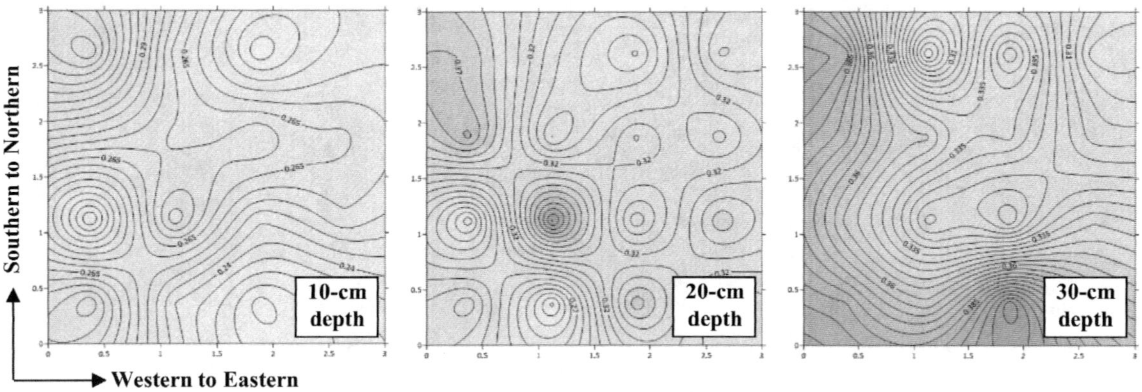


**Figure 2**. Spatial distribution of the volumetric water content of soil within the horizontal ridges of 30 x 30 x 30 cm dimension (G30) at the depth of 5, 10, and 20 cm.

Figure 3 shows the spatial distribution of the volumetric water content ($\theta$) of soil within the horizontal ridges of 30 x 40 x 30 cm dimension (G40) at the depth of 5, 10, and 20 cm. According to the figure, it can be seen that the $\theta$ increased within the depth of 5 to 10 cm and then slightly decreased in the depth of 20 cm, at which the locations were spatially correlated among those depths. For all the depths, the highest $\theta$ values (0.37 – 0.45 $cm^3\ cm^{-3}$) were spatially distributed around the eastern – northern, western – southern, and eastern – southern corner of the sub-plots, while the lowest $\theta$ values (0.22 – 0.27 $cm^3\ cm^{-3}$) were spatially distributed around the center of the sub-plots.

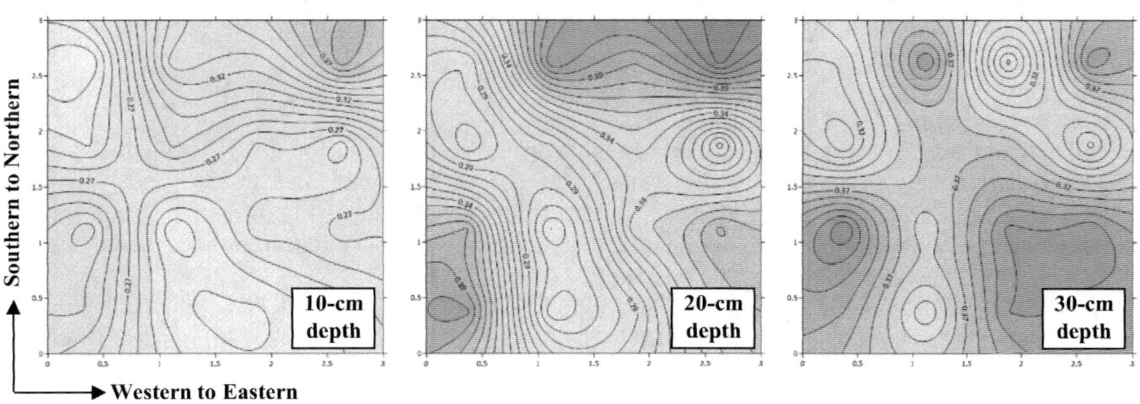


**Figure 3**. Spatial distribution of the volumetric water content of soil within the horizontal ridges of 30 x 30 x 40 cm dimension (G40) at the depth of 5, 10, and 20 cm.

For the horizontal ridge of 30 x 50 x 30 cm dimension (G50), the spatial distribution of the volumetric water content ($\theta$) of soil within the depth of 5, 10, and 20 cm were spatially correlated from one depth to another, at which the values increased with the depth increment (Figure 4). As shown in the figure, the highest $\theta$ values (0.410 – 0.500 $cm^3\ cm^{-3}$) were spatially distributed around the center to southern side of the sub-plots, while the lowest $\theta$ values (0.230 – 0.290 $cm^3\ cm^{-3}$) were spatially distributed along the northern – eastern side of the sub-plots, except those within the depth of 20 cm which were spatially distributed around the eastern side of the sub-plot at the coordinate of [1.5 – 3.0 cm, 1.5 – 2.5 cm].

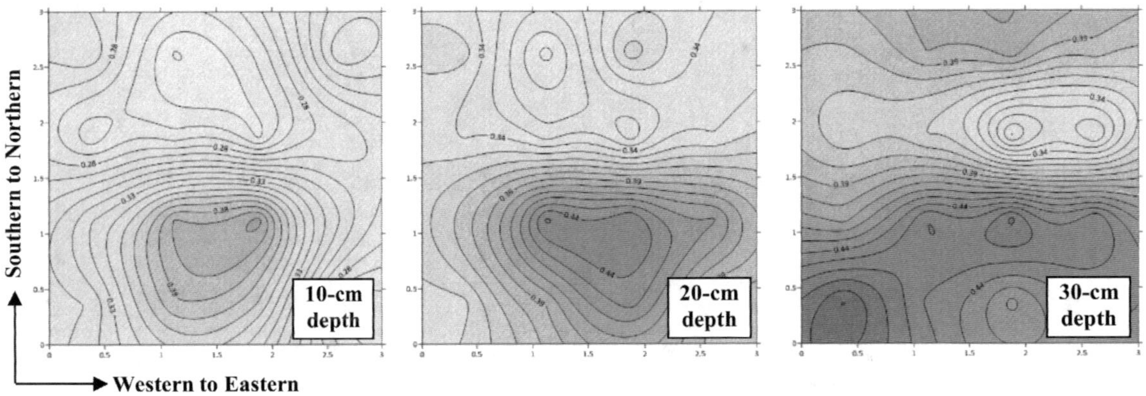


**Figure 4**. Spatial distribution of the volumetric water content of soil within the horizontal ridges of 30 x 30 x 50 cm dimension (G50) at the depth of 5, 10, and 20 cm.

The correlation of $\theta$ spatial distribution pattern among depths indicated that there were a linked of water movement as well as water distribution from one depth to another. The water tended to move and distribute within soil profile in all directions by either infiltration or evapotranspiration [11,12,3,19]. If we took a look more detail, the $\theta$ spatial distribution patterns within the upper depths (5 – 10 cm) were more evenly distributed compared to the lower depth (20 cm). This might be affected by the higher uniformity of soil structure at the upper depth than that at the lower depth [14,4,20].

Comparing those 3 dimensions of the horizontal ridges, it was found that the G50 dimension had more homogeneous in $\theta$ spatial distribution pattern within all depths than G30 and G40 dimensions. The more complex soil physical (e.g., structure, porosity) and biological (e.g., roots density and distribution, organic matters, microorganism/earthworm) properties existed within wider ridge dimensions [15,7,18,21,22] might corroborate the results.

### *3.2. Spatial variability of the volumetric water content of soil*

To understand the variability of spatial distribution pattern of certain collected data, the spherical model of semivariogram can be applied to the data. There is two main variables in the model, namely semivariance/sill (level or value of variation/variability) and range of influence (distance of variation/variablity), by which level/value and distance (degree of dependence/inter-correlated among data pairs) of variation/variability can be characterized respectively.

Figure 5 shows the spatial variability of the volumetric water content ($\theta$) of soil within the horizontal ridges of 30 x 30 x 30 cm dimension (G30) at the depth of 5, 10, and 20 cm. It can be seen from the figure that the $\theta$ data at these consecutive depths showed semivarogram trends with semivariance/sill of 0.00084, 0.00327, and 0.00130 $cm^3\ cm^{-3}$ and distance of 1.216, 0.629 and 1.049 m, respectively. This indicated that the $\theta$ data at the top and low profile were spatially correlated with similar level and distance of variation (similar variability), while those at the middle profile had slightly higher variability than others.

For the horizontal ridges of 30 x 40 x 30 cm dimension (G40), the spatial variability of the volumetric water content ($\theta$) of soil at the depth of 5, 10, and 20 cm is presented in Figure 6. According to the figure, the $\theta$ data at these consecutive depths had semivarogram trends with semivariance/sill of 0.00324, 0.00429, and 0.00239 $cm^3\ cm^{-3}$, and distance of 2.809, 1.216 to 0.624 m, respectively. This implied that the $\theta$ data were slightly different in level and distance of variation among the depths, by mean those had different spatial variability one to another. The $\theta$ data at the top profile showed higher spatial variability than those at middle and low profile.

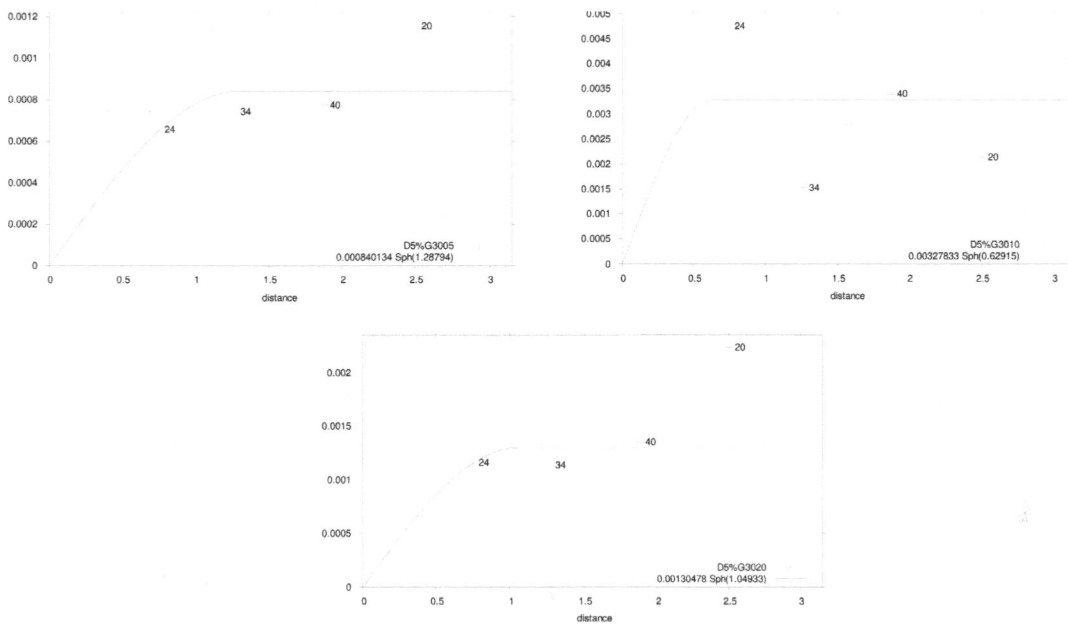


**Figure 5**. Spatial variability (semivariogram) of the volumetric water content of soil within the horizontal ridges of 30 x 30 x 30 cm dimension (G30) at the depth of 5, 10, and 20 cm.

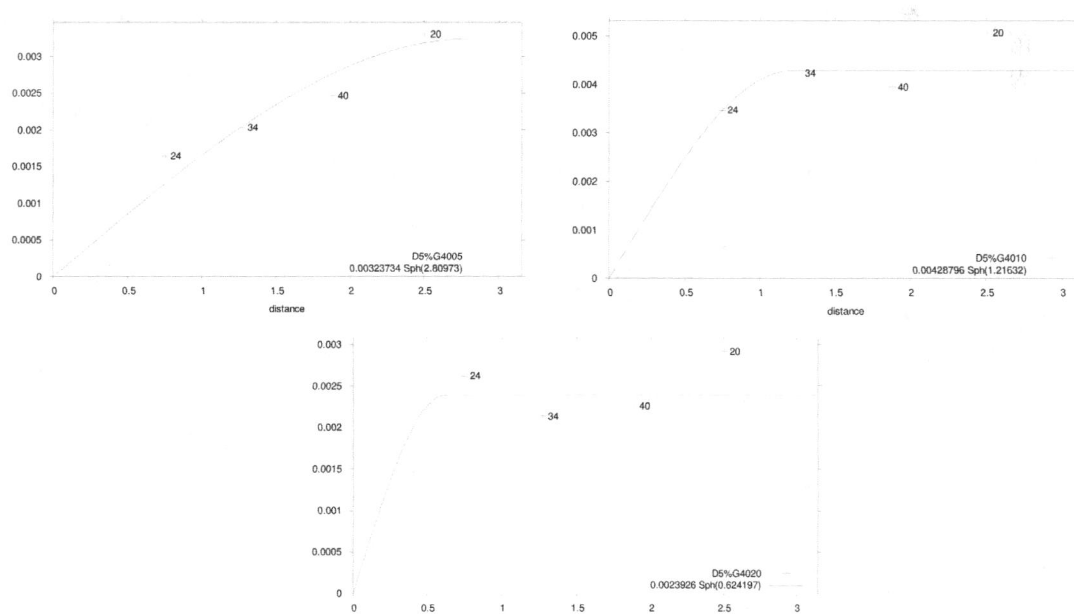


**Figure 6**. Spatial variability (semivariogram) of the volumetric water content of soil within the horizontal ridges of 30 x 40 x 30 cm dimension (G40) at the depth of 5, 10, and 20 cm.

Figure 7 shows the spatial variability of the volumetric water content of soil within the horizontal ridges of 30 x 50 x 30 cm dimension (G50) at the depth of 5, 10, and 20 cm. As shown in the figure, the $\theta$ data at these consecutive depths had semivarogram trends with semivariance/sill of 0.00321, 0.00254, and 0.00331 $cm^3\ cm^{-3}$, and distance of 1.487, 1.547 to 1.513 m, respectively. This revealed that the $\theta$ data were spatially correlated with similar level and distance of variation, by mean those had similar spatial variabilities among the depths.

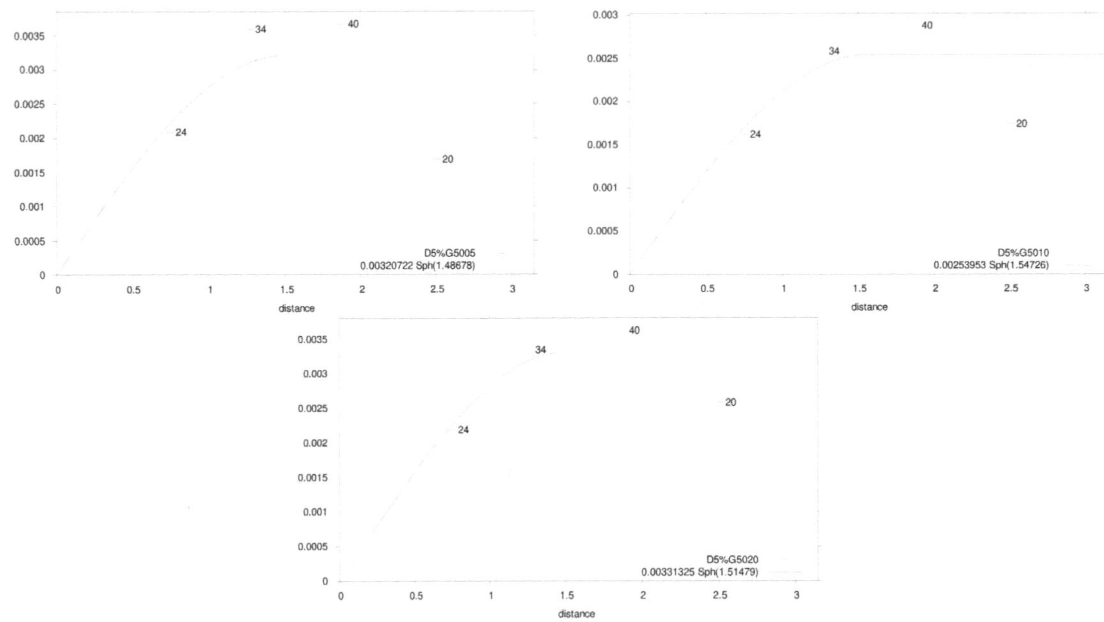


**Figure 7**. Spatial variability (semivariogram) of the volumetric water content of soil within the horizontal ridges of 30 x 50 x 30 cm dimension (G50) at the depth of 5, 10, and 20 cm.

More specifically, the slightly uniformity in $\theta$ spatial variability among the top-low and middle profile (G30, Figure 5) and among all the depths (G40, Figure 6) might be related to the complexity of soil physical (e.g., structure, porosity) and biological (e.g., roots density and distribution, organic matters, microorganism/earthworm) existed at each depth [12,7,20]. The G30 and G40 with relatively narrow ridges width might cause roots development as well as microorganism growth become more concentrated at their middle profiles than the top-low profiles [13,4,19]. This in turn might change the soil structure and water content at the intended profiles more frequent compared to others, thus affected their $\theta$ spatial variability.

On the other hand, the G50 with relatively wider ridges width might provide suitable space for roots as well as microorganisms to be better developed and uniformly distributed at all the depths. Accordingly, the soil structure, water content, and density tended to be stably existed at all the entire profiles [6,16,3,14,22]. This revealed that the ridge dimension had remarkable effect on the $\theta$ spatial variability.

## 4. Conclusion

The spatial distribution as well as spatial variability of the volumetric water content ($\theta$) of soil in the horizontal-ridge profile under various ridge dimensions was successfully characterized by applying a Spherical Model in geostastical analysis. Among the G30, G40, and G50 ridge-dimensions, the $\theta$ data at the depth of 5, 10, and 20 cm were in general spatially correlated and had similar spatial pattern from one depth to another, in which the G50 dimension showed more uniform in the $\theta$ spatial variability within all the depths compared to G30 and G40 dimensions. The G50 dimensions allowed the $\theta$ data at all the depths to be spatially distributed with relatively similar level and distance of variation, ranged from 0.00254 – 0.00321 $cm^3$ $cm^{-3}$ and 1.487 – 1.547 m, respectively. Therefore, the G50 dimension was considerably applicable for the sustainable potato cultivation in tropical highland with certain interval of drainage channel that might be determined from the above results.

**Acknowledgement**
This study was partly supported by Applied Research Grant of the Directorate General for Higher Education of The Ministry of Research, Technology, and Higher Education of Indonesia (Grant Number: Kept. 1636/UN23.14/PN.01.00/2018). The authors thank to Ms. Rarasati and Mr. Fauzi for the help during field and laboratory data collection.

**References**

[1] Badan Pusat Statistik. 2017. *Statistik Tanaman Sayuran dan Buah-buahan Semusim Indonesia 2016*. Badan Pusat Statistik Indonesia, Jakarta (in Indonesian).
[2] Soleh, M., Arifin, Z., Pratomo, G., Santoso, P., and Nitiawirawan, I.G. 2002. *Sistem Usahatani Tanaman Sayuran untuk Konservasi di Lahan Kering Dataran Tinggi Berlereng*. BPPT Jatim. pp. 1-13 (in Indonesian).
[3] Wijaya, K., Setiawan, B.I., and Kato, T. 2010. Spatio-temporal Variability of Soil Physical Properties in Different Potato Ridges Designs in Relation to Soil Erositon and Crop Production. *Proceeding of 2010 INWEPF-PAWEES Intl. Joint Symposium,* Jeju-South Korea, 27-29 October 2010.
[4] Wijaya, K., Kuncoro, P.H., and Arsil, P. 2019. Dynamics of soil physical and chemical properties within horizontal ridgesorganic fertilizer applied potato land. IOP Conf. Series: Earth and Environmental Science **255** (2019) 012024 (doi:10.1088/1755-1315/255/1/012024).
[5] Comegna, V., Damiani, P., and Somella, A. 2000. Scalling the saturated hydraulic conductivity of a vertic ustorthens soil under conventional and minimum tillage. *Soil and Tillage Research* 54: 1-9.
[6] Tominaga, T.T., Cassaro, F.A.M., Bacchi, O.O.S., Reichardt, K., Oliveira, J.C.M., and Timm, L.C. 2002. Variablity of soil water content and bulk density in a sugarcane field. *Aust. J. Soil Res.* 40: 605-614.
[7] Schwen, A., Bodner, G., Scholl, P., Buchan, G.D., and Loiskandl, W. 2011. Temporal dynamics of soil hydraulic properties and the water-conducting porosity under different tillage. *Soil and Tillage Research* 113: 89-98.
[8] Wijaya, K., Nishimura, T., and Kato, M. 2003. Estimation of dry bulk density of soils using amplitude domain reflectometry probe. *J. Jpn. Soc. Soil Phys.* 95: 63-73.
[9] Clark, I. 1979. *Practical Geostatistics*. Applied Science Publisher Ltd., London, UK.
[10] Isaaks, E.H. and Srivastava, R.M. 1989. *An Introduction to Applied Geostatistics*. Oxford University Press, Oxford, USA.
[11] Miyazaki, T. 1996. Bulk density dependence of air entry suctions and saturated hydraulic conductivities of soils. *Soil Sci.* 161, 84-90.
[12] Hillel, D. 1998. *Environmental Soil Physics*. Academic Press, San Diego, USA.
[13] Kirkham, M.B. 2005. *Principles of Soil and Plant Water Relations*. Elsevier Academic Press, USA.
[14] Wijaya, K., Nishimura, T., Setiawan, B.I., and Saptomo, S.K. 2010: Spatial variability of soil hydraulic conductivity in paddy field in accordance to subsurface percolation. *J. Paddy Water Environ*. 8, 113-120.
[15] Ciollaro, G. and Romano, N. 1995. Spatial variability of the hydraulic properties of a volcanic soil. *Geoderma* 65: 263-282.
[16] Wijaya, K., Nishimura, T., Kato, M., and Nakagawa, M. 2004. Field estimation of soil dry bulk density using amplitude domain reflectometry data. *J. Jpn. Soc. Soil Phys.* 97: 3-12.

[17] Zhang, J.H., Frielinghaus, M., Tian, G., and Lobb, D.A. 2004. Ridges and Contour Tillage Effect on Soil Erosion from Hillslope in the Sichuan Basin. China. *Journal of Soil and Water Conservation* (On-line). http://goliath.ecnext.com/.

[18] Zhou, L.M., Li, F.M., Jin, S.L., and Song, Y. 2009. How two ridges and the furrow mulched with plastic film affect soil water, soil temperature and yield of maize on the semiarid Loess Plateau of China. *Field Crops Research* 113, 41-47.

[19] Campbell, G.S. 1985. Soil Physics with Basic: Transport Model for Soil-Plant Systems. Elsevier, Amsterdam, p. 49-59.

[20] Etana, A., Larsbo, M., Keller, T., Arvidsson, J., Schjønning, P., Forkman, J., and Jarvis, N. 2013: Persistent subsoil compaction and its effects on preferential flow patterns in a loamy till soil. *Geoderma* 192, 430-436.

[21] Utami, F. 2012. Pengaruh Arah Guludan Lahan Terhadap Kadar Air Tanah dan Biomassa tanaman Kentang (*Solanum tuberasum L.*). *Skripsi.* Fakultas Matematika dan Ilmu Pengetahuan Alam, Institut Pertanian Bogor, Bogor (in Indonesian).

[22] Arya, L.M., Dierolf, T.S., Sofyan, A., Widjaja-Adhi, I.P.G., and van-Genuchten, M.Th. 1998: Field measurement of the saturated hydraulic conductivity of a macroporous soil with unstable subsoil structure. *Soil Sci.* 163(11), 841-852.

[23] Auerswald, K., Gerl, G., and Kainz, M. 2006. Influence of cropping system on harvest erosion under potato. *Soil and Tillage Research* 89, 22-34.

IOP Conf. Series: Earth and Environmental Science **406** (2019) 012026 doi:10.1088/1755-1315/406/1/012026

# THE EFFECT OF CHANGE IN WATER LEVEL ON THE ABUNDANCE OF LIPTINITE AND VITRINITE MACERALS IN UPPER KELESA FORMATION SHALE IN KUBURAN PANJANG, RIAU

**M. Heri Hermiyanto Zajuli and Taufik Ramli**
Pusat Survei Geologi, Badan Geologi, KESDM
Corresponding author: herisayazajuli@gmail.com; topikrml@gmail.com

**Abstract**

Research on changes in water levels to the abundance of liptnite and vitrinite maceral has focused on Kuburan Panjang area of the Sumai sub-Basin, Central Sumatra Basin. The subject of this research is shale rock from the Eocene-Oligocene Kelesa Formation. Series of rocks from bottom to top shows repetition between shale, sandstone and mudstone strata bounded by conglomerates at the top and bottom. The composition of organic material consists of vitrinite ranging from 0.20 - 5.0%, liptinite 0.60 - 4.70%, pyrite 0.20 - 16.00%, carbonate maceral 0.20 - 24.2%, and clay mineral which is the most dominant component, ranging from 71.60 - 98.00%. Based on the results of TOC analysis, the abundant organic material of the Kelesa Formation shale has a TOC value of 1.18% to 7.17% indicating the ability of shale as a source rock is very good. The presence of organic material in the study area shows that there is a cycle of enrichment of organic material from bottom to top. The smaller the ratio of pristane/phytane (anoxic condition, high water level), the smaller the content of liptinite and the higher the content of vitrinite. Otherwise, the greater the ratio (oxic condition, low water level), the higher the content of liptinite and the lower the content of vitrinite. Change in water level during sedimentation has effect on the quality of shale as a source of unconventional hydrocarbon energy in the Kuburan Panjang area.

Keywords: Kelesa Formation, Eocene-Oligocene, Central Sumatra Basin, Vitrinite, Liptinite

**INTRODUCTION**

IOP Conf. Series: Earth and Environmental Science **406** (2019) 012026 doi:10.1088/1755-1315/406/1/012026

Research on changes in water levels to the abundance of liptinite and vitrinite masals has focused on Eocene-Oligocene shale rocks in the Sumai Sub-Basin, Central Sumatra Basin, Riau. This research area occupies the southern part of the Central Sumatra Basin. Administratively, the research area is included in Rengat Regency, Riau Province (Figure 1). This area is included in a 1: 250,000 scale geological map, Rengat Sheet (8,10).

The objectives of this study are:

- To know the effect of fluctuations in changes of water level within the deposition process on the content of liptinite and vitrinite.
- To understand the relationship between the quality of hydrocarbon flakes and the deposition process and the mass characteristics of constituent organic material.

The benefit of this study is to provide an overview on the dynamics of precipitation, especially in relation to fluctuations and abundance of organic material in the upper shale layer of Kelesa Formation in the Kuburan Panjang, Riau.

Based on the regional geological map of Solok Sheet (8), the oldest Tertiary rocks revealed around the Nusa Riau, Manunggal and TBS coal fields, Taluk Kuantan Regency is Lower Telisa Formation. The formation is spread and extending towards southeast-northwest in the western boundary of the basin and covers, inharmoniously, Pre-Tertiary rocks of Kuantan Formation. In a total thickness of 300 meters, the Lower Telisa Formation is composed of quartz sandstone, clay bed, andesite breccia, glauconitic sandstone, pisolytic tuff, and mudstone containing seam of coal. Stratigraphically, the Lower Telisa Formation is likely to be comparable to Pematang Formation in Central Sumatra Basin and Lakat Formation in Lembar Rengat (Suwarna et.al., 1994) (Figure 2) which is consisting of a multi-material conglomerate, quartz sandstone, carbonate siltstone and claystone carbon, coal, tuff and siderite bintal. The two rock units are in the same basin, but the first unit is on the western edge of the basin and the second on the southern edge.

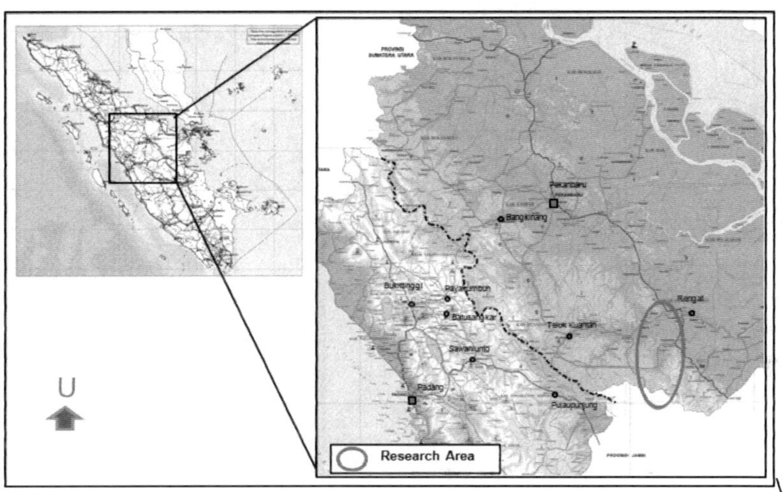


Figure 1. Sumai sub-basin research location, Central Sumatra Basin.

Furthermore, other Lower Members of Telisa Formation (Tmtl) in the coal mining area of PT Karbindo (Kiliranjau) consists of Light gray mudstone, brownish, greenish with conglomerate sandstone lenses; coal-carbon rocks, and brown/gray flakes. Some reseachers claim that the rock deposits in the area are Paleocene – Eocene – Oligocene Dike Groups. Some researchers contend that brown Eo-Oligocene flakes on the coal are deposited in a lake where organic-ric laminates are

IOP Conf. Series: Earth and Environmental Science **406** (2019) 012026 doi:10.1088/1755-1315/406/1/012026

formed during the dry season, whereas clay-rich laminates are formed during the rainy season. This means that the rock units in Kiliranjau area should be correlated to Kelesa Formasi (Teok) in Lembar Rengat consisting of conglomerate or sandstone conglomerate, claystone, siltstone, and coal deposited in fluviatile to lascustrine formations (Suwarna et.al., 1994) and Upper Sangkarewang Formation or Lower Sawahlunto in Ombilin Basin. In line with characteristics of lithology, age, and depositional area, the Shale unit in the research area can be compared to the Brown Shale Formation of Pematang Group (Heidrick and Aulia, 1993) or Eocene-Oligocene Kelesa Formation (Teok) (Suwarna, et.al., 1994).

## METHODOLOGY

### Laboratory Data

After activities in the field have been completed, the laboratory analysis of selected samples is performed in the Geological Survey Center laboratory and other institutions, including:

**a. Organic Petrographic Analysis**

Organic petrologic analysis was performed using a microscope on the basis of the optical method of reflected light with and without fluorescence. This examination aims to describe differences in types of organic material and mineral, measure the evolution of thermal ratings of organic material (level of maturity) based on the measurement of reflectance indexes on vitrinite, especially DOM (disseminated

IOP Conf. Series: Earth and Environmental Science **406** (2019) 012026 doi:10.1088/1755-1315/406/1/012026

organic material) to support the determination of depositional environment and rock maturity.

**b. Gas Chromatograph (GC) and Gas Chromatograph Mass Spectrometry (GC-MS) Analysis**

Bitumen fraction analysis is referred to, generally, as biomarker (biological marker) analysis. Compounds and classes of compounds found in oil and bitumen are called biomarker. Biomarker is useful to know the origin of organisms where bitumen or oil is located, as a parameter of diagenesis during the process of accumulation of organic material which is related to the maturity level of a rock. Determination of maturity with bitumen fraction can use the results of gas chromatographic analysis and GC-MS in the form of normal alkanes, sterane, and triterpane.

Some biomarkers used in source rock analysis are normal alkane, sterane, and triterpane. Biomarker is compound derived from pioneering biogenic molecule, so-called molecular fossil. Basically, certain precursor or molecule can be identified, but sometimes it can only be limited to the precursor to a few types. This molecular fossil cannot be used as index fossil for certain organisms.

***Normal Alkane***

Normal alkanes are the simplest series of hydrocarbons, having a straight chain structure caused by saturated hydrocarbon molecules. Normal alkanes arises in bitumen and oil due to the presence in fat and algae and the catagenic formation of long chain compounds such as fatty acids and alcohols.

Waples (11) explains that most of normal alkanes in higher plants have odd numbered carbon atoms especially on carbon atoms 23, 25, 27, 29 and 31, whereas marine algae produce normal alkanes having a maximum distribution only on carbon atoms 17 or 22, according to the current species. The appearance on the chromatography diagram shows a very sharp distribution and tends to have no odd or even numbers on the carbon atom (Table 1). Most of sediments receive normal alkane contribution from two sources, land and sea, so the chromatographic diagram will reflect a mixture of the two.

Table 1. Four important classes of biomarkers and their prazats (Waples, 1985)

| Biomarker | | Prazat |
|---|---|---|
| Normal Alkane | Normal Alkane (> 22) | higher plant wax |
| | Normal Alkane (< 22) | Algal fat |
| Isoprenoid | Isoprenoid (< 20) | A variety of chlorophylls |
| | Isoprenoid (> 20) | Fat or chlorophyll from hipersaline algae |
| Triterpane | | Triterpenoid |
| Sterane | | Steroid |

***Isoprenoid***

Isoprenoid is compound consisting of carbon atoms merged in certain rules with the methyl group ($CH_3$) bound to each of fourth carbon atom. The common isoprenoids used are pristane (C19) and fitane (C20). Pristane is an isoprenoid having 19 carbons, while fitane has 20 carbons (Figure 2).

IOP Conf. Series: Earth and Environmental Science **406** (2019) 012026 doi:10.1088/1755-1315/406/1/012026

According to Waples (1985), isoprenoid is a good indicator to detect the biogenic origin of bitumen and hydrocarbon, however, is it very limited to certain organisms. Isoprenoid is found in only specific depositional environment. The ratio of pristane to phytane can be used as an indicator to detect the oxygen level at the time of diagenesis; a high pristane/fitane ratio characterizes the environment under oxidation condition (Table 2), while according to Illich (3) the high ratio reflects association with sediment affected by terrestrial environment.

| Structure | Number of Carbon Atoms | Name |
|---|---|---|
| 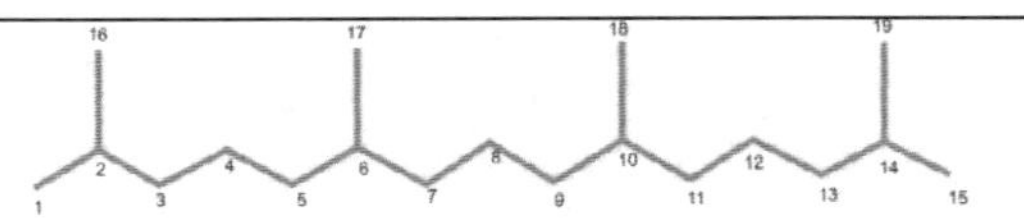 | 19 | pristane |
| 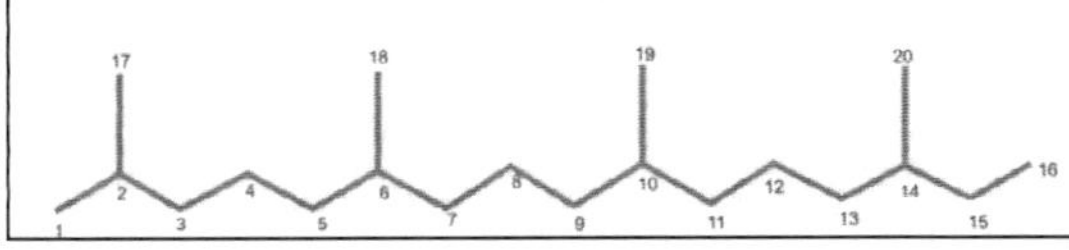 | 20 | fitane |

Figure 2. Structure of pristane and phytane isoprenoids.

Table 2. Comparison of pristane and phytane as indicators of depositional environment (Waples, 1985).

| **Type of sediment** | **Pristane/fitane** |
|---|---|
| Anoxic marine sediment | < 1 |
| Oxic marine sediment | 1 – 3 |
| Coal | > 3 |

## RESULTS

### Stratigraphy of Research Area

The study area is located in the Sumai Sub-Basin, Central Sumatra Basin which is, administratively, included in the Rengat Regency area. Geologically, the study area is included in Geological Map at Scale 1: 250,000 of Lembar Rengat and Lembar Solok (Figure 3).

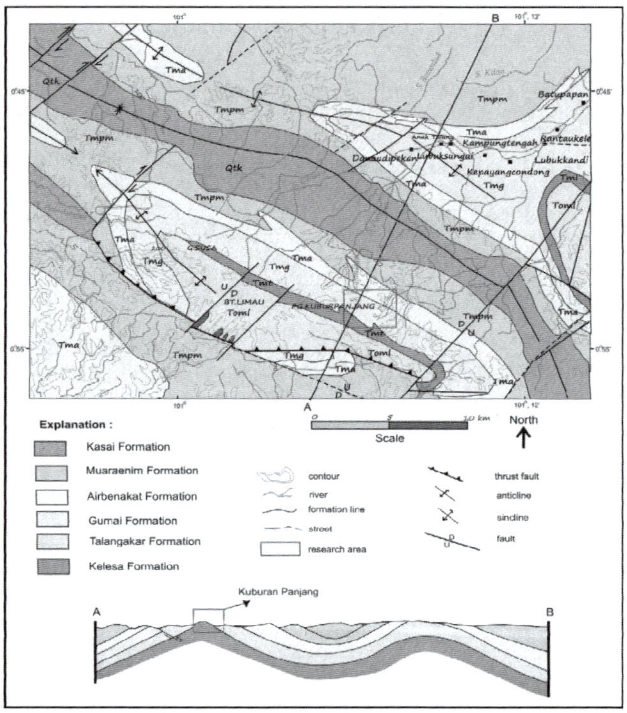


Figure 3. Regional geological map and geological cross section of Kuburan Panjang, Sumai Sub_basin, Riau (modification [8,10]).

The rocks composing Kelesa Formation (Figures 3 and 4) in the study area are composed of conglomerate, sandstone, mudstone and shale. Overall in Kuburan Panjang, the thickness of rocks composing Kelesa Formation on the Punti Kayu River is approximately 80 meters. Series of rocks from bottom to the top shows the repetition between layers of shale, sandstone and mudstone bounded by conglomerates at the top and bottom.

Figure 4. Exposure of the upper Kelesa Formation showing shale rock, conglomerate on the Punti Kayu river.

The most dominant layer is shale having a thickness about 27 meters. Shale, dark brown, grayish, soft, plastic, lamination alternated with mudstone, flaggy, gray, compact, having well coat of 1-1.5 cm.

Sandstone, brown to gray, moldy, consolidated, fine granulous, fragmentary composition was dominated by quartz and feldspar. Conglomerate, polymic, compact, bad sortation, halfway angled, halfway spherical, sand matrix, fragment was dominated by basalt, quartzite and andesite. Mudstone is in interlude with very fine sandstone, brown sandstone with light gray insert, platy-flaggy, soft-hard, well laminar, parallel layers.

Stratigraphically, there is a cycle of repetition between interlude of sandstones and shale and mudstone. Starting from the bottom, we found interlude of sandstone with sandstone dominated by sandstone, to the top we found conglomerate which is boundary of a facies at shale facies having a thickness about 9 meters. In the middle section is a cross section between sandstone and mudstone and shale. The alternation shows that there is a change in current that causes repeatedly precipitation of smooth and grainy material. On the top of cross section (Figure 5) we found a depositional facies from the series of quite thick shale, approximately, 12 meters. This thick shale deposition requires a precipitation system with a calm current persisted for a long time. Based on Nichols (5), this precipitation is, generally, occurring in deeper part of freshwater lake. Based on the material, the author interprets the depositional environment of this unit is a lake (lacustrine) environment.

Broadly speaking, the precipitation system from bottom to top is a cycle from the deposition of fluviatile to lacustrine. It is a gradual change in facies in the river system towards lacustrine. Lacustrine can be formed when there is an large abundance of water, so that the area that was originally a river is turned into a lake. Specifically for the fluviatile environment in the center of cross section, we found thin repetition between sandstone and fine rock (shale and mudstone) constituting a product or particularly rapid change in current. This can happen at the curve of river or meandering river.

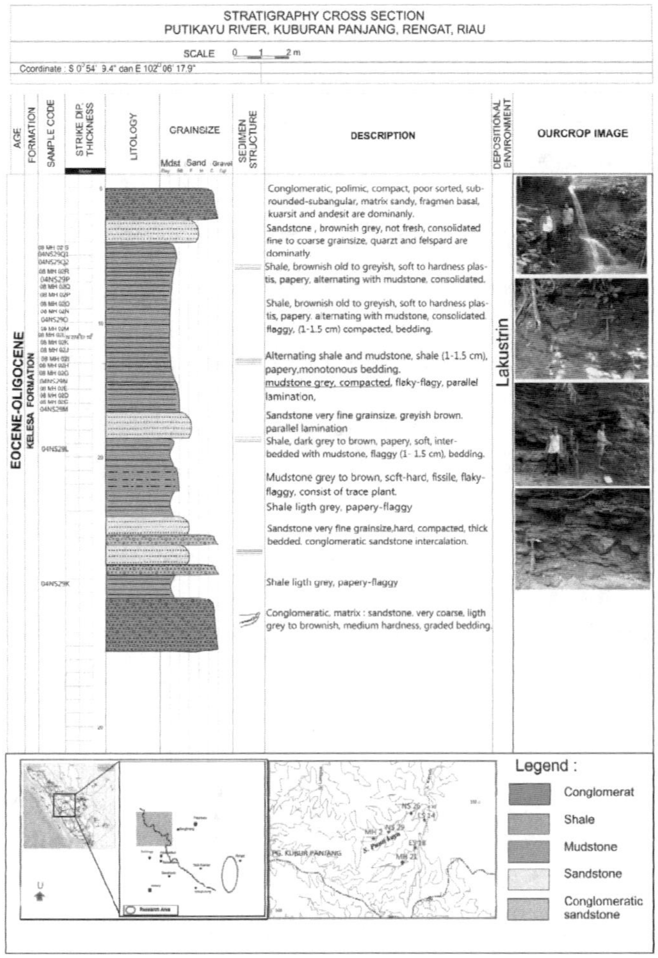


Figure 5. Stratigraphic cross section of the upper Kelesa Formation on the Puntikayu River

## Organic Matter

The results of organic petrographic analysis show that the upper Kelesa Formation shale is arranged by organic material such as vitrinite and liptinite. Vertically (Figures 6 ), there is fluctuation in the presence of liptinite and vitrinite.

IOP Conf. Series: Earth and Environmental Science **406** (2019) 012026 doi:10.1088/1755-1315/406/1/012026

Liptinite appear to be more dominant than vitrinite which are, vertically, present in all rock samples.

The results of TOC analysis and "Rock Eval Pyrolysis" (REP) show, vertically, the very fluctuative content of carbon, sometimes suggesting an abundant attendance, sometimes a little (Table 3). Comparison between TOC and Index Hydrogen shows a vertical similarity of pattern in stratigraphic cross section. When TOC is large, the value of Index Hydrogen is large, as well. This shows that the content of carbon has significant effect on the value of Index Hydrogen.

Table 3. Results of organic petrographic analysis in Kuburan Panjang, Rengat, Central Sumatra Basin

| No. | Sample | Litologi | Dv | V | Re | Btm | Sp | Sb | Cu | Alg | Exu | E | Cl | Carb | Py | MM | Rvmin | Rvmax | Rv |
|---|---|---|---|---|---|---|---|---|---|---|---|---|---|---|---|---|---|---|---|
| 1 | MH 02 S | Shale, olv gy | 0.00 | 0.00 | 0.00 | 2.90 | 0.00 | 0.00 | 0.00 | 0.60 | 0.00 | 0.00 | 96.00 | 0.00 | 0.50 | 96.50 | – | – | – |
| 2 | NS. 29$Q_1$ | Shale, brngy, papery | 3.00 | 3.00 | 1.00 | 0.00 | 0.40 | 0.00 | 0.00 | 0.00 | 0.00 | 1.40 | 87.60 | 2.00 | 6.00 | 95.60 | 0.34 | 0.48 | 0.43 |
| 3 | NS. 29$Q_2$ | Shale, gy-dkgy, papery | 1.00 | 1.00 | 3.00 | 0.00 | 0.00 | 0.00 | 0.00 | 0.00 | 0.00 | 3.00 | 88.00 | 3.00 | 5.00 | 96.00 | 0.21 | 0.33 | 0.27 |
| 4 | MH 02 R | Shale, dk gy | 0.20 | 0.20 | 0.00 | 3.40 | 0.00 | 0.00 | 0.00 | 1.20 | 0.00 | 4.60 | 95.00 | 0.00 | 0.20 | 95.20 | – | – | – |
| 5 | NS. 29P | Shale, gy, silty, papery | 0.80 | 0.80 | 0.80 | 0.00 | 0.00 | 0.00 | 1.4 | 0.00 | 0.00 | 2.20 | 71.60 | 24.2 | 1.2 | 97.00 | 0.16 | 0.22 | 0.18 |
| 6 | MH 02 Q | Shale, brnsh gy | 0.00 | 0.00 | 0.00 | 3.20 | 0.00 | 0.00 | 0.00 | 1.50 | 0.00 | 4.70 | 95.00 | 0.00 | 0.30 | 95.30 | – | – | – |
| 7 | MH 02 P | Shale, olv gy | 0.00 | 0.00 | 0.00 | 2.80 | 0.00 | 0.00 | 0.00 | 1.40 | 0.00 | 4.20 | 95.00 | 0.20 | 0.60 | 95.80 | – | – | – |
| 8 | MH 02 O | Shale, olv gy | 0.00 | 0.00 | 0.00 | 2.50 | 0.00 | 0.00 | 0.00 | 1.80 | 0.00 | 4.30 | 92.00 | 0.50 | 3.20 | 95.70 | – | – | – |
| 9 | MH 02 N | Shale, med dk gy | 0.00 | 0.00 | 0.00 | 1.80 | 0.00 | 0.00 | 0.00 | 0.70 | 0.00 | 2.50 | 93.00 | 0.20 | 4.30 | 97.50 | – | – | – |
| 10 | NS. 29 O | Shale, brngy, papery | 1.00 | 1.00 | 1.00 | 0.00 | 0.00 | 0.00 | 0.00 | 0.00 | 0.00 | 1.00 | 92.00 | 1.00 | 5.00 | 98.00 | 0.31 | 0.56 | 0.43 |
| 11 | MH 02 M | Shale, olv gy | 0.00 | 0.00 | 0.00 | 1.40 | 0.00 | 0.00 | 0.00 | 0.90 | 0.00 | 2.30 | 96.00 | 0.30 | 1.40 | 97.70 | – | – | – |
| 12 | MH 02 L | Shale, olv gy | 0.00 | 0.00 | 0.00 | 1.50 | 0.00 | 0.00 | 0.00 | 1.10 | 0.00 | 2.60 | 96.00 | 0.20 | 1.20 | 97.40 | – | – | – |
| 13 | MH 02 K | Shale, lt olv gy | 0.30 | 0.30 | 0.00 | 1.30 | 0.00 | 0.00 | 0.00 | 0.80 | 0.00 | 2.10 | 97.00 | 0.00 | 0.60 | 97.60 | – | – | – |
| 14 | MH 02 J | Shale, olv gy | 0.00 | 0.00 | 0.00 | 0.00 | 0.00 | 0.00 | 0.00 | 0.00 | 0.00 | 0.00 | 98.00 | 0.20 | 1.80 | 100.00 | – | – | – |
| 15 | MH 02 I | Shale, med k gy | 0.00 | 0.00 | 0.00 | 3.20 | 0.00 | 0.00 | 0.00 | 1.40 | 0.00 | 4.60 | 95.00 | 0.00 | 0.40 | 95.40 | – | – | – |
| 16 | MH 02 H | Shale, med gy | 1.50 | 1.50 | 0.00 | 0.60 | 0.00 | 0.00 | 0.00 | 0.00 | 0.00 | 0.60 | 96.00 | 1.30 | 0.60 | 97.90 | – | – | – |
| 17 | MH 02 G | Shale, olv gy | 0.70 | 0.70 | 0.00 | 2.50 | 0.00 | 0.00 | 0.00 | 1.30 | 0.00 | 3.80 | 94.00 | 0.00 | 1.50 | 95.50 | – | – | – |
| 18 | NS. 29N | Shale, gy, papery | 3.00 | 3.00 | 0.80 | 0.00 | 0.00 | 0.00 | 0.8 | 0.00 | 0.2 | 1.20 | 78.40 | 1.00 | 16.0 | 95.40 | 0.24 | 0.29 | 0.36 |
| 19 | MH 02 E | Shale, olv gy | 1.20 | 1.20 | 0.00 | 2.70 | 0.00 | 0.00 | 0.00 | 0.80 | 0.00 | 3.50 | 94.00 | 0.00 | 1.30 | 95.30 | – | – | – |
| 20 | MH 02 D | Shale, olv gy | 0.80 | 0.80 | 0.00 | 3.00 | 0.00 | 0.00 | 0.00 | 1.70 | 0.00 | 4.70 | 93.00 | 0.00 | 1.50 | 94.50 | – | – | – |
| 21 | NS.29M | Shale, brn.dkgy, fissile | 5.00 | 5.00 | 0.20 | 0.00 | 0.80 | 0.00 | 0.00 | 0.00 | 0.6 | 1.20 | 78.80 | 10.00 | 5.00 | 93.80 | 0.34 | 0.49 | 0.43 |
| 22 | NS.29 I | Shale, gy-brngy, silty | 3.60 | 3.60 | 2.00 | 0.00 | 0.00 | 0.00 | 0.00 | 0.00 | 0.00 | 2.60 | 85.80 | 2.40 | 5.60 | 93.80 | 0.30 | 0.40 | 0.35 |
| 23 | NS. 29K | Shale, gy-brngy, papery | 1.00 | 1.00 | 1.80 | 0.00 | 0.00 | 0.00 | 0.6 | 0.00 | 0.00 | 2.40 | 83.60 | 3.00 | 10.00 | 96.60 | 0.29 | 0.43 | 0.35 |

Information:

| | | | | | | | | | |
|---|---|---|---|---|---|---|---|---|---|
| V | : Vitrinite | Btm | : Bituminite | Cu | Cutinite | E | : Liptinite | Py | : Pyrite |
| Dv | : Detrovitrinite | Sp | : Sporinite | Alg | Alginite | Cl | : Clay | MM | : Mineral Substance |
| Re | : Resinite | Sb | : Subernite | Exu | Exudantinite | Carb | : Carbonate | Rv | : Vitrinite Reflectant |

Comparison of vertical abundance patterns between liptinite maceral and TOC shows a similarity less than 100%, but suitable pattern. This rather similar pattern can reflect that the total organic carbon from the upper Kelesa Formation shale is

IOP Conf. Series: Earth and Environmental Science **406** (2019) 012026 doi:10.1088/1755-1315/406/1/012026

dominated by abundance of liptinite. The abundance of liptinite affects shale to the type of kerogene, that is, shale tends to be dominant to include in type I and type II kerogene. The presence of vitrinite is closely related to type III kerogene.

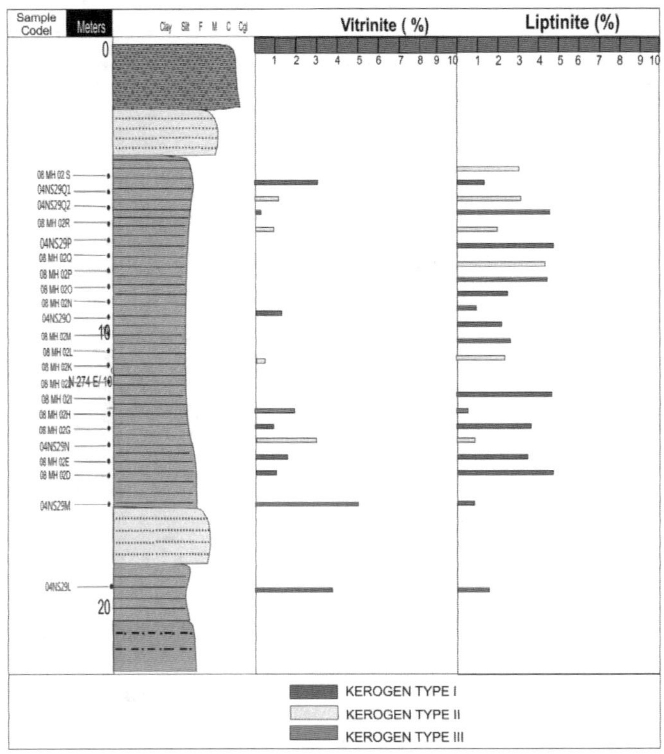


Figure 6. The content of vitrinite and liptinite macerals in shale layer in Kuburan Panjang area.

The richness of organic material in shale is not only deposited in the marine environment, but also non-marine and even the transition environment. The lacustrine facies environment is most in quantity [4] . For example, in North America the source of shale gas 1/3 is coming from shale deposited in the marine environment, while 2/3 from the transition and terrestrial environments, including lacustrine (12). Shale in the area of the study site was deposited in the lacustrine dosource inclining to have abundant content of organic material, so that the organic material is strongly influenced by organic material from the land and the lacustrine itself.

. Figure 7 shows the relationship between TOC with liptinite and vitrinite, suggesting liptinite has a more dominant content than vitrinite.

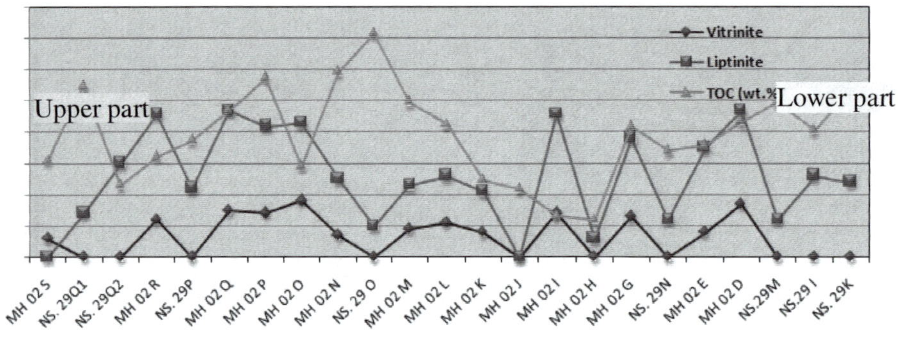


Figure 7. The relationship between TOC, Liptinite, and Vitrinite in Upper Kelesa Formation in Kuburan Panjang area

**Depositional Environment**

In this location, the continuous outcrops were found along Puti Kayu River, the tributary of Keruh River having its upper reach at Bukit Susah. The lithology compiling the rock formation in Putikayu River is dominated by shale with inserts of siltstone and sandstone which are considered as synrift deposit having identical characteristics to Sangkarewang Formation and Brani Formation in Ombilin Basin, and overlap conglomerates are incompatible which are considered late synrift deposit havingidentical deposits similar to the Sangkarewang Formation and Brani Formation in Ombilin Basin, and conglomerates which overlap incompatible which are considered to be synrift deposits having identical characteristics to Sawah Tambang Formation in Ombilin Basin.

The shale has brownish to blackish grayish, rather hard characteristics, showing papery structure (Figure 8), and there were fish fossils (Figure 9). There are inserts of grayish brown to blackish sandstone, very fine sand, rather hard, medium sorting, closed packaging, conglomerate in nature, cross bedding sedimentary structure and brownish gray to blackish siltstone, rather hard, and moderate parallel lamination sedimentary structure in the shale. Based on the characteristics of rocks found in the field, it is interpreted that the papery-coated shale is a lacustrine deposit, while conglomerate sandstone found as inserts from the shale is distal alluvial fan deposit associated with border faults during rifting and contaminating with shale. Based on the character of lithology and its spread in the field, we found the mechanism of deposition is taking place under calm current condition with a suspension current system. In addition, the absence of planktonic or bentonic foraminifera, bioturbation and characteristics of non-carbonate indicates the terrestrial environment under anoxic condition. Anoxic condition can be found in quite deep and cold freshwater environments. Based on the data above and the results of stratigraphic cross section measurement, the unit is predominantly arranged by quite thick shale. This is as stated by Nichols (7) by using a depositional model in lacustrine environment. Based on the statement of Nichols (5), this sedimentation is, generally, taking place in deeper parts of freshwater lake. Therefore, the writer interprets the unit depositional environment as lacustrine environment.

Figure 8. The structure of papery sediment on shale in Kuburan Panjang area.

The conglomerate found was quite thick and was seen incompatible overlapping on shale packages with insertions of sandstone and siltstone as erosional field was found. Conglomerate has blackish-gray characteristics, polomic dominated by components of basal fragments, poor sorting, components have halfway circular to angular form, matrix in the form of sandstone. This rock package is interpreted as fluvial braided channel sediment.

Historical sedimentation of rock package is initiated when there is rifting in the central basin of mountains characterized by precipitation of lacustrine sediment running around alluvial fan (Figure 10). When the intensity of the reduction of graben is decreased, the depositional environment is turning into fluvial channel (Figure 11).

IOP Conf. Series: Earth and Environmental Science **406** (2019) 012026 doi:10.1088/1755-1315/406/1/012026

Figure 9. Fish tail fossil found in sh

Indeed, based on the appearance o have identical resemblance to the rock pa the central mountain basin system. How researchers, the rock packages revealed in Kuburan Panjang are those in a back arc basin system.

Change in water level can be interpreted under the oxidic and anoxic conditions during the process of sedimentation. The conditions can be detected by the ratio of several parameters, i.e., normal alkane isoprenoid. Isoprenoid commonly used is pristane/phytane (Pr/Ph). By evaluating normal alkane data, we will get information on an organic facies system. The quality of shale is related to the potential of shale as an unconventional source of hydrocarbon.

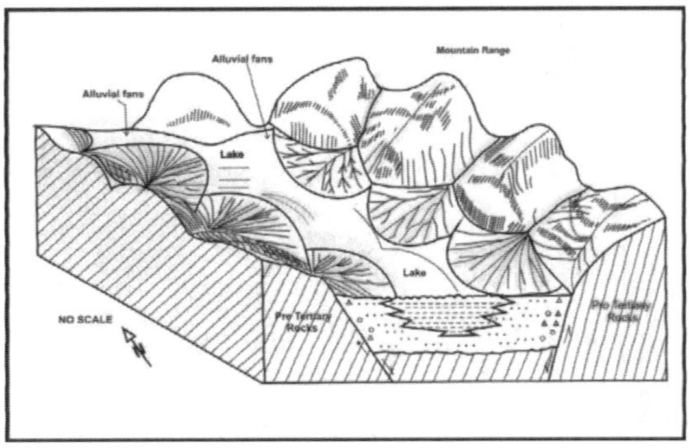

Figure 10. The shale sedimenation mechanism model with insertion of sandstone in Kuburan Panjang (cross section on the red line), modification of middle stage rifting sedimentation model in Ombilin Basin (Koesoemadinata and Matasak, 1981).

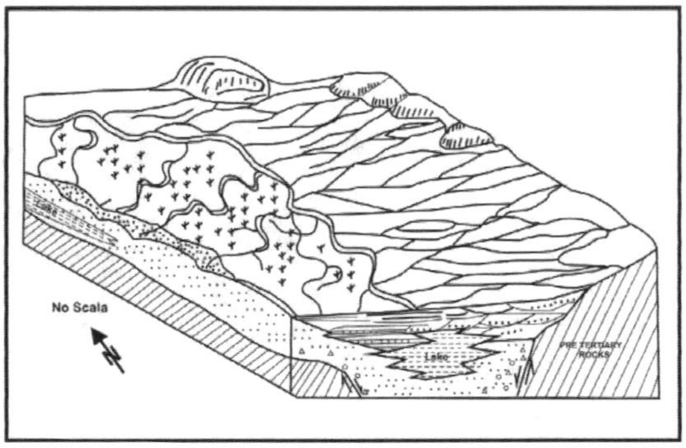


Figure 11. The conglomerate sedimentation mechanism model in Kuburan Panjang (cross section on the red line), modification of middle stage rifting sedimentation model in Ombilin Basin (Koesoemadinata and Matasak, 1981).

**Lacustrine Water Level Changes**

Table 4 illustrates the results of Gas Chromatography (GC) and Gas Chromatography - Mass Spectrometer (GCMS) analysis on shale of Kelesa Formation in the study area. Depositional environment can be interpreted based on the ratio of several isoprenoid parameters, i.e., pristane/phytane (Pr/Ph), pristane/nC17 and phytane/nC18. Powell and McKirdy, (6) stated the ratio of Pr/Ph with value of 5 - 11, the depositional environment of organic material is coming from non-marine under high wax conditions, while the low wax of marine source rock is ranging from 1 – 3. Peters and Moldowan (6) stated that the high ratio of Pr/Ph shows the depositional environment under oxidation conditions, and the lower content of salinity. The ratio of Pr/Ph tends to be high (> 1) in an oxidation environment, such as turf swamps due to the transformation of phytol into phytanoic acid is followed by decarbonization into pristane.

In the reduction environment, Pr/Ph has low ratio (<1) due to the transformation of phytol into dehydrophyol and reducing pristane. In addition, the ratio of Pr/nC17 < 0.5 indicates oil is coming from marine source rock, while Pr/ nC17 > 1 indicates oil coming from source rock deposited in nonmarine area (6).

Table 4. The results of Gas Chromatography (GC) and Gas Chromatography - Mass Spectrometer (GCMS) analysis

| GC And GCMS Analysis | | Sample Code | | | | | | | | | | | | | | | | |
|---|---|---|---|---|---|---|---|---|---|---|---|---|---|---|---|---|---|---|
| | | 14 MH 2 C | 14 MH 2 D | 14 MH 2 E | 14 MH 2G | 14 MH 2 H | 14 MH 2 I | 14 MH 2 J | 14 MH 2 K | 14 MH 2 L | 14 MH 2M | 14 MH 2N | 14 MH 2O | 14 MH 2P | 14 MH 2Q | 14 MH 2R | 14 MH 2S |
| Rasio n-alkana | pristana/fitana | 2.92 | 2.95 | 3.37 | 3.30 | 2.84 | 4.42 | 3.83 | 2.16 | 3.91 | 3.45 | 4.71 | 3.41 | 3.59 | 5.29 | | |
| | pristana/$nC_{17}$ | 5.73 | 6.16 | 8.70 | 6.79 | 6.66 | 8.20 | 7.27 | 10.01 | 4.55 | 13.90 | 7.93 | 7.64 | 10.07 | 9.83 | | |
| | fitana/$nC_{18}$ | 2.44 | 2.81 | 2.21 | 2.23 | 2.74 | 3.10 | 2.95 | 1.44 | 1.57 | 4.87 | 2.71 | 7.22 | 2.57 | 2.08 | | |
| | *Carbon Preference Index* (CPI) | 1.27 | 1.16 | 1.20 | 1.18 | 1.27 | 1.27 | 1.21 | 1.22 | 1.09 | 1.12 | 1.16 | 1.20 | 1.11 | 1.13 | | |
| Rasio Triterpana | $C_{30}$ Moretana/$C_{30}$ Hopana | 0.15 | 0.20 | 0.23 | 0.21 | 0.24 | 0.29 | 0.20 | 0.20 | 0.21 | 0.19 | 0.35 | 0.21 | 0.20 | 0.22 | 0.29 | 0.37 |
| | $C_{29}$/$C_{30}$ Hopana | 0.39 | 0.59 | 0.83 | 0.78 | 0.91 | 0.99 | 0.66 | 0.40 | 0.77 | 0.66 | 0.54 | 0.66 | 0.68 | 0.61 | 0.60 | 0.47 |
| Rasio sterana | 20S/(20S+20R)$C_{29}$ | 0.42 | 0.15 | | | | | | | | | | | | | | |
| | $C_{27}$ sterana | - | - | - | - | 59.24 | - | 42.30 | 42.00 | - | - | - | - | - | 44.53 | 45.14 | 46.77 |
| | $C_{28}$ sterana | - | - | - | - | 40.05 | - | 55.97 | 56.19 | - | - | - | - | - | 54.25 | 53.37 | 50.84 |
| | $C_{29}$ sterana | - | - | - | - | 0.71 | - | 1.74 | 1.81 | - | - | - | - | - | 1.22 | 1.50 | 2.39 |
| Rasio Aromatik | DNR-1 | 1.64 | 1.16 | 2.25 | 0.56 | 1.34 | 0.69 | 0.92 | 2.42 | 2.81 | 2.58 | 1.72 | 0.57 | 2.11 | 1.20 | 1.34 | 2.11 |
| | TNR-1 | 1.01 | 0.98 | 0.97 | 0.97 | 0.97 | 0.99 | 1.05 | 1.05 | 0.97 | 0.97 | 0.89 | 0.84 | 0.92 | 0.91 | 0.91 | 0.97 |
| | MPI-1 | 0.38 | 0.47 | 0.33 | 0.44 | 0.40 | 0.61 | 0.44 | 0.30 | 0.35 | 0.41 | 0.52 | 0.45 | 0.38 | 0.52 | 0.50 | 0.38 |
| | MPI-2 | 0.40 | 0.49 | 0.36 | 0.45 | 0.43 | 0.72 | 0.49 | 0.31 | 0.33 | 0.44 | 0.53 | 0.44 | 0.42 | 0.45 | 0.50 | 0.39 |
| Sterana/Hopana | | | | | | 0.05 | | 0.05 | 0.03 | | | | | | 0.05 | 0.06 | 0.04 |
| Diasterana/sterana | | | | | | 0.24 | | 0.27 | 0.30 | | | | | | 0.23 | 0.31 | 0.34 |

Shale in Kuburan Panjang has Pr/nC17 ranging between 4.55 - 13.90 which tends to be hydrocarbon source rock originating from source rock deposited in land area (nonmarine). The ratio of n-alkane shown by the ratio of pristane/phytane in Kelesa Formation shale ranges from 2.16 to 6.07 is classified as having a high value, that is, approximate 3 and > 3. The condition indicates that shale is deposited in the area under oxidation conditions and the lower content of salinity or the lower effect

IOP Conf. Series: Earth and Environmental Science **406** (2019) 012026 doi:10.1088/1755-1315/406/1/012026

of marine. Notwithstanding, the fluctuation in pristane/fitane value can indicate fluctuation in water level.

## DISCUSSION

Figure 12 shows the relationship between the ratio of pristane/phytane and the content of vitrinite and liptinite suggesting a similarity; that is, when the ratio of pristane/phytane is high, the content of liptinite maceral is large, while vitrinite shows the opposite condition; that is, when the ratio of pristane/phytane is high, the content of vitrinite is low. The ratio of pristane/phytane illustrates fluctuation in the ratio of pristane and phytane. The ratio is related to the oxidation zone, the high oxygen zone that occurs when water level (lacustrine) is low, so it can be illustrated that when the water level is going down, the supply of organic material is rising, especially liptinite resulting in the formation of shale in a high Index Hydrogen which means it has characteristics of I and II type kerogen. The higher the ration of pristane/phytane, the higher the oxidation condition or when sedimentation occurs under high oxygen conditions. In the MH 02 K model, the ratio of pristane/fitane is lowest, it means that at that time there was a reduction in condition or anoxic; that is, a condition with the lower content of oxygen. The case may be taking place when lacustrine water level has fluctuation under high water level condition. As the fluctuation in lacustrine water level go up, it seems the ratio is high, heading to high oxygen or oxic conditions, or, in other words, the water level is getting lower.

IOP Conf. Series: Earth and Environmental Science **406** (2019) 012026 doi:10.1088/1755-1315/406/1/012026

Einsele (1) suggested difference in production of organic material in the basin under oxic and anoxic conditions. The oxic condition has a tendency to produce sedimentary rocks and the lower carbon organic content, 0.5 - 1%, as compared to anoxic conditions ranging from 2 - 10%. The potential of hydrocarbons will also be affected by the conditions; that it, the potential of hydrocarbons will be more likely to produce oil and gas 2 - 4% under anoxic conditions, while under oxidic conditions it produces only around 0.01%.

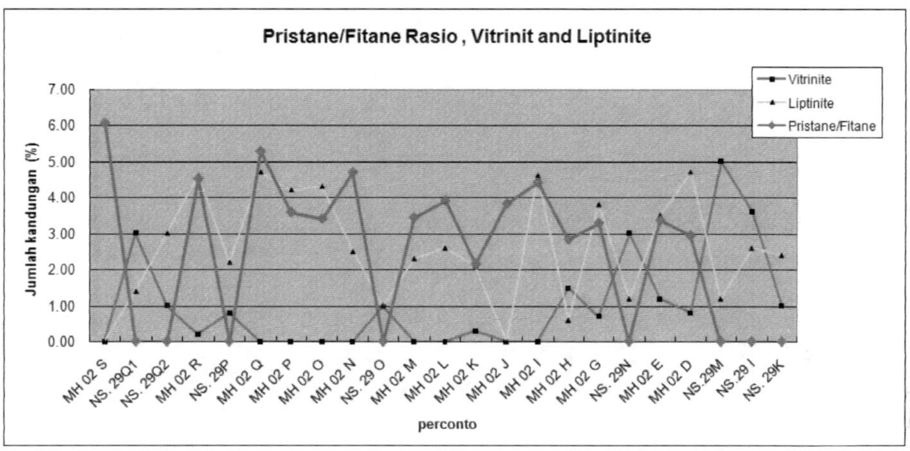


Figure 12. Relationship between pristane/phytane ratio and vitrinite and liptinite Upper part in Kelesa Formation shale. Lower part

Figure 13 shows the relationship between the content of carbon (TOC )and the ratio of pristane/fitane. When MH 02 D is in motion to MH 02 E (Y), the value of TOC decreases but the value of pristane/phytane rises, then MH 02 I, MH02 J, MH 02 K and MH 02P moved to MH 02 Q (X) and the value of TOC increases but the value of pristane/phytane decreases and MH 02 S is in the opposite position of TOC and the ratio of pristane/phytane. Based on the fact, it seems the content of TOC is

IOP Conf. Series: Earth and Environmental Science **406** (2019) 012026 doi:10.1088/1755-1315/406/1/012026

related to fluctuation in water level or a process of sedimentation under oxic and anoxic conditions.

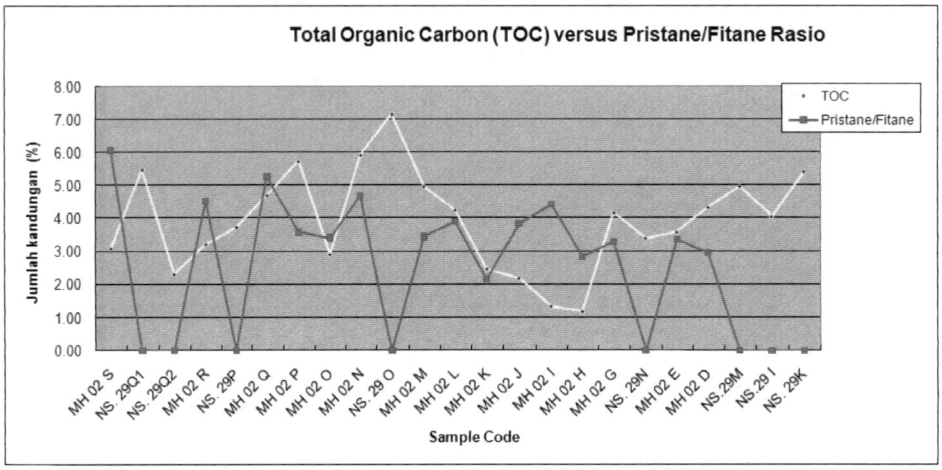


Figure 13Upper part relationship between TOC and pristaLower partatio showing the relationship between oxidic-anoxic conditions and abundance of TOC.

X X Y

Figure 14 shows the Relationship between TOC and Potential Yield (PY) in the shale layer in Kuburan Panjang from the bottom (NS 29K) to the top (MH 02S). Although different scale, the distribution of TOC and PY has a similar pattern, the TOC is in the range of 1.18% to 7.17%, while PY ranges from 1.28 to 70.92. In MH 02 H and NS 29Q2, both TOC and PY show the lowest position. MH 02 G, MH 02 I, NS 29 O and NS 29 Q1 describes that when TOC is rising, the PY will be increasing, as well.

The quality of shale can be indicated by TOC and PY which can describe potential shale as a source of energy, either conventional or unconventional. High

IOP Conf. Series: Earth and Environmental Science **406** (2019) 012026 doi:10.1088/1755-1315/406/1/012026

TOC and PY are potential shale indicators as sources of hydrocarbon energy. The relationship between shale quality and change in water level can be seen in Figures 13 and 14. If we look at the graph between TOC, PY and pristane/phytane ratio, it seems a similar distribution pattern. When we look at MH 02 K, the ratio of pristane/phytane is lower as shown by TOC and PY.

Figure 14 shows MH 02 H, MH 02 I and MH 02 J (zone A) have the same pattern between the ratio of pristane/phytane, TOC and PY; that is, increase to decrease indicating one cycle of fluctuation. Meanwhile, from MH 02M to MH 02O (zone B) and from MH 02 P to MH 02 Q (zone C) shows the same pattern, i.e., the ration of pristane/phytane, TOC and PY have the same increase. In general, there is a relationship between the three cases; that is, the ratio of pristane/phytane shows change in water level with TOC and PY as potential indicators of potential shale as sources of hydrocarbon.

Based on change in water level and potential shale as a source of hydrocarbon, it is revealed that there is a relatively significant relationship between the two things.

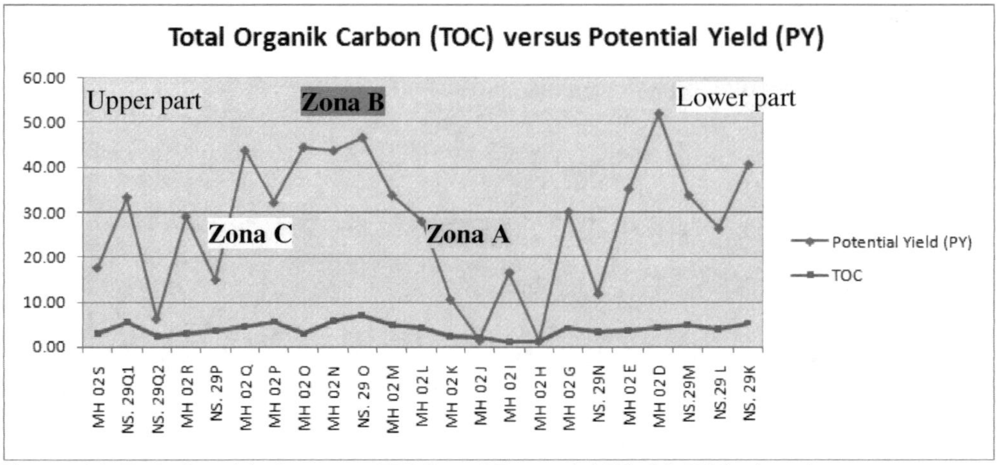


Figure 14. Relationship between TOC and Potential Yield (PY) showing the relationship between TOC and PY in each zone

This evidence is important to give an idea of what shale might be a potential source of hydrocarbon. Change in water level is attributable to climate factors during the process of sedimentation. The ratio of pristane/fitane tends to be high under oxic environmental conditions; otherwise, the TOC will be going down. However, when the ratio of the pristane/phytane is lower, the content of TOC will be going high under anoxic conditions. The conditions reflects that when the water level (lacustrine) is going down, the supply of organic material have relatively decrease and when the water level is going up, the organic material will have relatively increase.

Based on the results of analysis, the content of liptinite and vitrinite macerals is influenced by change in water level. When the ratio of pristane/phytane shows a small price, the water level will be high or anoxic condition. The abundance of vitrinite will be larger than the content of liptinite under noxic conditions. However, when the oxic orr oxygen conditions are high as indicated by high ratio of pristane/phytane, the content of liptinite shows very good, so the content of liptinite will be abundant under oxic conditions. The abundant content of liptinite may be due to organic formation and deposition in the lacustrine. Even though the water condition is low, the supply of oxygen is sufficient for the development of organisms in the lacustrine environment.

Instead, the anoxic conditions are characterized by low oxygen content and high water level. This can be interpreted organic material have not well development under those conditions due to the lower supply of oxygen so that liptinite is not abundant, while vitrinite which is a maceral originating from tall plants outside of the environment itself is formed due to transport from other places under high water level conditions.

## CONCLUSION

Stratigraphically, the abundance of organic material has fluctuation from bottom to top. The abundance of organic material may be visible at the top of stratigraphic cross section in the study area. The abundance is characterized by a change in oxidation conditions. The abundance of organic material is dominated especially by liptinite and vitrinite. Change in water level in the process of sedimentation has effect on the quality of shale as a non-conventional energy source of hydrocarbon in Kuburan Panjang. It is based on difference in potential from good to perfect in the shale layer, the varying patterns of pristan/phytane ratios between shale layers showing pattern of change in water level, similar distribution patterns between the ratio of pristane/fitane, TOC and PY, and the smaller the ratio of pristane/fitane (anoxic condition, high water level), the smaller the content of liptinite and, the higher the content of vitrinite; conversely, the greater the ratio (oxic condition, low water level), the higher the content of liptinite and the lower the content of vitrinite. In general, the development of layers from the bottom to the top gives a picture of change in water level from high to low or from a reduction condition (low oxygen) to oxidation (high oxygen) condition.

**ACKNOWLEDGMENT**

The author would like to thank profusely to the Editorial Board, so that the writing and data can be processed into scientific paper article.

**REFERENCES**

1. Einsele, G., 1992, Sedimentary Basin; Evolution, Facies, and Sediment Budget, Springer-Verlag Berlin Heidelberg, New York.
2. Eubank, R. T. and Makki A. C. 1981. "Structural Geology of the Central Sumatra Back Arc-Basin, 10th Annual Convention, Vol I. IPA Proceedings". Jakarta.
3. Illich, R.A. (1983): Pristane, phytane, and lower molecular weight isoprenoid distributions in oil, Bulletin of the American Association of Petroleum Geologists, 67, 385-393.
4. Ju YW., Wang, G., Bu, H., Li, Q., Yan, Z., 2014, China Organic Rich Shale Geologic Features and Special Shale Gas Production Issues, Journal of Rock Mechanics and Geotechnical Engineering 6, 196- 207.

5. Nichols, G., 2009, Sedimentology and Stratigraphy Second Edition, John Wiley and Son, New York.
6. Peters, E., and Moldowan, J. (1993): The biomarker guide: Interpreting molecular fossils in petroleum and ancient sediments, Englewood Cliffs, Prentice Hall, 363.
7. Peters, E., Walters, C., and Moldowan, J. (2005): The Biomarker Guide, Cambridge University Press, Chapter 4, 72-118.
8. Silitonga, P.H. and Kastowo, 1995. Geology of Lembar Solok, Sumatra, scale of 1:250,000. Geology Research and Development Center, Bandung.
9. Susanto, E., Suwarna, N. and Hermiyanto, H., 2004. Research of Fuel Fossil and Paleontologist, Study of Oil Shale, in Central Sumatra. Internal Report, Geology Research and Development Center, Bandung.
10. Suwarna, N., Budhtrisna, T., Santosa, S., and Andi Mangga, S., 1994. Geological Map of the Rengat Quadrangle, Sumatra, scale 1: 250,000. Geological Research and Development Center, Bandung.
11. Waples, D.W., 1985, Geochemistry in Petroleum Exploration. Brown and Ruth Labotaries, Inc. Denver, Colorado.
12. Zhang, DW., Li YX., Zhang JC., Qiao DW., Jiang WL., And Zhang JF., 2012, National Survey and Assessment of Potential Shale Gas Resources in China, Beijing: Geologic Publishing House in Chinese.

# Effect of Sugar on Nutrient Composition and Shelf Life of Red Guava Jams

**Choirun Nissa[1], Fahmi Arifan[2], Rico Febrianto[2], William Aditya[2], Ida Hayu Dwimawanti[3], Reyno Pramudyono Widyasmara[4]**

[1]Nutrition Department, Faculty of Medicine, Diponegoro University, Jl. Prof Soedharto, Tembalang, Semarang, 50239, No Telp/Fax: (024) 7460058

[2]Chemical Engineering Department, DiponegoroUniversity, Jl. Prof Soedharto, Tembalang, Semarang, 50239, No Telp/Fax: (024) 7460058

[3]Public Administration Department, DiponegoroUniversity, Jl. Prof Soedharto, Tembalang, Semarang, 50239

[4]Publik Health Study Program, Hakli College of Health Sciences, Jl. Dr. Ismangil No.27 Bongsari, Semarang, Telp / Fax: (024) 7601844

nissaeyong@gmail.com, fahmiarifan80@gmail.com, ricofebrianto98@gmail.com, william.ayodyasena@gmail.com, idahade@gmail.com, rpwidyasmara@gmail.com

**Abstract.**Red guava as fruit jams normally preserved using added sugar to prolong its shelf life. Heating process and different sugar content could affect in nutrient content and expiration date. Hence, the aim of this study was to evaluate the nutrient content and shelf life of jams based on two different sugar content. Two hundred (jam A) and two hundred and fifty grams of sugar (jam B) was added in 800 grams of guava then cooked at 70°C within an hour. Protein and fat was measured using Kjeldahl and soxhlet, while water and ash content was analyzed using thermogravimetry and dry ash method. Carbohydrate was calculated using by difference method. Shelf life determination was obtained using organoleptic test, containing taste, aroma, and visual sensory. In result, jam A have a higher protein (0.334%), fat (1.567%), and water (29.21%) than jam B (0.306%; 1.092%; 27.32%). On the other hand, ash (0.34%) and carbohydrate (68.549%) in jam A are lower than B (0.37% and 70.912 respectively). Overall, jam A have a higher protein, fat, water and carbohydrate content than jam B. A higher water content potentially indicate a lower shelf life which seen in jam A. In conclution, terms of nutritional composition, jam A have higher content than jam B, but have lower shelf life.

**Keywords**: red guava, jam, shelf life, nutrition composition, organoleptic

IOP Conf. Series: Earth and Environmental Science **406** (2019) 012027 doi:10.1088/1755-1315/406/1/012027

## Introduction

Food is anything that comes from biological sources and water, whether it is processed or not, which is intended as food or drinks for human consumers, including food additives and other materials used in the process of preparing, processing and the making of food or drinks. Food processing is food or drinks that is processed by particular methods with or without additives.

One of biological sources that can be used as a food resources is red guava fruit. Red Guava *(Psidiumguajava L.)* is a part of the Myrtaceae family, where guava has been planted and developed in various countries in the world. Red guava is famous for its distinctive taste, scent, nutritional value as a fruit and is a potential source of *phytochemicals*[5].The characteristics of red guava fruit, namely oval-shaped with green or yellow color, with red or pink guava grains depending on the variety [11]. Red guava grows as a green tree with a height up to 2-11 meters. The plants grow quickly and will bear fruit within 2-4 years from the beginning of the seeding period. Red guava production alone is estimated to reach 500,000 tons worldwide. Red guava itself is used to be processed into drinks, syrup, ice cream, jams, and *jelly*[7]. Based on an analysis conducted by McCook-Russell [11], red guava has quite a lot of nutrients. These contents include fat, protein, ash, fiber, calcium, polyphenols, anti-oxidant and absolutely red guava fruit is also rich of vitamin C.

In Indonesia alone, red guava is quite abundant. Based on data from the Central Bureau of Statistics [2], the production of red guava reached 200,495 tons where Central Java became the highest-producing areas. More specifically, in Temanggung, red guava fruit is quite abundant found in Kaloran and Bejen. However, the red guava is still only used for direct sales so that it has a fairly low economic value. In addition, because red guava is rich with vitamin C, the fruit tends to easy to rot when it is ripe, which is 2 weeks after the ripe phase. This decay can occur because of the vitamin C in red guava will be oxidized irreversibly by the enzyme *Ascorbate oxidase. Ascorbate oxidase* is a copper-containing enzyme which will oxidize and degrade vitamin C and increase microbial activity in fruit [10]. More clearly, the enzymatic activity is influenced by the storage process. The storage process in the open air, places exposed to sunlight, and places exposed to pathogens and chemicals will affect the enzymatic activity that results in the decay of red guava. Therefore, it needs the correct storage processes. The storage process can be in the form of processing red guava to other products that are more durable.

IOP Conf. Series: Earth and Environmental Science 406 (2019) 012027 doi:10.1088/1755-1315/406/1/012027

However, the processing of red guava has not varied. Nevertheless, one of the processing solutions is by processing red guava into jam.

Jam is food that is cooked using the meat/juice of fruits or vegetables which are then converted into jelly-like form. In general, jam is made using only one type of fruit with the characteristics of a good jam is to have a soft and even texture, favorable color and good fruit taste [3]. Jam production can use various types of fruit, but in general the fruit that is used contains pectin. *Pectin* is a sugar/*polysaccharide* compound that makes jam to have a soft but thick texture. One of them is in red guava. Red guava is suitable for jam because it contains quite a lot of *pectin*. The pectin in red guava is around 1.4% -1.6% of the total content. Regarding the content of pectin in various fruits, red guava has a fairly high content of pectin. Furthermore, the red guava pectin content is close to the pectin content in fruits which are generally processed into jams, such as strawberries, blueberries, grapes, etc. The advantages of processing besides red guava jam have good economic value, red guava jam products have also a long shelf life, and have a delicious and nutritious taste. In this study using the variable of sugar amount with the expected results is to know the nutrition of red guava jam and its durability.

**Research Methods**

The material used to make red guava jam is 800 grams red guava fruit, sugar for variable A is 200 gram and variable B is 250 gram, 3 cinnamons, and 3 cloves. The steps to make red guava jam is, first the juice in red guava is taken using a blender and separated from the seeds, then the juice is heated on a pan to a temperature of 70°C for one hour, when it has reached the temperature, ingredients such as sugar, cinnamon and cloves are added and stirred evenly. Stirring is carried out until the red guava jam reaches a consistency like jam in general. After that, red guava jam is put into aluminum foil as a storage container. Then, to find out the nutrition contained and endurance in red guava jam, a proximate test was performed. This proximate test includes a protein content test, a fat content test, a water content test and an ash content test. In the protein content test, the test uses the Kjeldahl method where the sample will undergo desctruction using concentrated sulfuric acid and catalysts. In the fat content test, the test is carried out using the Soxhletation method where the sample will be extracted using a petroleum ether solvent. In the ash content test, the test is carried out by the method of drying ash in which the sample will be made up on the flame and blended in a furnace at 550°C until s complete ashing. In the water content test, this test uses a thermogravimetric method in which the sample will be dried to a certain constant weight [12]. For carbohydrate content can be obtained from the calculation results, which are:

$$\%Carbohydrate = 100\% - \%(Protein + Fat + Water + Ash)\ [12]$$

For red guava jam durability can be seen from the water content and the endurance supported through direct observation of changes with organoleptic tests. Organoleptic test is a test based on the body's sensory stimulus to an object, especially food. This organoleptic test aims to determine the quality of products based on direct consumer acceptance [1]. With organoleptic testing, the quality of the jam can be known to determine whether the jam is still suitable for consumption or not. Organoleptic tests were carried out for 4 weeks in the second month with an assessment of 1-10 in the categories of taste, color, and scent. The greater the value in each category shows that the quality in each category is still maintained. Proximate test in this study was conducted at UPT Laboratory Diponegoro University. Organoleptic tests were carried out in the Department of Chemical Engineering, Diponegoro University.

**Results and Discussion**

The results of the proximate test include protein content tests, fat content tests, ash content tests, and water content tests. For carbohydrate levels, these levels can be obtained through the calculation of the difference between the levels 100% with the amount of percent protein, fat, ash and water. For dissolved solids, it can be obtained through a percentage of protein, fat, ash and carbohydrate. These results can be seen in the following Table 1.

Tabel 1 Proximate Test Result Data

| Variable | Protein (%) | Fat (%) | Ash (%) | Water (%) | Carbohydrate (%) | Dissolved Solids (%) |
|---|---|---|---|---|---|---|
| A Jam | 0,334 | 1,567 | 0,34 | 29,21 | 68,549 | 70,79 |
| B Jam | 0,306 | 1,092 | 0,37 | 27,32 | 70,912 | 72,68 |

Datafrom Indonesian National Standard [14] states that the minimum dissolved solid content requires for fruit jam is 65%. When compared with A jam and B, both of them have successively higher levels of dissolved solids than the requirements of the Indonesian National

Standard. Thus, A jam and B have met the requirements of the Indonesian National Standard as a fruit jam.

There are found differences in levels of protein, fat, ash, water and carbohydrates in the A jam and B. This can be caused by the amount of sugar added to the jam. The addition of different level sugars will cause different results as well. The more addition of sugar will affect the reduction which is also more on the levels of protein, fat, ash and water. This is also supported by a temperature increase in the processing of red guava fruit into jam [9]. However, carbohydrates have increased levels because sugar contains glucose where glucose is one of the constituent compounds of carbohydrate polymers [13]. The addition of sugar will result in more glucose in the jam. So, from the results we can get that the jam which experiences more sugar additions will cause a decrease in levels of protein, fat, ash, water but also causes an increase in carbohydrate levels.

From the results of the proximate test, we also found out the water content of each jam variable. This content is related to jam microbial activity and the durability. The water content test results can be seen in Table 1 above.

Based on Table 1, we found out the results of the water content of the two jam variables. The A jam has a higher water content than B jam. A jam has 29.21% water content and in B jam has a 27.32% water content. The water content has a relation with microbial activity and the durability of jam itself. The higher the water content is, then microbial activity will increase. According to the research results from Bussiere[4], the water content in jams generally ranges between 20% -40% where in that range there are *Mycotoxigenicaspergilli* microbes. *Mycotoxigenic aspergilla* microbes are a type of microbe that plays a role in the breakdown and destruction of components in food. Furthermore, with a water content in range of 20% -40%, jam has an estimated shelf life of approximately 4 months [4]. Furthermore, the water content plays an important role in the growth period, especially in the exponential phase of the bacterium *Mycotoxigenic aspergilla*. The exponential phase is the phase in which bacteria undergo the most active processes of growth and activity before entering the peak and death phases. This bacterium has an exponential phase of 1,5 months [6]. To find out if there is a change in the quality of jam during storage, an organoleptic test is performed. The collection of organoleptic assessment data was carried out for 4 weeks in order to find out if there were changes when the

*Mycotoxigenic Aspergilla* bacteria entered an exponential growth phase. Organoleptic assessment data can be seen in Table 2 below.

Tabel 2 Organoleptic Assessment Results

| Category | A Jam | | | | B Jam | | | |
|---|---|---|---|---|---|---|---|---|
| | Week 1 | Week 2 | Week 3 | Week 4 | Week 1 | Week 2 | Week 3 | Week 4 |
| Color | 9 | 9 | 8 | 8 | 9 | 9 | 9 | 8 |
| Taste | 8 | 8 | 8 | 7 | 8 | 8 | 8 | 8 |
| Scent | 8 | 8 | 7 | 7 | 8 | 8 | 8 | 7 |

Based on the requirements from Indonesian National Standard (SNI) [14], good quality jam have a normal aroma, taste, and color. Based on the above organoleptic data, A jam and B jam have quality that is still maintained during the storage period. In more detail, it can be seen that the quality of B Jam has more consistent endurance than A Jam. This can occur because in the A jam the sugar has less water content. The lower water content indicates less microbial and bacterial activity so that it has a longer life. So that, during the 4 weeks test period, the quality of guava jam B was maintained.

## Conclusion

The result, the B jam contains less protein, fat, water, and ash than the jam A due to reduction, but it has more carbohydrate levels because sugar contains glucose as a constituent part of carbohydrates. In durability, the B jam has less water so it has less microbial activity as a result, the jam has better endurance. Therefore, the B jam has better endurance and has higher carbohydrate content, but for protein, fat, water and ash content, the A jam is much higher.

## References

[1] Ana, A., Subekti, S., Hamidah, S., Komariah, S. 2017. "*Organoleptic Test Patisserie Product Based on Consumer Preference*". IOP Publishing: IOP Conf. Series: Materials Science and Engineering 180.

[2] BadanPusatStatistik (BPS). 2017. "*StatitstikTanamanBuah-BuahandanSayuranTahunan, tahun 2017*". Indonesia.

[3] Berolzheimer, Ruth. 1969. "*Culinary Arts Institute Encyclopedic Cookbook*". United State: Chicago, Culinary Arts Institute, Page 830.

[4] Bussiere G and Serpelloni M. 1985. "*Confectionery and Water Activity Determination of $a_w$ by Calculation*". In: Simato D and Multon JL, (eds.) Properties of Water in Foods in

Relation to Quality and Stability. Dordrecht, The Netherlands: MartinusNijhoff Publishers. Page 626-645.

[5] De Souza, Tercio da Silva et al. 2018. "*Chemotype diversity of Psidiumguajava L.*", Phytochemistry 153, Page 129-137.

[6] Gibson, Angela M., Baranyi, Jozsef, Pitt, John I., Eyles, Michael J., Roberts, Terry A. 1994. "*Predicting Fungal Growth: The Effect of Water Activity on Aspergillusflavus and Related Species*". International Journal of Food Microbiology Vol. 23, Issues 3-4, Pages 419-431.

[7] Jimenez-Escrig, A., Rincon, M., Pulido, R., danSaura-Calixto, F. 2001. "*Guava Fruit (Psidiumguajava L.) as a New Source of Antioxidant Dietary Fiber*". Journal of Agricultural and Food Chemistry, 49, Page 5489-5493.

[8] Joshi, Homi. 2016. "*Nutritional Evaluation of Different Varieties of Guava (Psidiumguajava L.) and Their Preserved Products*". Pakistan: Department of Food and Nutrition, Punjab Agricultural University.

[9] Khan, Aftab Ahmed et al. 2016. "*Influence of Sugar Concentration on Physicochemical Properties and Sensory Attributes of Sapodilla Jam*", Pakistan: Institute of Agricultural Sciences, Quaid-e-Azam Campus, University of the Punjab.

[10] Lee, Seung K., Kader, Adel A. 2000. "*Preharvest and Postharvest Factors Influencing Vitamin C Content of Horticultural Crops*". Journal of Postharvest Biology and Technology, 20, Page 207-220.

[11] McCook-Russell, Kayanne P. et al. 2012. "*Nutritional and Nutraceutical Comparison of Jamaican Psidiumcattleianum (Strawberry guava) and Psidiumguajava (Common guava) Fruits*". Journal of Food Chemistry vol.134, Page 1069-1073.

[12] Musfiroh, I., Indriyati, W., Muchtaridi, Yudhi, S. 2010. "*AnalisisProksimatdanPenetapan Kadar β-KarotendalamSelaiLembarangTerungBelanda (CyphomandrabetaceaSendtn.) denganMetodeSpektrofotometriSinarTampak*", Bandung: FakultasFarmasi, UniversitasPadjajaran.

[13] Prentice, Ann. 2015. "*Carbohydrates and Health*". London: Scientific Advisory Committee on Nutrition, The Stationery Office.

[14] Standard Nasional Indonesia (SNI). 2008. SNI 3746:2008. Indonesia.

IOP Conf. Series: Earth and Environmental Science **406** (2019) 012028 doi:10.1088/1755-1315/406/1/012028

# A SPATIO-TEMPORAL ANALYSIS ON THE COMPOSITION AND ABUNDANCE OF PHYTOPLANKTON IN SEGARA ANAKAN LAGOON AREA

**Rose Dewi[1, 3*], MuhammadZainuri[2,4], SutrisnoAnggoro[3], TjahjoWinanto[1]**

[1]Fisheries and Marine Science Department, Jenderal Soedirman University
[2] Department of Oceanography, Fisheries and Marine Science Department,Diponegoro University
[3] Doctoral Programof Coastal Resources Management,
Departmentof Fisheries and Marine Science,Diponegoro University
[4] Centre of Excellent Mitigation of Natural Disaster and Coastal Rehabilitation, Diponegoro University
*Corresponding author: rose.83unsoed@gmail.com

**Abstract.** SegaraAnakan Lagoon (LSA) is a dynamic estuary with humid tropical climate, of which water fluctuates as the result of water hydrodynamics and changes in monsoon. The LSA area is influenced by natural factors (estuary of freshwater flow of Citanduy Watershed and high sedimentation) and various anthropogenic factors. Such conditions may influence the water's ecological conditions. This research aims at examining such ecological changes as phytoplankton's biological indicators (composition and abundance), since phytoplankton is able to naturally respond to environmental changes. The method is conducted with spatially in-situ measurement on phytoplankton abundance in the seven areas of SegaraAnakan Lagoon with different characteristics: (1) TPI and Port of Majingklak; (2) Pelawangan Barat LSA bordering the Indian Ocean; (3) Locks of Cikonde and Cimeneng Rivers; (4) Mangrove Forest; (5) KampungLaut Community Settlement; (6) Conservation Area "Mina Wisata" Ujung Alang; (7) Floating Net Culture. A temporal approach (time series) is then conducted for 1 year based on the representativeness of monsoon: West, Transition I, East and Transition II. Phytoplankton abundance is quantitatively analyzed using Lackey Drop Microtransect Counting method, and for ease of spatio-temporal interpretation, the results are displayed in the form of thematic map using Arc Gis 10.2 software. The analysis results show that spatially, the phytoplankton abundance in the seven areas ranges from 4474.80 ± 3408.25 – 8712.00 ± 6444.89 indv/L. Meanwhile, with the temporal analysis, the highest abundance takes place in the Transition I season of 9956.57 ± 5300.23 indv/L, and the lowest is in West season of 4129.71 ± 3139.88 indv/L. In the whole spatio-temporal period, the phytoplankton composition is dominated by Division Chrysophyta, and then respectively by Divisions Chlorophyta, Cyanophyta, Pyrophyta and Euglenophyta. The composition and abundance of freshwater, estuary and sea phytoplankton are spatio-temporally, evenly distributed in the seven areas, which is expectedly resulted from the Flushing Time mechanism from freshwater flow of Citanduy Watershed and the influence of water hydrodynamics of Indian Ocean at Pelawangan Barat entering the lagoon area.

**Keywords**: Monsoon, Phytoplankton, SegaraAnakanLagoon, Spatio–Temporal

## Introduction

Cilacap Regency covers a coastal area of 225,361 km$^2$. According to Act No. 27 of 2007 j.o Act No. 1 of 2014 on the Management of Coastal Area and Small Islands, coast is a transitional area between land and sea ecosystems and is under influence of changes in land and sea.This coast includes the SegaraAnakan Lagoon (LSA) area, which is shielded by Nusa Kambangan Island from Indonesian Ocean Watersand surrounded by river mouths, particularly Citanduy Watershed, thus land's influence is highly dominant with sedimentation process. Land conversion forvarious anthropogenic activities in its surrounding causes high process of sedimentation and expansion of mangrove vegetation in the lagoon area. Such conditions cause the depth and size of lagoon water to decrease [1]. Dahuri [2] states that various anthropogenic activities whichcause erosion of an ecosystem, such as land clearing for settlement, agriculture and industry without terracing around the river,may cause sediment on surface flow (surface run off) to increase. Such sediment will enter into the bodies of rivers and go to the lagoon, coastal area and the sea. As reported by [3;4], sediment derived from land use for anthropogenic activities contains organic matter. The contents of Nitrogen (N) and Phosphate (P) organic matter are strongly bound by water molecules, resulting in high concentration in the sediment. There is an ecological change in the quality of water (nutrient loading) or enrichment of organic matter from the perspective of biological indicators, which are the composition and abundance of phytoplankton. The reason is that phytoplankton has the ability to naturally respond to ecological changes [5. 6] and [7] add that phytoplankton may tolerate unstable condition in water, such as a high increase in organic matter, and quickly respond to various forms of pollutant entering body of water, as marked with a change in type dominance, change in community structure, biomass of phytoplankton and fluctuation of chlorophyll content [8; 9].

Moreover, Pintado, *et al.*,[11] state that weather (rainfall pattern) as the controlling factor in nutrients fluctuation significantly influences organic matter fluctuation. The lagoon area highly depends on closed and open waters and annual (rainy and dry) seasons,which will influence the structure of community and abundance of phytoplankton. Seasons are the cause of variation in primary productivity in the water [5].The biological parameter serving as the initial indicator directly influenced by such changing season conditions in the water is phytoplankton, which does not only serves an important role in the flow of energy of primary producer, but may also serve as the bio-indicator of fertility level of a water with observation of the composition and abundance of phytoplankton [12; 13]. Therefore, this research aims at studying the ecological change in the LSA area as the biological indicator of phytoplankton (composition and abundance) both spatially, by comparing the difference in characteristics of the research's station location, and temporally (time series) of the Indonesian Waters influenced by the monsoon pattern.

**Materials and Method**

The analysis on the composition and abundance of phytoplankton in the LSA Area in this research is conducted in the temporal period (time series) in 2016, in which the measurement is conducted based on monsoon representativeness. Pursuant to the statement of Nontji [14], that based on the data of the Meteorology and Oceanography Body, the Indonesian waters are influenced by monsoon,consisting of: West Season: December – February (Representativeness of sampling in February and December); Transition Season I: March – May; East Season : June - August; Transition Season II: September –November.

Such reference is taken for the time of phytoplankton sampling in-situ with spatial approach in the research's seven stations with ecological difference in characteristics (Table.1).

- The phytoplankton is sampled vertically using plankton-net 25 μm by screening 100 liters of wateron the surface of lagoon water at depth of 0.5 - 1 m (APHA, 2005).
- The samples are preserved using formalin up to 4% [14] for 10 drops and identified using binocular light microscope with 40 times magnification.The sample phytoplankton is observed under microscopeby having them dropped onto the object glass and covered with cover glass. The calculation is conducted with 20 times of field of vision and repeated 3x. It is then identified using: *identifying Marine Phytoplankton*, Tomas, C.R [15]; Planktonologi, Sachlan, [16] and *Marine Plankton, a practical guide*, Newel and Newel [33]. The abundance of phytoplanktonis calculated using the formula based on the *Lackey Drop MicrotransectCounting*method [17].

**Table. 1 Characteristics of Spatial Approach of Phytoplankton Sampling in SegaraAnakan Lagoon Area, Cilacap Regency**

| | Characteristics | Ordinate | |
|---|---|---|---|
| 1 | TPI and Port of Majingklak | 108°48'02.4"BT | 07°40'27.6"LS |
| 2 | Gate (*Pelawangan Barat*) of LSA directly bordering the Indian Ocean | 108°46'56.7"BT | 07°41'59.0"LS |
| 3 | Locks of Cikonde and Cimeneng Rivers | 108°49'47.9"BT | 07°40'34.6"LS |
| 4 | Mangrove Forest Area | 108°51'36.5"BT | 07°41'44.9"LS |
| 5 | KampungLaut Community Settlement | 108°52'14.0"BT | 07°42'19.5"LS |
| 6 | Conservation Area "Mina Wisata"UjungAlang, KampungLautSubdistrict | 108°52'45.4"BT | 07°42'55.4"LS |
| 7 | Floating Net Culture Area | 108°48'56.0"BT | 07°41'01.0"LS |

- The data obtained are displayed in the table and analyzed descriptively and reviewed based on the literature [18]. For ease of spatio-temporal interpretation, the result of calculation of the composition and abundance of phytoplankton is displayed in the form of a thematic map using Arc Gis 10.2 software with output in the form of image format (Jpeg) equipped with a pie chart. The big and small sizes in figure show the quantity of abundance of phytoplankton, while color variation in the pie chart shows the composition (divisions of phytoplankton).

**Results and Discussion**

The research sampling results in 82 types of phytoplankton from 5 (five) divisions, which are 54 types of Chrysophyta, 13 types of Chlorophyta, 8 types of Cyanophyta, 3 types of Pyrophyta and 4 types of Euglenophyta.The quantity of abundance of phytoplankton with the spatial approach (characteristics of the research's 7 stations) and temporal approach (monsoon) is presented in (Table 2).

Division Chrysophyta dominate smost of the seasons and stations, of class Bacillariophceae (diatom), particularly: *Gyrosigmakuetzingii*.;*Cyclotella*, sp.; *Neidium Affine, Cymbellanaviculiformis*; *Melosira granulate, Chaetoceros*, sp.; *Rhizosoleniarobusta*; *Coscinodiscuslinetus*; *Thalassiosirarotula, Navicula,* sp. and various species of genus *Nitzschia*. The reason is, according to Basmi [32], that diatoms are an important (adaptive)species in phytoplankton population and are the key to every collection of planktonic algae (identification). The other phytoplankton found in the LSA includes Division Chlorophyta: various species of genus *Scenedesmus*, *Closteriumlunula*, *Coelastrumastroideum* and *Pediastrum simplex*; Division Cyanophyta: *Oscillatoriaprinceps, Lyngbyabergei, Anabaena hallensis*;DivisionPyrophyta (dinoflagellata): *Peridinium*, sp.;and Division Euglenophyta : *Trachelomonas*, sp.; Phacus, sp.

**Table.2 Measurement of Abundance of Phytoplanktonin the Water of SegaranAnakan Lagoon Area with Spatio-Temporal Approach**

| Abundance ofPhytoplankton(indv/L) | | | |
|---|---|---|---|
| Spatial Approach | | Temporal Approach | |
| 1 | 8355.60 ± 3049.46 | West Season (February) | 6873.43 ± 2296.90 |
| 2 | 4474.80 ± 3408.25 | Transition Season I | 9956.57 ± 5300.23 |
| 3 | 6058.80 ± 1860.57 | East Season | 6618.86 ± 3650.68 |
| 4 | 8712.00 ± 6444.89 | Transition Season II | 5685.43 ± 2075.74 |
| 5 | 7286.40 ± 4137.68 | West Season (December) | 4129.71 ± 3139.88 |
| 6 | 5940.00 ± 4356.00 | | |
| 7 | 5742.00 ± 1945.04 | | |

There is variation of types of freshwater, estuaryand sea phytoplankton obtained spatio-temporally throughout the research which, according to Hawkes (1978) in Suryanti [19], is expectedly because of the influence of hydrodynamic factor in the water: direction and speed of flow, which influences the composition and abundance of plankton. Such factors play a role in the process of migration and distribution of plankton as an organism with passive movement.

***Abundance of Phytoplankton and spatial approach***

The highest abundance of phytoplankton (Table2) during the research takes place in the mangrove forest area (station 4) of8712.00 ± 6444.89 indv/L which, according to Ridd*et al.* [20] ,water around mangrove forest plays a key role in the nutrients cycle since it is rich of nutrients. That the abundance of plankton is high is because the station is known as an area with fertile organic matter producing characteristics. Layborn and Parry [21] add that mangrove ecosystem may potentially generate productive organic matter through leave brown waste. After decomposition process by bacteria and protozoa, the next cycle produces macronutrients (N and P) which are the key to water fertility. The degree of macronutrient contents in the water will then influence the abundance and chlorophyll-a content of phytoplankton [8;9].

Meanwhile, the lowest abundance takes place in (station 2),which is the gate (*Pelawangan Barat*) of LSA directly bordering the Indian Ocean, of only 4474.80 ± 3408.25 indv/L. The reason is expectedly that the growth of phytoplankton depends on fluctuation of nutrient elements and hydrodynamics in the water [22]. Moreover, Ji [23] states that an estuary area is highly influenced by the tidal dynamics. The hydrodynamic factor in the water of station 2 expectedly causes nutrients in this stationto be distributed and enter into the lagoon (other stations) besides runoff of nutrients from Citanduy Watershed into the lagoon. This is supported by Chua's statement (1970) in Sediadi&Sutomo [24] that in tropical waters, phytoplankton's cell division process occurs quickly and the quantity is abundant towards land as the result of nutrients intake from land through river and, on the contrary, tends to decline offshore. This causes the station in side the lagoon to have higher abundance of phytoplankton than station 2.

***Abundance of Phytoplankton and Temporal Approach***

The result of five temporal abundance measurements (monsoon) as outlined in (Table.2) throughout year 2016 shows that the highest abundance of phytoplankton takes place in Transition Season I of 9956.57 ± 5300.23 indv/L with sampling in May. According to Nontji [14],during the transition season, there will be uncertain but generally relatively calm wind power and direction pattern. This transition season will determine the distribution of phytoplankton's chlorophyll, both spatially and temporally. That the abundance in transition season I is high is expectedly caused by upwelling (increase of water mass), causing nutrients from water's inner layer to be lifted upwards and fertilize (enrich with nutrients) the surface layer. The upwelling causes primary productivity to increase, as reflected with an increase in abundance and chlorophyll of phytoplankton.

Meanwhile, the lowest abundance takes place in West Season (December) or rainy season of 4129.71 ± 3139.88 indv/L. The reason, according to Panggabean [25],is that in tropical region, eutrophication is likely to occur in rainy season. Eutrophication will occur in the water with weathering of organic matters derived from river mouth by various anthropogenic activities, causing an increase in nutrients entering into coastal water. In eutrophic condition,only certain types of phytoplankton may adapt well and respond quickly with blooming [26]. Besides such factor, the low abundance in West Seasonin the research is expectedly because of the Flushing Time mechanism derived from the freshwater flow of Citanduy Watershed supported with high rainfall. Based on BMKG's data,the rainfall in the LSA area in December ranges from 483-958 mm. Dyer [27] further explains that Flushing Time is the time a water needs for replacement of freshwater (flushing) in lagoon area.Such condition causes the abundance of phytoplankton to be distributed towards the sea (Indian Ocean). Pednekar [28] adds that in tropical region, run off influenced by rainy reason will play a role in the spatio-temporal distribution of types of phytoplankton.

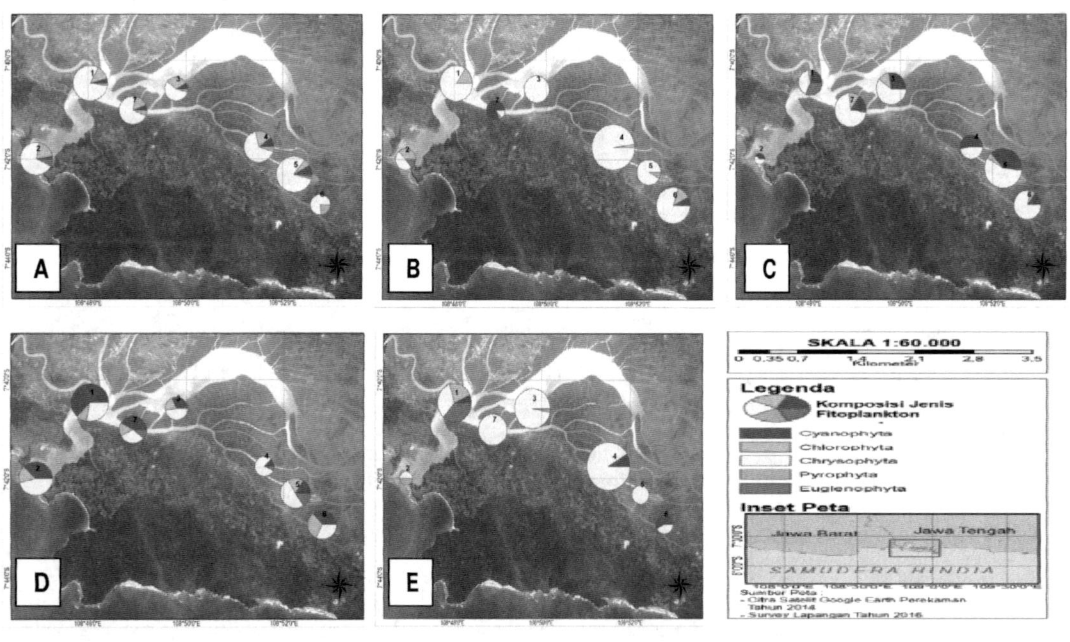


**Figure .1 Spatio-Temporal Composition of Phytoplanktonin the SegaraAnakan Lagoon area in 2016, (A) West Season (February); (B) Transition Season 1; (C) East Season; (D) Transition Season 2; (E) West Season (December)**

***Composition of Phytoplankton with Spatio-Temporal Approach***

The composition of phytoplankton during the research shows that Division Chrysophyta with abundance of 31957.2 ± 15795.75 indv/L dominates most of the stations (spatial) and season (temporal), particularly with class Bacillariophceae (diatom).Division Chlorophyta (4831.2 ± 2443.51 indv/L) increases in abundance, particularly in West Season (February) atstation (1,2,3, 4) and in transition season II (2,6,7), and although found in most of the seasons, but with low abundance. Division Cyanophyta (7603.2 ± 6113.36 indv/L) increases in abundancefrom transition season I (7), east seasontotransition season II (at all stations). Division Pyrophyta (673.2 ± 892.09 indv/L) tends to adapt well with an increase of abundance of phytoplankton, particularly in West Season in December (2, 3) and February (1,3,4,5,6,7) until transition I (5),but from east season to transition season II, there is no abundance of Division Pyrophyta found. Division Euglenophyta (1504.8 ± 665.58 indv/L) increases in abundance from east season (1), transition season II (5) toWest Season (1), although not in all stations.

Spatio-temporal changes in the composition and abundance of phytoplankton are the form of adaptation to the change in ecological condition in the water, since phytoplankton plays an important role in water that, besides serving as the primary producer, it may also respond to environmental changes [29]. Odum [30] adds that the dynamic of abundance and community structure of phytoplankton are influenced by its form of adaptation to physical and chemical factors in the water, particularly the availability of nutrients and phytoplankton's ability to utilize them. Phytoplankton community is controlled by dominant species, showing such species' power against other species. Dominant species and lost species will cause important changes in phytoplankton community ecologically. Some plankton species may be abundant (blooming) in eutrophic water, while the other species are sensitive to organic waste and will be lost. The composition of species of plankton may also mark the quality of water mass [31].

The identification finds some types of phytoplankton are able to produce toxin which may potentially be (HABs) Harmful Algae Blooms in West Season (rainy season) to transition season I, particularly at station around anthropogenic activities. So far, however, there is no clear threshold of quantity of phytoplankton during blooming. Such condition is commonly only marked with a change in domination of type, change in community structure and biomass of phytoplankton [8; 9].Moreover,Mulyasari, *et al.*, [10] state that phytoplankton blooming condition may occur when the quantity of abundance is higher than the average quantity of phytoplankton annually.

**Conclusion and Suggestion**

***Conclusion***
The fluctuation of abundance and composition of phytoplanktonin the LSA Area is influenced by difference in ecological characteristics of the research'sstations (spatial) or temporally. The monsoon pattern, supported with hydrodynamic of the water (tide, flow), Upwelling mechanism, Flushing Time and macro nutrients in the water, will spatially influence forming of the research stations' characteristics. Such condition leads to a change in dominance of type, change in community structure and biomass of phytoplankton as the form of adaptation ofthe change in the ecological condition of the water, as marked with change in the composition and abundance of phytoplankton.

***Suggestion***
Based on type identification, with the finding of phytoplanktoncapable of producing toxin which may potentially be (HABs),the Local Government, related parties and surrounding communityneed to prepare a lagoon management strategy to maintain ecological stability in the LSAArea, particularly in relation to land conversion for various anthropogenic activities, since they are expected to increase macronutrientswhich may lead to dominance of HABs type phytoplankton, although willnot be easy that there are ecological differences in the lagoon from the upstream (freshwater) to the downstream (seawater).With its geological factors, the lagoon is a very active hydrology zone which is highly sensitive to environmental changes.

**References**

[1] Setyawan,W.B. 2010[b]KonservasibagiestuariSegaraAnakan. ProsidingNasionalBiologi.UniversitasJenderalSoedirman. ISBN 978-979-16109-4-0
[2] Dahuri, R. 2003. *KeanekaragamanHayatiLautAset Pembangunan Berkelanjutan Indonesia.*GramediaPustakaUtama. Jakarta.
[3] Dsikowitzky,L, I. Nordhaus, TC. Jennerjahn, P. Khrycheva, Y.Sivatharshan, E.Yuwono&J. Schwarzbauer. 2011. Anthropogenic organic contaminants in water, sediments and benthic organisms of the mangrove-fringed SegaraAnakan Lagoon, Java, Indonesia. *Marine Pollution Bulletin* 62.851–862.
[4] Jennerjahn, B. Nasir, I. Pohlenga. 2009. Spatio-temporal variation of dissolved inorganic nutrients related to hydrodynamics and land use in the mangrove-fringed SegaraAnakan Lagoon, Java, Indonesia. *Reg Environ Change* 9:259–274

[5] Coutinho, M.T; A. C. Brito; P. Pereira; A.S. Gonçalves& M.T Moita. 2012. A phytoplankton tool for water quality assessment in semi-enclosed coastal lagoons: Open vs closed regimes. *Journal Estuarine, Coastal and Shelf Science*. 110: 134-146

[6] Yunev,O.A; J. Carstensen; S. Moncheva; A. Khaliulin; G. Ærtebjerg&S.Nixon. 2007. Nutrient and phytoplankton trends on the western Black Sea shelf in response to cultural eutrophication and climate changes. *Journal Estuarine, Coastal and Shelf Science*. 74: 63-76

[7] Dewi,R., M. Zainuri& T. Hardijati. 2010. Ujioptimalisasiintensitascahayaterhadapkandunganklorofila,bpadasistemkulturDunaliellasalinadan Chlorella vulgaris. ProsidingNasionalBiologi.UniversitasJenderal Soedirman.ISBN978-979-16109-4-0

[8] Vadrucci, M.R., E. Stanca., C. Mazzioti, N. Ungaro& A. Basset. 2013. Ability of Phytoplankton trait sensitivity to tighlightantropogenic pressures in Mediteraneanlagoons:A size spectra sensitivity index (ISS-phyto). Ecological Indicators. 34: 113-125.

[9] Ayuningsih, M.S, I.B. Hendrart& P.W.Purnomo. 2014. Distribution and Abundance of Phytoplankton and Chorophyll-a in the Sekumbu Bay JeparaRegency: Relationship with Nitrate and Phosphate Content in Water. *Diponegoro Journal OfMaquares. ManajementOf Aquatic Resources*. Vol. 3, No.2. 138-147

[10] Mulyasari, R. Peranginangin&Th. D. Suryaningrum, A. Sari. 2003. PenelitianMengenaiKeberadaanBiotoksinpada Biota danLingkunganPerairanTeluk Jakarta.*JurnalPenelitianPerikanan Indonesia* Vol. 9 No. 5. 39-64

[11] Pintado, J.G; M.M. Mena, G.G. Barberá;J. Albaladejo and V.M. Castillo. 2007. Anthropogenic nutrient sources and loads from a mediterranean catchment into a coastal lagoon: Mar Menor, Spain .*Science of the total environment*. 373: 220-239

[12] Nielsen, S.L, G.T Banta &M.F Pedersen. 2004. Estuarine Nutrient Cycling: The Influence of Primary Producers.The Fate of Nutrients and Biomass. Aquatic Ecology Series.Vol.2. Kluwer Academic Publishers. Printed in the Netherlands.. ISBN:1-4020-2638-2. Pp.303 Hlm.

[13] Jannah, R. & Z.A, Muchlisin. 2012. Komunitasfitoplankton di daerahEstuariaKrueng, Aceh. *JurnalIlmu-ilmuPerairan, PesisirdanPerikanan*.Vol.1, no.3.

[14] Nontji, A. 2008. Plankton Laut. Jakarta: LembagaIlmuPengetahuan Indonesia. LIPI Press. Jakarta. ISBN: 978-979-799-085-5. 331 Hlm.

[15] Tomas, Carmelo,R,. 1997. *Identifiying Marine Phytoplankton*. Academic Press, United States Of America.ISBN 0-12-693018-X(Pbk:Alk.paper)

[16] Sachlan, M, 1982. Planktonologi.FakultasPerikanandanIlmuKelautanUniversitasDiponegoro.

[17] American Public Health Association (APHA). 2005. Standard Methods ForThe Examination Of Water And Waste water. 21stEdition.EditedBy:Andrew. DEaton ,Lenore.SClesceri, Eugene. WRice, Arnold. E Greenberg.Centennial Edition. American Public Health Association, American Water Work Association. Water Environment Federation.ISBN0-87553-047-8,ISSN55-1979.

[18] Sugiyono. 2007. MetodePenelitianPendidikan. Bandung: Alfabeta.

[19] Suryanti. 2008. Kajian Tingkat Saprobitas di Muara Sungai MorodemakpadasaatPasangdanSurut.UniversitasDiponegoro. Semarang. *JurnalSaintekPerikanan*. 4 (1) 2 : 76 – 83.

[20] Ridd PV, Wolanski E &Mazda Y. 1990. Longitudinal diffusion in mangrove fringed tidal creeks. Estuarine Coastal and Shelf Science 31:541-544.

[21] Layborn, J. & Parry. 1992. Protozoan Plankton Ecology. Chapman & Hall, 2-6 Boundary Row, London. p. 1-30.

[22] Rokhim, K., A. Arisandi&I. W. Abida. 2009. AnalisaKelimpahanFitoplanktondanKetersediaanNutrien (NO3 dan PO4) di PerairanKecamatanKwanyarKabupatenBangkalan.*JurnalKelautan*. 2 (2) :1907-9931.

[23] Ji,Z.G.2008.HydrodynamicsandWaterQuality.ModellingRivers,LakesandEstuaries A JohnWiley&Sons,Inc., Publication.

[24] Sediadi, A. &Sutomo.1990. DiatomaeMargaChaetocerosdanRhizosolenia di perairanWaisarisaPulauSeram, SuatuStudiPendahuluanDalamPerairan Maluku danSekitarnya.LIPI. Jakarta. Hal 121-126

[25] Panggabean, L.M.G. 1994. "Red Tide" di Indonesia.: PerlukahDiwaspadai? *Oseana*. 19(1):33-38

[26] Livingston, R.J. 2001. Eutrophication Processes in Coastal Systems. Origin and Succession of Plankton Blooms and Effect on Secondary Production in Gulf Coast Estuaries.. CRC Press LLC. Marine Science Series Boca Raton.Printed in the United State of America.ISBN.0-8493-9062-1. Pp 327.

[27] Dyer, K.R. 2000. Estuaries: A physical introduction (2 edition), John Wiley and Sons, New York,195pp

[28] Pednekar, S.M, S.G.P, Matondkar&V. Kerkar. 2012. Spatio temporal Distribution of Harmfull Algal FlorainThe Tropical Estuarine Complex of Goa, India. The Scientific World. 2012:11. P.doi: 10.1100/2012/596276

[29] Pirzan, A. M & P. Rani. 2008. HubunganKeragamanFitoplanktondenganKualitas Air Di PulauBaulung, Kab. Tahalar, Sulawesi Selatan. *Biodiversitas*.Vol 9 No. 3. (217-221). ISSN: 1412-033X

[30] Odum, E.P. 1971. Fundamental of Ecology.(3rd edition).W.B. Saunders Co.

IOP Conf. Series: Earth and Environmental Science **406** (2019) 012028 doi:10.1088/1755-1315/406/1/012028

[31] Krebs, J. C. 1978. Ecology The Experimental Analysis of Distribution and Abundance. Third Edition.Harper and Row Publisher. New York.
[32] Basmi, J. 1999. Planktonologi: Bioekologi Plankton Algae. FakultasPerikanandanIlmuKelautan, InstitutPertanian Bogor.
[33] Newell, G.E & R.C Newell. 1993. Marine Plankton, APracticalGuide. Hutchinson of London.244 hal.

# Fish conservation status in eastern part of segara anakan Cilacap Indonesia

**Agatha Sih Piranti*, Nuning Setyaningrum, Diana Retna USR, Erwin R Ardli, Dwi Sunu Widyartini, Ilalqisny Insan**
Fakultas Biologi Unsoed Purwokerto, Indonesia
*Corresponding Email : agatha.piranti@gmail.com

**Abstract.** Segara Anakan is an estuarine ecosystem located on the southern coast of Central Java Province which is very fertile with high biodiversity. However, the current condition is thought to have experienced a decline in biodiversity caused by sedimentation, overexploitation and water pollution problems so that it is feared that it will disrupt the preservation of organisms including fish. The purpose of this study was to determine the diversity of fish found in Segara Anakan Cilacap, and to know the conservation status of each species of fish caught in Segara Anakan in eastern Cilacap. This research was conducted using a survey method with cluster random sampling technique at 6 stations (3 stations on the Sapuregel River and 3 stations on the Kembangkuning River) with 3 replications each during highest tide. Fish collection is done by using a net and gillnet or gillnet that is installed during low tide and carried out during high tide. The diversity of fish in the waters of Segara Anakan is 36 species from 25 families. There are 16 species from 15 families in the Donan River and 34 species from 25 families in the Sapuregel River. The diversity of fish in the Sapuregel River is higher than in the Donan River. The status of fish in the eastern Segara Anakan waters management area is mostly in the LR (Least Concern) or low risk status and 1 NT (Near Threatened) species or threatened with extinction namely Sidat Fish (*Anguilla bicolor*)

Keywords: Segara Saplings, fish, conservation status, diversity

## 1. Introduction

Segara Anakan is a brackish water lagoon ecosystem with the largest mangrove forest in Java located on the southern coast of Central Java Province, precisely in Cilacap Regency or in the north of Nusakambangan Island at coordinates 07 ° 34'29.42 "LS-07 ° 47'32.39" LS and 108 ° 46'30.12 "East – 109 ° 03'21.02" East. Segara sapling ecosystems are very fertile ecosystems with high biodiversity. The area of Segara Anakan reaches approximately 34,018 ha [1]. Segara Anakan is an estuary that is a meeting between the Citanduy River, Cibeurem, Cikujang, Cikonde, Kayu Mati, Ujung Alang, Dangkal, Kembang Kuning, Sapuregel, and Donan. The Segara Anakan area is divided into 3 management areas, namely, the western part, the central part and the eastern part. The western part is a waters area which includes Citanduy River, Cibeureum River and Cikonde River. The eastern part covers the Donan River and the Sapuregel River, and the middle part is the Kembangkuning River which connects the two regions.

The Eastern Segara Anakan area is a shipping lane for industry and is used for other activities [1]. Industrial activities include petroleum processing industry with oil refining activities, crude oil loading and unloading in the area and oil processing, Cilacap Industrial Zone (KIC), cement industry, and residential areas so that the potential for the entry of solid material into the waters. Changes in water conditions can increase the chemical compounds so that it can trigger increased nutrient concentrations that will affect water quality and fish production.

Estuary ecosystem is a river estuary that has the potential to accumulate sedimentation originating from material carried by the river flow from upstream which can be in the form of clay, mud and mixed with organic material [1]. There are 3 (three) main problems in Segara Anakan which are possible to cause the threat of fish diversity in Segara Anakan, such as sedimentation (water quality problems), decline in biological resources (illegal logging, conversion, overexploitation) as well as population growth problems. If the problem is not immediately addressed, the possibility of ecosystems and organisms in the ecosystem becomes threatened. In 2000, fisheries production in the lagoon reached 488 tons consisting of 41% shrimp, 39% fish, 13% crabs and 7% other biota [5]. The 2015 data was based on research by Nurfiarini *et al* (2015) that the diversity of fish in puppies consisted of 45 families and 87 species dominated by the families of Ambassidae, Engraulidae, Leognathidae, Mugilidae, Atherinidae, and Bagridae. If it is not quickly overcome, this situation clearly disturbs the balance of the communities that inhabit it, especially fish. What is likely to happen is the reduction in fish populations inhabiting the area. Even though the Segara Anakan mangrove area has contributed to coastal fishery production in a year of more than 62 billion rupiah (Budiman 2007).

Menezes and Camaraschi (1999) stated that the presence and distribution of fish in rivers is influenced by biotic and abiotic factors such as: temperature, pH, salinity, substrate type, feed availability and reproductive strategy. Based on reports from the Citanduy Project Department of Public Works, sediments carried by the three major rivers namely the Citanduy river are 5 million $m^3$, the Cimeneng river is 0.4 million $m^3$, and the Cikonde river is 1.2 million $m^3$ annually due to sedimentation that is constantly feared by the surface water of the Citanduy river and other rivers that flow into Segara Anakan continue to increase and cause flooding around the river area. In addition, siltation due to sedimentation will accelerate the destruction of ecosystems in Segara Anakan. Also reported by Ardli & Wolff [2] that the lagoon area of Segara Anakan currently was ± 600 ha. Under these conditions it is feared that it will disrupt the sustainability of fish. Based on the description above, the objectives of the study were to know the diversity of fish found in Segara Anakan Cilacap, and to know the conservation status of each species of fish caught in Segara Anakan in eastern Cilacap. By knowing the conservation status of each of these fish species, efforts can be made to protect, conserve and utilize these fish resources to ensure their existence, availability and sustainability for current and future generations.

## 2. Method

This research was conducted using a survey method with cluster random sampling technique at 6 stations (3 stations on the Sapuregel River and 3 stations on the Kembangkuning River) with 3 replications each during highest tide (Figure 1). Data of fish caught in the Sapuregel River were based on research data published by Setyaningrum [11].

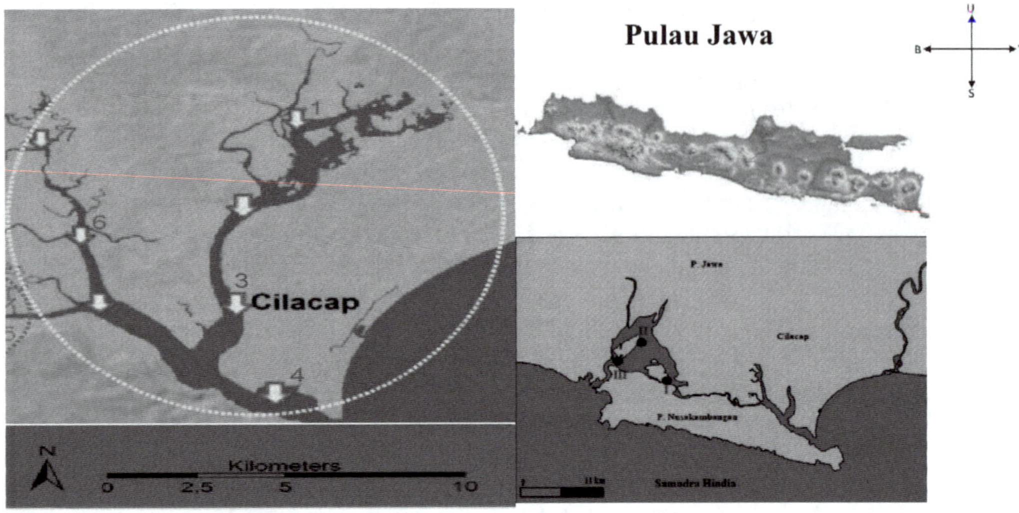


Gambar 3.1 Researh location in Eastern Part of Segara Anakan, Cilacap

Fish collection was done by using a net and gillnet or gillnet. The mesh size used is based on the mesh diameter of 1-2 cm, while the gill nets used are 0.5 - 4 inch in diameter with a net length of 30 m and width of 18 m. Installation of nets is done at low tide and then removal of nets is done at high tide. The fish caught are preserved using 4% formalin and labeled. Furthermore, the handling of fish is carried out in the Aquatic Biology Laboratory, where formalin is washed with running water, then 75% alcohol is replaced as a permanent preservative. Furthermore, the fish were identified by referring to Kottelat et al. [9], Nelson (2006), Marine Portal Identification and Fish Base.

The composition of fish species was analyzed qualitatively and used to differentiate fish species based on identification. Relative abundance is used to see the density of fish species at the study site. The relative abundance of species is obtained by counting the number of individuals of each species of fish per total individual species caught. To find out the conservation status of each organism, the diversity data of each organism is compared with the conservation status in the Red List of International Union for Conservation of Nature (IUCN Red List).

## 3. Result and Discussion

### 3.1 Fish Diversity in the Sapuregel River

The number and species of fish caught in the Sapuregel River were 1,446 individuals representing 23 families and 34 species (Table 1). Sapuregel River is included in the mangrove ecosystem so it is still classified as fertile waters for fish habitat.

Fish caught in the rainy season are more dominant than the dry season. This is because during the rainy season the water level rises and a number of fish in the estihari habitat which is eurihalin are brought into the river, especially during high tide conditions. Fishing is done by installing a net at low tide then the net is lifted at high tide. So that the fish caught are affected by tidal conditions. During high tide, the physical chemistry conditions of waters in rivers that are estuary tend to be the same as mangrove waters due to tidal currents [10].

Table 1. Fish Diversity in Sapuregel River [11].

| No | Familia | No | Spesies | Kemarau | | Penghujan | |
|---|---|---|---|---|---|---|---|
| | | | | ∑ ind | KR | ∑ ind | KR |
| 1 | Gobidae | 1 | *Pseudogobus javanicus* | - | 0,00 | 207 | 16,83 |
| 2 | Mugilidae | 2 | *Valamugil seheli** | 42 | 19,44 | 70 | 5,69 |
| 3 | Cygnoglossidae | 3 | *Cygnoglossus microlepis* | - | 0,00 | 259 | 21,06 |
| 4 | Tetraodontidae | 4 | *Tetraodon kretamensis** | 1 | 0,46 | 4 | 0,33 |
| | | 5 | *Lagocephalus sp* | - | 0,00 | 1 | 0,08 |
| 5 | Leiognathidae | 6 | *Leiognathus oculus** | 5 | 2,31 | 14 | 1,14 |
| 6 | Carangidae | 7 | *Caranx ignobilis* | - | 0,00 | 29 | 2,36 |
| | | 8 | *Trachurus symmetricus* | - | 0,00 | 4 | 0,33 |
| | | 9 | *Alectis indicus* | - | 0,00 | 1 | 0,08 |
| 7 | Ambassidae | 10 | *Ambassis kopsii* | - | 0,00 | 3 | 0,24 |
| | | 11 | *Parambassis sp* | - | 0,00 | 1 | 0,08 |
| 8 | Drepanidae | 12 | *Drepane longimana* | - | 0,00 | 6 | 0,49 |
| 9 | Scianidae | 13 | *Nibea sp** | 4 | 1,85 | 6 | 0,49 |
| 10 | Sillagnidae | 14 | *Sillago sp** | 1 | 0,46 | 1 | 0,08 |
| 11 | Belanidae | 15 | *Xenentodon canula* | - | 0,00 | 1 | 0,08 |
| 12 | Oxuderudae | 16 | *Taenoides cirratus* | - | 0,00 | 4 | 0,33 |
| | | 17 | *Sicyopterus sp* | - | 0,00 | 8 | 0,65 |
| 13 | Terapontidae | 18 | *Terapon jarbua* | - | 0,00 | 1 | 0,08 |
| | | 19 | *Terapon theraps* | - | 0,00 | 1 | 0,08 |
| | | 20 | *Mesopristes cancellatus** | 14 | 6,48 | 1 | 0,08 |
| 14 | Engraulidae | 21 | *Stolephorus indicus** | 124 | 57,41 | 524 | 42,60 |
| 15 | Ophichthidae | 22 | *Lamnostoma mindora* | - | 0,00 | 1 | 0,08 |
| | | 23 | *Ophichthus aphostistos** | 1 | 0,46 | 57 | 4,63 |
| | | 24 | *Ophichthus frontalis* | 4 | 1,85 | - | 0,00 |
| 16 | Placycephalidae | 25 | *Platycephalus indicus* | - | 0,00 | 4 | 0,33 |
| 17 | Serranidae | 26 | *Epinephelus sp** | 4 | 1,85 | 5 | 0,41 |
| 18 | Scatophagidae | 27 | *Scatophagus argus* | - | 0,00 | 4 | 0,33 |
| 19 | Lutjanidae | 28 | *Lutjanus russelii* | - | 0,00 | 12 | 0,98 |
| 20 | Paralichthydae | 29 | *Pseudorhombus elevatus** | 1 | 0,46 | 1 | 0,08 |
| | | 30 | *Pseudorrhombus arsius* | 1 | 0,46 | - | 0,00 |
| 21 | Muraenidae | 31 | *Gymnothorax polyuranodon* | 2 | 0,93 | - | 0,00 |
| | | 32 | *Uropterygius concolor* | 4 | 1,85 | - | 0,00 |
| 22 | Eleotridae | 33 | *Eleotris acanthopoma* | 4 | 1,85 | - | 0,00 |
| 23 | Anguillidae | 34 | *Anguilla bicolor* | 4 | 1,85 | - | 0,00 |
| | | | Kelimpahan | 216 | | 1230 | |
| | | | Total kelimpahan | | 1446 | | |
| | | | Jumlah jenis | 16 | | 28 | |

The most caught species of *Stolephorus indicus* (anchovy) are able to adapt well in the Sapuregel River in both the dry and rainy seasons. These results are consistent with research by Suprastini et al., [12] *Stolephorus indicus* is found and adapted well in the Segara Anakan area of Cilacap. Furthermore, the Valamugil seheli species (mullet fish) have eurihalin properties that can live in a wide range of salinity, because they can live in freshwater, brackish and sea that are associated with reefs, often entering estuaries and rivers. Catfish mulls are migratory so they migrate to the river to reproduce (fish base). This is consistent with the results of research Latupapua (2011); Redjeki [10]; Wahyudewantoro [14] that mullet fish, anchovy, glodok fish caught in mangrove waters are mostly juvenile and juvenile phases.

Kottelat *et al* [9] that *Ophichthus aphostistos* (sea eel) was caught a lot especially in the rainy season even though this fish is a habitat in the muddy waters and belongs to the snake eels (Aguilliformes) group. In general, the types of fish caught in both seasons (dry and rainy) are able to adapt to the Sapuragel river habitat and have a relatively high abundance.

The relative abundance of fish in the Sapuregel River is highest in the families of Engraulidae (dry and rainy), Mugillidae (dry) and Cynoglossidae, Gobiidae (rainy) (Figure 2). The types of fish in the waters are part of the community only a few species are predominant. Among the 23 fish families found in the Sapuregel River, two families that are quite dominating

IOP Conf. Series: Earth and Environmental Science **406** (2019) 012029 doi:10.1088/1755-1315/406/1/012029

both in the dry and rainy seasons are the Engraulidae and Mugillidae families. The Engraulidae family and also the Gobiidae are mostly caught in the Sapuregel River, species of fish belonging to the Engraulidae family are quite a lot such as the *Stolephorus indicus* (Teri fish) and the Gobiidae family *Pseudogobius javanicus* (Glodog fish). Glodog fish, belonging to the Gobiidae family, have not been used as food in Indonesia [13]. Based on [3] when juvenile fish, Gobiidae is a herbivor that eats more diatoms and filamentous algae, while as adults switch to crustaceans, polichaeta and land insects.

Mugillidae family were found in Sapuregel River mainly during dry season. They are migratory species from the sea and are the largest group in estuaries in both subtropical and tropical regions [15]. Whereas, the Family of Cygnoglossidae dominated in the Sapuregel river area during rainy season. This family is eurihalin species but they were abundant in the Sapuregel River was caused by tides. They were migratory species from the sea and carried into the river.

Based on the IUCN Red list, the conservation status of each fish caught in the Sapuregel River is listed in Table 2. Based on the IUCN Red List, the conservation status for each fish in the Sapuregel River showed that there were 8 species of fish have not been evaluated yet (Table 3), 17 species in LC status (Table 4), 2 species of DD (Table 5) and 1 species in NT status (Table 6).

Table 2. Concervation Status of fish in Sapuregel River

| No | Familia | Spesies | Status Konservasi |
|---|---|---|---|
| 1 | Gobidae | *Pseudogobus javanicus* | NE |
| 2 | Mugilidae | *Valamugil seheli** | NE |
| 3 | Cygnoglossidae | *Cygnoglossus microlepis* | LC |
| 4 | Tetraodontidae | *1. Tetraodon kretamensis** | NE |
| | | *2. Lagocephalus sp* | LC |
| 5 | Leiognathidae | *Leiognathus oculus** | LC |
| 6 | Carangidae | *1. Caranx ignobilis* | LC |
| | | *2. Trachurus symmetricus* | LC |
| | | *3. Alectis indicus* | LC |
| 7 | Ambassidae | *1. Ambassis kopsii* | NE |
| | | *2. Parambassis sp* | LC |
| 8 | Drepanidae | *Drepane longimana* | NE |
| 9 | Scianidae | *Nibea sp** | LC |
| 10 | Sillagnidae | *Sillago sp** | LC |
| 11 | Belanidae | *Xenentodon canula* | LC |
| 12 | Oxuderudae | *1. Taenoides cirratus* | DD |
| | | *2. Sicyopterus sp* | LC |
| 13 | Terapontidae | *1. Terapon jarbua* | LC |
| | | *2. Terapon theraps* | LC |
| | | *3. Mesopristes cancellatus** | LC |
| 14 | Engraulidae | *Stolephorus indicus** | NE |
| 15 | Ophichthidae | *1. Lamnostoma mindora* | NE |
| | | *2. Ophichthus aphostistos** | NE |
| | | *3. Ophichthus frontalis* | LC |
| 16 | Placycephalidae | *Platycephalus indicus* | DD |
| 17 | Serranidae | *Epinephelus sp** | LC |
| 18 | Scatophagidae | *Scatophagus argus* | LC |
| 19 | Lutjanidae | *Lutjanus russelii* | LC |
| 20 | Paralichthydae | *1. Pseudorhombus elevatus** | NE |
| | | *2. Pseudorrhombus arsius* | LC |
| 21 | Muraenidae | *1. Gymnothorax polyuranodon* | LC |
| | | *2. Uropterygius concolor* | LC |
| 22 | Eleotridae | *Eleotris acanthopoma* | LC |
| 23 | Anguillidae | *Anguilla bicolor* | NT |

Notes :
NE = Not Evaluation, LC = Least Concern, DD = Data Deficient, NT = Near Threatened

Fish with not evaluation (NE) status meant that the species have not yet been evaluated for the possibility of extinction and distribution criteria, there were Gobidae, Mugilidae, Tetraodontidae, Ambassidae, Drepanidae, Engraulidae, Ophichthidae, Paralichthydae.

Table 3. Species of fish in Sapuregel River with NE (Not Evaluation) Status

| No | Familia | Spesies | Status Konservasi |
|---|---|---|---|
| 1 | Gobidae | *Pseudogobus javanicus* | NE |
| 2 | Mugilidae | *Valamugil seheli** | NE |
| 3 | Tetraodontidae | *Tetraodon kretamensis** | NE |
| 4 | Ambassidae | *Ambassis kopsii* | NE |
| 5 | Drepanidae | *Drepane longimana* | NE |
| 6 | Engraulidae | *Stolephorus indicus** | NE |
| 7 | Ophichthidae | *1. Lamnostoma mindora* | NE |
| | | *2. Ophichthus aphostistos** | NE |
| 8 | Paralichthydae | *Pseudorhombus elevatus** | NE |

Keterangan : NE = Not Evaluation

There are 17 species of fish that have Least Concern (LC) status. Least Concern status is the IUCN category given for species that have been evaluated but do not fall into any category.

Table 4. Species of fish in Sapuregel River with LC (*Least Concern)* Status

| No | Familia | Spesies | Concervation Status |
|---|---|---|---|
| 1 | Cygnoglossidae | *Cygnoglossus microlepis* | LC |
| 2 | Tetraodontidae | *Lagocephalus sp* | LC |
| 3 | Leiognathidae | *Leiognathus oculus** | LC |
| 4 | Carangidae | *1. Caranx ignobilis* | LC |
| | | *2. Trachurus symmetricus* | LC |
| | | *3. Alectis indicus* | LC |
| 5 | Ambassidae | *Parambassis sp* | LC |
| 6 | Scianidae | *Nibea sp** | LC |
| 7 | Sillagnidae | *Sillago sp** | LC |
| 8 | Belanidae | *Xenentodon canula* | LC |
| 9 | Oxuderudae | *Sicyopterus sp* | LC |
| 10 | Terapontidae | *1. Terapon jarbua* | LC |
| | | *2. Terapon theraps* | LC |
| | | *3. Mesopristes cancellatus** | LC |
| 11 | Engraulidae | *Ophichthus frontalis* | LC |
| 12 | Serranidae | *Epinephelus sp** | LC |
| 13 | Scatophagidae | *Scatophagus argus* | LC |
| 14 | Lutjanidae | *Lutjanus russelii* | LC |
| 15 | Paralichthydae | *Pseudorrhombus arsius* | LC |
| 16 | Muraenidae | *1. Gymnothorax polyuranodon* | LC |
| | | *2. Uropterygius concolor* | LC |
| 17 | Eleotridae | *Eleotris acanthopoma* | LC |

Keterangan : LC = Least Concern

There were 2 species with DD concervation status, they were *Taenoides cirratus* and *Platycephalus indicus* (Table 5). It meant that the information about that fish were inadequate to estimate the risk of extinction based on population distribution and status.

Table 5. Species of fish in Sapuregel River with DD (*Data Deficient)* Status

| No | Familia | Spesies | Concervation Status |
|---|---|---|---|
| 1 | Oxuderudae | *Taenoides cirratus* | DD |
| 2 | Placycephalidae | *Platycephalus indicus* | DD |

There was one species of fish that had NT status, that was species of Eel fish (*Anguilla bicolor)* belonging to Familia of Anguilidae (Table 6). Eel fish have unique habitat characteristics, because they are able to adapt in fresh, estuary and sea waters. In its life cycle the fish undergoes six phases, namely eggs, pre-leptocephalus, leptocephalus, glass eels, adults and broodstock. Eel has katadromus properties, which live in fresh waters and when they will spawn (reproduce) eels migrate very far into the sea. To reach the river mouth, the eel must be able to follow the river flow. After reaching the river mouth, the eel must oppose the waves to arrive in the middle of the sea and must adapt to changes in salt content.

Table 6. Species of fish in Sapuregel River with NT (*Near Threatened)* Status

| No | Familia | Spesies | Concervation Status |
|---|---|---|---|
| 1 | Anguillidae | *Anguilla bicolor bicolor* | NT |

The Eel fish (*Anguilla bicolor bicolor*) is important commodities, and the demand for fish availability in the global market was increasing drastically nowadays. This demand impact to high eel exploitation. Reduced catches of eel be caused by high rates of explotation, declining environmental quality, sedimentation, mining, and other development impacts. The life cycle of eel is very complex, one of the interesting stages is the movement between Leptocephalus that live in the sea into glass eel which migrates into fresh waters. Eel have a habit of spawning in warm seas at a depth of about 400 m (Matsui, 1982). Glass Eel will enter the river mouth at night when the tide is high and salinity at the river mouth is low. The high level of catching results in the increasingly threatened availability of glass eel in nature. Initially excessive exploitation was only glass eel, but starting from 2010, it became more diverse, that was from glass eel to fingerling [6]. As a result of these threats the status of eel fish (*Anguilla bicolor*) became Near threatened (NT)

### 3.2 Fish Diversity in the Donan River

The number and species of fish caught in the Donan River during the study were 534 individuals representing 16 families and 16 species. In the dry season, the fish caught consisted of 6 families, namely the Gobiidae family (48.5%), Sillaginidae (30.3%), Mugilidae as much as 12.1% and other families (Cichlidae, Serranidae, and Carangidae), respectively, only around 3%. In the rainy season 11 families were obtained, but as many as 90% of the fish caught were Engrauilidae (90.2%), namely *Stolephorus indicus* (Table 7). *Stolephorus indicus* is a fish that lives in groups that are found in coastal waters entering the river mouth and to tolerate brackish water [8].

IOP Conf. Series: Earth and Environmental Science **406** (2019) 012029 doi:10.1088/1755-1315/406/1/012029

Table 7. The number and species of fish caught in the Donan River during rainy and dry season

| No | Familia | Spesies | Dry | | Rainy | |
|---|---|---|---|---|---|---|
| | | | ∑ ind | KR % | ∑ ind | KR % |
| 1 | Tetraodontidae | *Lagocephalus lagocephalus* | - | - | 2 | 0,4 |
| 2 | Terapontidae | *Mesopristes cancellatus* | - | - | 1 | 0,2 |
| 3 | Leiognathidae | *Secutor indicius* | - | - | 1 | 0,2 |
| 4 | Leiognathidae | *Leiognathus equulus* | - | - | 5 | 1,0 |
| 5 | Cygnoglossidae | *Cygnoglossus microlepis* | - | - | 1 | 0,2 |
| 6 | Paralichthyidae | *Pseudorrhombus arsius* | - | - | 3 | 0,6 |
| 7 | Ambassidae | *Ambassis vachelli* | - | - | 11 | 2,2 |
| 8 | Scatophagidae | *Scatophagus argus* | - | - | 1 | 0,2 |
| 9 | Gobiidae | *Pseudogobius javanicus* | 16 | 48,5 | 13 | 2,6 |
| 10 | Engrauilidae | *Stolephorus indicus* | - | - | 452 | 90,2 |
| 11 | Polynemidae | *Eleutheronema tetradactylum* | - | - | 4 | 0,8 |
| 12 | Cichlidae | *Oreochromis niloticus* | 1 | 3,0 | - | - |
| 13 | Mugilidae | *Valamugil seheli* | 4 | 12,1 | 7 | 1,4 |
| 14 | Serranidae | *Epinephelus sp* | 1 | 3,0 | - | - |
| 15 | Sillaginidae | *Sillago sihama* | 10 | 30,3 | - | - |
| 16 | Carangidae | *Caranx ignobilis* | 1 | 3,0 | - | - |
| | Total | | 33 | | 501 | |

Fish caught in the rainy season was more diverse (12 families) than the dry season (6 families). This was because during the rainy season the estuarine water level rises with increasing the debit of river water, so the composition of fish found were a mixture of species of freshwater fish and seawater but it was still tolerably to brackish conditions.

Based on observations in the 2 river of Donan and Sapuregel obtained of a diverse fish compositions. Sapuregel River has more diverse composition compared to Donan River. The composition of a species of fish is generally influenced by habitat conditions and water quality of waters. When referring to water quality (Table 8) in the two rivers, the results were not significantly different, because the water quality in the two locations still met the quality standards for the life of organisms in mangrove or estuarine areas. Therefore differences in the composition of fish species between the two rivers could be affected technically when sampling, for example sampling time, sampling operation techniques, which then resulted in differing composistion of fish caught between location.

Table 8. Water Quality Measurement in Donan and Sapuregel River

| Parameter | Unit | Sampling Location | | Water quality standard (Mangrove ecosystem)* |
|---|---|---|---|---|
| | | Donan River | Sapuregel River | |
| pH | unit | 6.6 | 6.8 | 7 – 8.5 |
| Water Temperature | °C | 30.5 | 28.1 | 28 – 32 |
| Salinity | ppt | 30.5 | 29.3 | < 34 |
| TSS | $mg.l^{-1}$ | 115.1 | 83.4 | 80 |
| DO | $mg.l^{-1}$ | 3.2 | 3.4 | >5 |
| BOD | $mg.l^{-1}$ | 4.4 | 4.3 | 20 |

Note * = Kepmen LH Nomor 51 Tahun 2004 : Water Quality for Biota Life

IOP Conf. Series: Earth and Environmental Science **406** (2019) 012029 doi:10.1088/1755-1315/406/1/012029

Water quality will influence the fish diversity and there was a relationship between habitat and environmental variables [4]. The measurement of Total Suspended Solid (TSS) concentrations in the Sapuregel and Donan Rivers have beyond the standard for biota life determined at 80 $mg.l^{-1}$. The high concentration of TSS is caused by the two locations experiencing the insertion of solid material carried from the river due to human activities such as agriculture, industry and settlements in the catchment area. Dissolved oxygen is one of the main elements in the metabolic processes of organisms, especially for the process of respiration. Dissolved Oxygen (DO) in water has a very important role for the survival of marine organisms (Dewanti *et al.* 2018). Besides oxygen is also needed for the oxidation of organic and inorganic materials. The main source of oxygen in a waters comes from a process of diffusion of free air and photosynthesis of organisms that live in these waters. The results of DO measurements at the study site showed low DO concentrations ranging from 3.2 to 3.4 mg.l-1 lower than those required for biota life

Fish has a very important role in human life because of its position in the web of life [9], and has a very high potential in commercial terms. The comparison of fish composition and abundance between river of Donan and Sapuregel are presented in Table 9.

Table 9. The comparison of fish composition and abundance between river of Donan and Sapuregel

| No | Familia | Species | Local Fish Name | S. Donan | | S. Sapuregel | |
|---|---|---|---|---|---|---|---|
| | | | | ∑ ind | KR % | ∑ ind | KR % |
| 1 | Gobiidae | *1. Pseudogobus javanicus* | Glodok | 29 | 5.4 | 207 | 14.3 |
| 2 | Tetraodontidae | *2. Lagocephalus lagocephalus* | Bunteg pisang | 2 | 0.4 | | |
| | | *3. Lagocephalus sp* | | | | 1 | 0.1 |
| | | *4. Tetraodon kretamensis* | | | | 5 | 0.3 |
| 3 | Mugilidae | *5. Valamugil seheli* | Belanak | 11 | 2.1 | 112 | 7.7 |
| 4 | Terapontidae | *6. Mesopristes cancellatus* | Rekrekan | 1 | 0.2 | 15 | 1.0 |
| | | *7. Terapon jarbua* | | | | 1 | 0.1 |
| | | *8. Terapon theraps* | | | | 1 | 0.1 |
| 5 | Cygnoglossidae | *9. Cygnoglossus microlepis* | Lengdai | 1 | 0.2 | 259 | 17.9 |
| 6 | Leiognathidae | *10. Secutor indicius* | Petek baworan | 1 | 0.2 | | |
| | | *11. Leiognathus equulus* | Petek | 5 | 0.9 | 19 | 1.3 |
| 7 | Sillaginidae | *12. Sillago sihama* | Bojor | 10 | 1.9 | 2 | 0.1 |
| 8 | Carangidae | *13. Caranx ignobilis* | Kue | 1 | 0.2 | 29 | 2.0 |
| | | *14. Trachurus symmetricus* | | | | 4 | 0.3 |
| | | *15. Alectis indicus* | | | | 1 | 0.1 |
| 9 | Ambassidae | 16. *Ambassis vachelli* | Peprek | 11 | 2.1 | | |
| | | *17. Ambassis kopsii* | | | | 3 | 0.2 |
| | | *18. Parambassis sp* | | | | 1 | 0.1 |
| 10 | Paralichthyidae | 19. *Pseudorrhombus arsius* | Tapel buruk | 3 | 0.6 | 1 | 0.1 |
| | | *20. Pseudorhombus elevatus* | | | | 2 | 0.1 |
| 11 | Drepanidae | *21. Drepane longimana* | | | | 6 | 0.4 |
| 12 | Scianidae | *22. Nibea sp* | | | | 10 | 0.7 |

| | | | | | | | |
|---|---|---|---|---|---|---|---|
| 13 | Belanidae | *23. Xenentodon canula* | | | | 1 | 0.1 |
| 14 | Oxuderudae | *24. Taenoides cirratus* | | | | 4 | 0.3 |
| | | *25. Sicyopterus sp* | | | | 8 | 0.6 |
| 15 | Polynemidae | *26. Eleutheronema tetradactylum* | Kuro | 4 | 0.7 | | |
| 16 | Cichlidae | *27. Oreochromis niloticus* | Nila | 1 | 0.2 | | |
| 17 | Serranidae | *28. Epinephelus sp* | Kerapu | 1 | 0.2 | 9 | 0.6 |
| 18 | Engrauilidae | *29. Stolephorus indicus* | Teri mancung | 452 | 84.6 | 648 | 44.8 |
| 19 | Scatophagidae | *30. Scatophagus argus* | Kiper | 1 | 0.2 | 4 | 0.3 |
| 20 | Ophichthidae | *31. Lamnostoma mindora* | | | | 1 | 0.1 |
| | | *32. Ophichthus aphostistos** | | | | 58 | 4.0 |
| | | *33. Ophichthus frontalis* | | | | 4 | 0.3 |
| 21 | Placycephalida e | *34. Platycephalus indicus* | | | | 4 | 0.3 |
| 22 | Lutjanidae | *35. Lutjanus russelii* | | | | 12 | 0.8 |
| 23 | Muraenidae | *36. Gymnothorax polyuranodon* | | | | 2 | 0.1 |
| | | *37. Uropterygius concolor* | | | | 4 | 0.3 |
| 24 | Eleotridae | *38. Eleotris acanthopoma* | | | | 4 | 0.3 |
| 25 | Anguillidae | *39. Anguilla bicolor* | | | | 4 | 0.3 |
| | | Number of Individual | | 534 | 100 | 1446 | 100 |
| | | *Number of Spesies* | | | 16 | | 34 |
| | | *Number of Familia* | | | 15 | | 25 |

The dominant fish species in the Donan and Sapuregel rivers with a relative abundance of> 10%) are *Pseudogobus javanicus*, *Cygnoglossus microlepis*, and *Stolephorus indicus*. The abundance of *Pseudogobus javanicus* and *Cygnoglossus microlepi*s is higher in the Sapuregel River higher than in the Donan River, whereas *Stolephorus indicus* is more abundant in the Donan River.

Table 10. Fish Conservation Status in Donan River with LC (Least Concern) status

| No | Nama famili | Nama Spesies | Status Konservasi |
|---|---|---|---|
| 1 | Tetraodontidae | *Lagocephalus lagocephalus* | LC |
| 2 | Terapontidae | *Mesopristes cancellatus* | LC |
| 4 | Leiognathidae | *Leiognathus equulus* | LC |
| 5 | Cygnoglossidae | *Cygnoglossus microlepis* | LC |
| 6 | Ambassidae | *Ambassis vachelli* | LC |
| 7 | Scatophagidae | *Scatophagus argus* | LC |
| 8 | Gobiidae | *Pseudogobus javanicus* | LC |
| 9 | Engrauilidae | *Stolephorus indicus* | LC |
| 10 | Serranidae | *Epinephelus sp* | LC |
| 11 | Sillaginidae | *Sillago sihama* | LC |
| 12 | Carangidae | *Caranx ignobilis* | LC |

Based on Tables 10 and 11 can be seen from the conservation status of the IUCN Red List (Froese & Pauly, 2017) fish caught in the Donan River include: 12 low risk status species or LC (Least Concern) and 5 NE species (not evaluated).

Table 11. Fish Conservation Status in Donan River with NE (Not Evaluated) status

| No | Nama famili | Nama ilmiah | Status Konservasi |
|---|---|---|---|
| 1 | Leiognathidae | *Secutor indicius* | NE |
| 2 | Paralichthyidae | *Pseudorrhombus arsius* | NE |
| 3 | Polynemidae | *Eleutheronema tetradactylum* | NE |
| 4 | Cichlidae | *Oreochromis niloticus* | NE |
| 5 | Mugilidae | *Valamugil seheli* | NE |

One of the 5 species with NE status was *Oreochromis niloticus* (Tilapia). This fish is an introducing fish that is very invasive and disturbs various ecosystems, especially those located in the tropics. *Oreochromis niloticus* is an aquatic species that is often cultivated and possibly carried by floods and released into rivers to estuarins. If the diversity of fish is compared between the two locations, it can be seen that both the number of families and the number of species are more in the Sapuregel River than in the Donan River (Table 12).

Table 12. Comparison Of Fish Community Structure In Sapuregel dan Donan River

| No | Parameter | Donan River | Sapuregel River |
|---|---|---|---|
| 1 | Number of familia | 18 | 35 |
| 2 | Number of spesies | 18 | 34 |
| 3 | Conservation Status | | |
| | - NE (Not Evaluation ) | 5 | 9 |
| | - LC (Least Concern) | 11 | 22 |
| | - DD (Data Deficient) | | 2 |
| | - NT (Near Threatened) | | 1 (*Anguilla bicolor* |
| 4 | Dominating fish | *Stolephorus indicus (84,6%)* | 1. *Stolephorus indicus (44,8%)* |
| | | | 2. *Cygnoglossus microlepis (17,9%)* |
| | | | 3. *Pseudogobus javanicus (14,3%* |

## 4. Conclusion

The diversity of fish in the waters of Segara Anakan is 36 species from 25 families. There are 16 species from 15 families in the Donan River and 34 species from 25 families in the Sapuregel River. The diversity of fish in the Sapuregel River is higher than in the Donan River. The status of fish in the eastern Segara Anakan waters management area is mostly in the LR (Least Concern) or low risk status and 1 NT (Near Threatened) species or threatened with extinction namely Sidat Fish (Anguilla bicolor)

**Acknowlegement**

Thanks to UNSOED the Directorate General of Higher Education Research and Technology of Indonesia for funding this project through the Institutional Reseach (RISIN) Scheme with contract Number of Kept.3715/UN23.14/PN.01.00/2018 March 29, 2018.

**References**

[1]. Ardli, E. R. & Wolff, M., 2008. Quantifying habitat and resource use changes in the Segara Anakan Lagoon (Cilacap, Indonesia) over the past 25 years (1978-2004). *Asian Journal of Water, Environment and Pollution*, Volume 5, pp. 59-67.

[2]. Ardli, E. R. & Wolff, M., 2009. Land use and land cover change affecting habitat distribution in the Segara Anakan lagoon, Java, Indonesia. *Regional Environmental Change*, Volume 9, p. 235–243.

[3]. Bob-Manuel, F.G. 2011. Food and feeding ecology of the Mudskipper Periopthalmus koelreuteri (PALLAS) Gobiidae at Rumuolumeni Creek, Niger Delta, Nigeria. Agric. Biol. J. North America 2(6): 897.

[4]. Hossain, M.S., N. G. Das., S. Sarker ., M. Z. Rahaman. 2012. Fish Diversity And Habitat Relationship With Environmental Variables At Meghna River Estuary,Bangladesh. Egyptian Journal Of Aquatic Research (38): 213–226.

[5]. Dudley, R. G., 2000. SegaraAnakan Fisheries Management Plan., s.l.: SegaraAnakan Conservation and Development Project..

**[6].** Haryono dan Wahyudewantoro, G, 2016. Pemetaan Habitat Ruaya Benih Ikan Sidat (*Anguilla bicolor*) dan Potensinya di Pantai Selatan Jawa. *Omni-Akuatika* Vol. 12 No. 3, 2016 : 47 - 58

[7]. Hossain, M.S., N. G. Das., S. Sarker ., M. Z. Rahaman. 2012. Fish Diversity And Habitat Relationship With Environmental Variables At Meghna River Estuary,Bangladesh. Egyptian Journal Of Aquatic Research (38): 213–226.

[8]. Kumar, M. A., G. Padmavati, and S. Venu. 2015. Food and Feeding Dynamics of *Stolephorus commersonnii* (Lacepede, 1803) (Family: Engraulidae) from South Andaman. Journal of Marine Biology Volume 2015l: 1-8

[9]. Kottelat, M.;A.J. Whitten, S. N. K. and S. W. (1996). Fresh Fishes of Western Indonesia and Sulawesi. Hongkong.

[10]. Redjeki, R. 2013. Komposisi dan Kelimpahan Ikan di Ekosistem Mangrove di Kedungmalang, Jepara. Ilmu Kelautan 18(1):54-60.

[11]. Setyaningrum, N., Piranti, A,S., Carmudi, Widyartini, D.S.I, Rahayu, D.R.U.S, & Adli, E.R., 2018. Fish Diversity in Sapuregel River of Segara Anakan Cilacap Central Java. *Paper Presented in South-East Asian + Conference in Biodiversity and Technology 2018*. Purwokerto, Indonesia 5-7 November 2018

[12]. Suprastini, E.R.Ardli, A. Nuryanto. 2014. Diversitas Dan Distribusi Ikan Di Segara Anakan Cilacap. Scripta Biologica 1(2):147–151.

[13]. Wahyudewantoro,G. 2009. Keanekaragaman Fauna Ikan Ekosistem Mangrove Dikawasan Taman Nasional Ujung Kulon, Pandeglang-Jawa Barat (Biodiversity of Fish Fauna Mangrove Ecosystem at Ujung Kulon National Park, Pandeglang-West Java]. Berita Biologi 9(4):379-386.

[14]. Wahyudewantoro, G. 2012. Ragam Ikan Mangrove Di Muara Sungai Bojong Langkap Dan Sungai Ciperet, Segara Anakan-Cilacap. Zoo Indonesia 21(1): 9-15.

[15]. Zahid, A & C.P.H. Simanjuntak, M.F. Rahardjo, Sulistiono. 2011. Iktiofauna ekosistem estuari Mayangan, Jawa Barat. Jurnal Iktiologi Indonesia, 11(1):77-85.

[16]. Yuwono, E., Jennerjahn, T.C., Nordhaus, I., Riyanto, E.A, Sastranegara, M. H. & Pribadi, R. 2007. Ecological status of Segara Anakan, Java, Indonesia, a mangrove-

fringed lagoon affected by human activities. Asian Journal of Water, Environment and Pollution, 4(1), pp. 29-3.

# Technology engineering of rearing red tilapia saline (*oreochromis niloticus*) fed on artificial diet enriched with protease enzymes in an eroded brackish water pond

**Istiyanto Samidjan[1*], Safar Dody[2], Diana Rachmawati[1]**

[1] Department of Aquaculture, Faculty of Fisheries and Marine Sciences, Diponegoro University, Jl. Prof. Sudarto Tembalang-Semarang, Indonesia
[2] Research Center for Oceanography,Indonesian Institute of Sciences, Indonesia
*Corresponding author: istiyanto_samidjan@yahoo.com

***Abstract.*** Technology engineering of red tilapia saline (*O niloticus*) was reared in brackish water pond fed on artificial diet enriched with protease enzymes in an eroded brackish water pond. The purpose of this study was to know of technology engineering of red tilapia saline reared in brackish pond that is exposed to effect eroded brackish water pond and the effect of adding protease enzymes in artificial feed to accelerate the growth and survival of red tilapia saline fish. The fish samples in this study used red tilapia saline with and average weight of 1.25 ± 0.09 g / fish with stocking 10 fish / m$^2$. This research used experimental method with Completely Randomized Design, 4 treatments and 3 replications. The treatments in this study were the addition of papain protease enzyme in the artificial feed with different dose ie A (0 mg / kg of feed), B (350 mg / kg of feed), C (700 mg / kg feed), D (1050 mg / kg feed ). The parameters to be determed include relative growth rate (RGR), feed utilization efficiency (EPP), protein efficiency ratio (PER), feed conversion ratio (FCR), survival rate (SR), and water quality.

**Keywords:** Growth, survival, red tilapia saline, protease enzyme

## 1. Introduction

Red tilapia fish farming (*O. niloticus*) intensively in ponds in hilink program using artificial feed enriched with protease papain enzymes to accelerate growth and survival. The protease papain enzyme added in the diet can hydrolyze the protein so as to improve the efficiency of feed utilization and growth.

The fish Red tilapia was one of the cultivars that have good prospect to be developed because much favored by society [1,2]. Chervinski [3] adds that red tilapia was also tolerant of low or high temperatures and was euryhalin. According to Hassan et al. [4] and Chervinski [3] red tilapia which is nourished in the sea has advantages such as faster growth, more compact meat, smell and taste more tasty.

The most perodedlems faced in the cultivation of red tilapia intensively include slow growth of fish resulting from the use of vegetable materials as a source of protein in commercial food, such as soybean meal meal containing anti-nutritional substances or phytic acid. The high content of phytic acid in soybean meal resulted in the utilization of vegetable protein in the feed less than optimal. According to Cao et al. [5], Hossain and Jauncey [6] phytic acid content in soybean flour reached 3.88% or 59.9% of total phosphorus. According to Kumar et al. [7], phytic acid can bind minerals such as calcium, magnesium, zinc, copper, iron and potassium, thus interfering with the absorption and digestion of fish

Eroded is a condition of sea water in tidal conditions that inundate parts of the river and land. Eroded phenomenon that occurred in approximately the last 10 years was a very high eroded condition, the impact was not only the river that overflows the water, but some of the land area in the coastal areas included settlements, yard, rice fields, maritime industry, community activity space, offices, also participated. Many human activities are disturbed by this eroded phenomenon, even the most apprehensive impact of this eroded phenomenon was the changing socio-cultural conditions as well as the very basic pattern of environmental adaptation [8, 9].

Environmental changes due to the phenomenon of eroded was very visible on the coast, the pattern of geographic changes in a region was a very real impact, one of which was a productive pond area that can no longer diopersionalkan, pond land has been turned into a stretch of submerged sea water , so that changes in the physical properties of the camps and biological pond water also experienced a very significant change. Biota that usually live in pond waters have a habitat shift except that have strong physiological adaptation capability.

Eroded phenomenon not only affects biological biological changes, but also affects the damage to the construction of ponds as well as the design of the pond before. The missing ponds of the embankment due to high tides become a common sight in the coastal areas, so the ponds have been very difficult to strive for Eroded phenomenon. The disappeared ponds of the embankment due to high tides became a common sight in the coastal areas, so the pond has been very difficult to the cultivation area, this incident is very influential on the decreased of fish land production (pond).

This high eroded is one of the consequences of world climate change. Behavioral changes that tend to violate environmental moral ethics have been carried out by humans on earth, some of which are the burning of earth fossils that exceed the natural absorption capacity, the cutting of vegetation as a carbon sink absorbs more than the naturally balancing capacity, resulting in decreased decline buffer capacity in various ecosystems on earth. Increased geothermal was one of the negative impacts of a series of events. The increase in sea water volume was a reflection of the increase in geothermal that melts icebergs in the polar regions [8].

One attempt to overcome the above perodedlem was to add protease enzyme papain in the feed ingredients. Chung [10], argues that papain protease enzyme is one of the exogenous enzymes that are expected to inhibit anti-nutrients, especially phytic acid, so as to improve the efficiency of feed utilization and growth. Papain protease enzymes added in the diet can increase nutrient absorption and regulate nutrient excretion (such as phosphorus, nitrogen, and minerals) and can hydrolyze phytic acid (phosphate elemental reserves) in fish feed into inositol and phosphoric acid.

Research on the addition of papain protease enzyme to the feed has been performed on several species, and has been shown to have an effect on the increase of protein digestibility, feed utilization, and growth in rainbow trout (Oncorhynchus myksiss Walbaum) [11], in fish Salmon salar L) [12], on vaname shrimp (Lipopenaeus vannamei) [13], in tiger grouper *(Epinephelus fuscoguttatus*) [23] and on tilapia ( O.niloticus)[31]. Research on the addition of papain protease enzyme in red tilapia fish meal (*O. niloticus*) on pendederan I has not existed. Therefore, this study aims to evaluate the effect of the addition of papain protease enzyme to the efficiency of feed utilization, growth and survival of red saline tilapia on I pendederan.

The main problem in this research was the cultivation of red tilapia fish experiencing constraints resulting from the use of feed ingredients. One of them was the use of vegetable source of raw materials containing anti-nutrients in the form of phytic acid. This causes important minerals and proteins in the feed can not be absorbed properly by the body, so that the efficiency of feed utilization was not maximal in fish body. Efforts that can be made to reduce the content of phytic acid was the addition of protease enzyme papain in artificial feed that serves to hydrolyze phytic acid in vegetable standards, such as soybean meal meal. According to Rachmawati and Hutabarat [23], grain plants such as soybeans are one source of vegetable protein containing anti-nutritional substances such as phytic acid that can inhibit the absorption of nutrients by the body so that the efficiency level of nutrient utilization of feed is less than optimal. Efforts made to improve the utilization of nutrients is to add protease enzyme papain in the feed. Papain protease enzymes in the diet can increase nutrient uptake and regulate nutrient excretion (such as phosphorus, nitrogen, and minerals) and can hydrolyse phytic acid (phosphate elemental reserves) in fish feed into inositol and phosphoric acid. The destruction of these phytic acid anti-physique substances, then the metabolic processes such as breakdown of proteins and minerals complex in the body can run well. The scheme of this perodedlem solving approach can be seen in the intensive cultivation of red tilapia fish in ponds in hilink program using artificial feed enriched with protease papain enzymes to accelerate growth and survival. The protease papain enzyme added in the diet can hydrolyse phytic acid in fish feed to inositol and phosphoric acid, thereby increasing the efficiency of feed utilization and growth.

The purpose of this study is to examine the effect of adding papain protease enzyme in artificial feed and to know the best dose of papain protease enzyme in artificial feed on the efficiency of feed utilization, growth, and survival of red tilapia saline (*O. niloticus*). As well as finding the best dose of papain protease enzyme in artificial feed on the efficiency of feed utilization, growth and survival of red salt tilapia.

## 2. Materilas and Methods

### 2.1. All Time and place

This research was conducted on April-Agustus 2018 at Pond culture Marine and Fishery goverment of Pekalongan City.

### 2.2. Research Materials

The materials used in this study were test animals, test feed, container maintenance and protease papain enzymes. The research materials used were as follows.
Animal test

The test animals used in this study were red saline tilapia (*O. niloticus*) at stalk stage I which had an average weight of 1.07 ± 0.09 g / fish. Red tilapia fish used amounted to 120 fish derived from red tilapia fish saline marine and fishery goverment in Pekalongan City.
The test feed used in this study was pellet-shaped artificial feed with 30% protein content and minimum energy 260 kcal DE / kg referred to SNI [25], plus protease papain enzyme with dose A (0 mg / kg of feed), B ( 350 mg / kg of feed), C (700 mg / kg of feed), D (1050 mg / kg of feed.

### 2.3. Maintenance container

The container used during maintenance is a 1x1x1 size penculture in a 1200 m² pond area which was sown by the Marine and Fisheries Office of Pekalongan City. maintenance using seaweed biofilter system to keep water quality to stay good, then stocked red tilapia saline seed density 10 fish / m². referring to Rachmawati and Samidjan's research [9].
Papain protease enzyme

The protease papain enzyme used in this research plan is the papain protease enzyme of Natuphos 5000® brand in powder form manufactured by PT. BASF Indonesia. Papain protease enzyme used in the study.

### 2.4. Research methods

The method used in this research is an experimental method conducted at Pond culture Marine Department fisheries and Marine science used Randomized Complete Design (RCD) which was a design where the treatment was involved entirely randomly on the experimental units. This study used 4 treatments with 3 repetitions, namely:

The parameters to be determed include relative growth rate (RGR), feed utilization efficiency (EPP), protein efficiency ratio (PER), feed conversion ratio (FCR), survival (SR) and water quality. The treatments in this study modify Primary (2015), which states that the addition of a dose of the papain protease enzyme 500 mg / kg of feed is the best dose for the red saline tilapia.

### 2.5. Research procedure

The procedures in this study include the preparation stage and the implementation stage. The stages are as follows:

2.5.1. Preparation phase

Preparation of test animals

The test fish to be used are first selected based on size and weight, completeness of body organs and physical health, then red tilapia adapted to the media saline. Based on the research from Setiawati and Suprayudi [24], red tilapia grow optimally in 20 ppt. The process of adaptation on red tilapia fish seed is done for 4 days. Adaptation was done at 1 day interval from low salinity (fresh water) to high salinity (brackish water). The salinity on the media was gradually increased from 5ppt, 10ppt, 15ppt, to 20ppt. This method of adaptation of red tilapia refers to the method used in the Laboratorium Pekalongan University at Samaran, Pekalongan red tilapia fish and then inserted into penculture for 1 week for adaptation to feed and new environment.

The test feed used in this study uses the ingredients such as fish meal as a source of animal protein, soybean flour as a source of vegetable protein, corn flour, bran flour and wheat flour as a source of carbohydrate, fish oil and corn oil as a source of fat, minerals and vitamin mix as a source of vitamins, CMC as binder or adhesive and protease papain enzyme as input material to break up phytic acid molecule. Preparation of test feed conducted in this study, among others, to test proksimat feed ingredients, preparing formulations and making feed. Proximate test results can be seen in Table 1.

**Table 1.** Proximate Analysis of Raw Materials for Composers Used in Research (in% Dry Weight)

| Material | ash | lipid | fiber | Protein | BETN | Total |
|---|---|---|---|---|---|---|
| Fish meal (*) | 31,97 | 12,15 | 7,14 | 50,40 | 3,93 | 100,00 |
| Soybean meal (**) | 8,94 | 1,47 | 2,53 | 55,55 | 31,52 | 100,00 |
| Corn meal (*) | 0,27 | 0,68 | 0,03 | 0,45 | 98,57 | 100,00 |
| Ricemeal (*) | 10,83 | 1,22 | 12,04 | 14,71 | 61,21 | 100,00 |
| Wheat flour (*) | 1,31 | 12,76 | 0,22 | 12,88 | 72,82 | 100,00 |

Source: * Animal Feed Science Laboratory, Faculty of Animal Husbandry and Agriculture, Diponegoro University (2015)
** Laboratory of Physics and Environmental Chemistry, Center for Brackishwater Aquaculture (BBPBAP), Jepara

Based on the results of proximate analysis in Table 1. it can be seen the value of each raw material. The nutritional content of the proximate analysis results are then used to calculate the feed formulation. The feed feeding formulation can be seen in Table 2.

Table 2. Formulation and Proximate Analysis of Test Feed Used in the Study

| Types of Feeding Materials | Treatment*) | | | | |
|---|---|---|---|---|---|
| | A | B | C | D | |
| Papain Enzime | 0 | 0,035 | 0,070 | 0,105 | |
| Fish meal | 25,00 | 25,00 | 25,00 | 25,00 | |
| Soybean meal | 25,00 | 25,00 | 25,00 | 25,00 | |
| Corn meal | 13,30 | 13,50 | 13,60 | 13,77 | |
| Rice meal | 13,80 | 13,69 | 13,57 | 13,47 | |
| Wheat flour | 13,90 | 13,80 | 13,80 | 13,71 | |
| Fish oil | 1,50 | 1,50 | 1,50 | 1,50 | |
| Corn oil | 1,50 | 1,50 | 1,50 | 1,50 | |
| Vit Min Mix | 4,00 | 4,00 | 4,00 | 4,00 | |
| CMC | 2,00 | 2,00 | 2,00 | 2,00 | |
| Total (g) | 100 | 100 | 100 | 100 | |
| Proximate analysis | | | | | |
| Protein (% )* | 30,37 | 30,34 | 30,32 | 30,29 | 30,31 |
| Lipid (%)* | 8,44 | 8,43 | 8,43 | 8,41 | 8,44 |
| BETN (%)* | 40,54 | 40,60 | 40,62 | 40,66 | 40,63 |
| Energy (kcal) | 275,99 | 275,92 | 275,93 | 275,84 | 275,99 |
| Rasio E/P | 9,09 | 9,10 | 9,10 | 9,11 | 9,11 |

Information:
a. Calculated based on Digestible Energy according to Wilson [31] for 1 g of protein is 3.5 kcal / g, 1 g of fat is 8.1 kcal / g, and 1 g of carbohydrate is 2.5 kcal / g.
b. According to De Silva [14], the E / P value for optimal growth of fish ranges from 8-9 kcal / g.
c. (*): Animal Food Science Laboratory, Faculty of Animal Husbandry and Agriculture, Diponegoro University (2015)

dose ie A (0 mg / kg of feed), B (350 mg / kg of feed), C (700 mg / kg feed), D (1050 mg / kg feed ).

The feeding process begins by adding the protease papain enzymes according to the treatment of soy flour. The papain protease enzyme to be added to the soybean mixture beforehand was dissolved with 50 ml of water at 45°C. This refers to Fox et al. [17] in which the papain protease enzyme was dissolved at 45-60 °C, then added to soybean meal, stirred evenly and allowed to stand for ± 24 hours in an airtight container.
Feed making was done by weighing all the necessary materials. After that, mix all ingredients starting from the smallest amount of ingredients to the largest amount of ingredients. Pellet-shaped feed dough is dried by aerated air by putting it in the room. A complete feeding procedure is presented in Appendix 3.

Implementation of research

The red saline tilapia (*O. nilotius*) to be used in the research was first weighed to determine the initial weight of the fish. Red tilapia salt (*O. nilotius*) then incorporated into penculture in ponds penculture size 1x1x1 in the area of 1200 m² ponds were souvenirs belonging to the Office of Marine and Fisheries Pekalongan City. maintenance using seaweed biofilter system to keep water quality to stay good, then stocked red tilapia seed saline density 10 fish / m². referring to Rachmawati and Samidjan's research [9] were used dose 300 to 600 mg/kg adding the protease papain enzymes.

**Data collection**

**Relative Growth Rate (RGR)**

The relative growth rates in this study can be calculated using the Steffens [27] formula as follows:

$$RGR = \frac{Wt - Wo}{Wo \times t} \times 100\%$$

Information :

RGR = Relative growth rate (RGR)
W0 = Saline tilapia weight at the beginning of the study (g)
Wt = Saline tilapia weight at the end of the study (g)
t = Duration of study (days)

Efficiency of Feed Utilization (EPP)

The Value of Feed Utilization Efficiency (EPP) can be determined by Tacon [28] formula, as follows

$$EPP = \frac{Wt - Wo \times 100\%}{F}$$

Information :

EPP = Feed Utilization Efficiency (%)
W0 = Saline red tilapia biomass weight at the beginning of the study (g)
Wt = Saline red tilapia biomass weight at the end of the study (g)
F = Number of feed red tilapia consumed during the study (g)

**Protein Eficiency Ratio (PER)**

The value of the protein eficiency ratio (PER) was calculated using the Tacon [28] formula, as follows:

$$PER = \frac{(Wt-Wo)}{Pi} \times 100\ \%$$

Information :

PER = Protein Efficiency Ratio
W0 = Biomass weight of the test animal at the beginning of the study (g)
Wt = The weight of animal biomass test at the end of the study (g)

Pi = The amount of test feed consumed multiplied by the protein content of the test feed

**Food Convention Ratio (FCR)**

According to Steffens [27], FCR is calculated based on the following formula:

$$FCR = \frac{F}{(Wt+D)-Wo}$$

Information:

FCR = (Food Convertion Ratio) Feed conversion ratio;
F = Weight of food eaten (g)
Wt = The biomass weight of the salt tilapia fish at the end of maintenance (g);
D = The weight of dead tilapia fish (g);
Wo = The biomass weight of the red tilapia copy at the beginning of maintenance (g)

Survival Rate (SR)

The survival rate or Survival Rate (SR) was calculated to determine the survival rates of test animals during the study, the survival can be calculated based on the Effendi [15] formula:

$$SR = \frac{Nt \times 100\%}{No}$$

Information :

SR = Survival rate
N0 = Number of test animals at the beginning of the study (tail)
Nt = Number of test animals at the end of the study (tail)
D =% protein in food

**Data analysis**

Data analysis was done after the research was done, after got the data of SGR, RGR, PER, FCR, EPP, and SR at the end of the research. Data obtained then analyzed using analysis of variance (ANOVA). Prior to the ANOVA, the data were first tested for normality, homogeneity test, and adivitas test to know that the data are normal, homogenous and additive to be tested further ie the analysis of variance. After analysis of variance, if found a highly significantly difference (P <0.01) or significantly difference (P <0.05) then Tukey> S test to be able to know the differences between treatments [26].

## 3. Results and Discussion

### Results

**Relative growth rate (RGR)**

Based on the weight measurement done at the beginning and end of research conducted for 42 days, obtained relative growth rate data on red tilapia fish (*O. niloticus*). The average calculation of the relative growth rate was presented Fr Based on Table 3 above, the average value of relative growth rates at each treatment from highest was C treatment of 7.75 ± 0.10$^{a}$% / day and lowest was treatment A of 4.19 ± 0.01$^{c}$% / day. Based on data relative growth rate of red tilapia saline (*O. niloticus*). The daily growth rate data have been tested the distribution of normality test, homogeneity test, and aditivity test indicating that the data spread normal, homogeneous, and additive so that has fulfilled the requirement of analysis of variance. Result of analysis of variance to relative growth rate of red saline tilapia gave significant effect (P <0,01) with F count> F table (0,01) to relative growth rate on red tilapia fish in the medium of hygiene.

Differences in the effect of treatment with other treatments can be determined by Tukey'S test for relative growth rates of red tilapia significantly different From Tukey'S test results showed that the mean value of C-A, C-B treatment was very different, but not significantly different from C-D.

**Efficiency of protein feed utilization (EPP)**

Efficiency of feed utilization showed how much feed was used by fish. Based on the research indicated that the highest EPP (%) value was at C ($55.25 \pm 0.33^{a}$) and lowest A ($36.25 \pm 0.24^{c}$) (Table 3).

The result of EPP data variety analysis on red saline tilapia fish showed the addition of papain enzyme in artificial feed with different percentage gave highly significantly effect (P <0,01) with F count> F table (0,01) against EPP on red tilapia saline. Furthermore, to find out the mean of intermediate values between treatments, Tukey'S test on red tilapia fish showed that there was a significant difference between the C-B, C-A treatment and did not differ significantly with C-D.

**The protein efficiency ratio (PER)**

The ratio of protein efficiency was a value that showed the amount of fish weight resulting from the weight of the test feed protein consumed. Based on the research results, the value of protein efficiency ratio calculation presented in Table 3. Very significantly effect on Protein Efficiency Ratio (PER) Ratio (%) red tilapia saline (P <0.01 Furthermore, to determine the mean of intermediate values between treatments Tukey'S test on red tilapia fish showed that there was a marked difference between the treatment of CB, CA and did not different significantly with the CD.

The result of the analysis of various data of protein efficiency ratio on red tilapia saline fish showed the addition of protease papain enzyme in artificial feed with different dose gave a very highly significantly effect (P <0,01) with F count> F table (0 , 01) to the protein efficiency ratio of saline red tilapia. Furthermore, to find out the mean of intermediate values between treatments, Tukey'S test on red tilapia fish (O. niloticus) showed that there was a significant difference between the C-B, C-A treatment and did not differ significantly with C-D.\

**Feed conversion rate (FCR)**

The feed conversion ratio is an index of total feed utilization for growth or the number of grams of feed required by fish to produce 1 g of wet weight of fish (Stickney, 1979 in 9). Based on the result of research, showed that the lowest value of FCR at C ($1.25 \pm 0.02^{a}$) and highest A ($1.52 \pm 0.03^{c}$), the value of feed conversion ratio presented in Table 3.

Very significantly effect (P <0.01) on the Feed Conversion Rate Ratio (FCR) of Red Tilapia saline. Furthermore, to find out the mean of intermediate values between treatments Tukey'S test on red tilapia fish that there are significant differences between CB, CA and CD treatments.

Based on the above table, the average feed conversion ratio in each treatment A ($1.52 \pm 0.03^{c}$), B ($1.47 \pm 0.17^{b}$), C ($1.25 \pm 0.02^{a}$), and D ($1.34 \pm 0.09^{b}$).

The result of data analysis of feed conversion ratio on red tilapia fish showed the addition of protease papain enzyme in soybean meal on artificial feed with different dose gave a highly significantly effect (P <0,01) with F value> F table (0,01) to feed conversion ratio on red tilapia fish. From Tukey'S test results on saline red tilapia showed that there were significant differences between the treatments of C-B, C-A and C-D.

**Survival rate**

Based on the results of the study, it was found that the survival value of red saline *O. niloticus* showed the highest yield on C treatment ($93.33 \pm 5.77^{a}$). The results of the calculation of survival are presented in Table 3. The same supercript in the column showed no significant difference.

The result of variance analysis has significant effected on the survival of red tilapia saline (P <0,05). Furthermore, to find out the mean of intermediate values between treatments, Tukey'S test on red tilapia fish showed that there were significantly differences between the treatments of C-B, C-A and C-D. Based on Table 3 above, the analyzed variance showed a significantly effect (P <0.05), whereas the highest survival outcome was C treatment of $93.33 \pm 5.77^{a}$%, then the lowest was D ($83.33 \pm 5.77^{b}$%), B ($76.67 \pm 5.77^{b}$)%, and lowest A ($70 \pm 10^{c}$%). The results of the analysis of multiplication data on survival red tilapia showed that the addition of papain protease enzyme in the artificial feed with different doses gave a significant effect (P< 0,05) with F count ≤ F table (0,05) on the survival of red tilapia saline fish. in Table 3.

Table.3. Absolute growth of weight. Survival Rate , Conversion ratio on the cultivation of red tilapia fish in ponds on the program Agromina politan Pekalongan City.

| | Treatment*) | | | |
|---|---|---|---|---|
| | A (0 mg/kg feed) | B (350 mg/kg feed) | C(700 mg/kg feed) | D(1050 mg/kg feed) |
| 1.Relative growth rate (RGR) (%/day) | 4.19±0.01$^{c}$ | 5.97±0.15$^{b}$ | 7.75±0.10$^{a}$ | 6.50±0.07$^{b}$ |
| 2. EPP value (%) | 36.25±0.24$^{c}$ | 43±1.13$^{b}$ | 55.25±0.33$^{a}$ | 46.31±0.38$^{b}$ |
| 3.Protein efficiency ratio (PER) (%) | 1.23±0.01$^{c}$ | 1.43±0.05$^{b}$ | 2.76±0.01$^{a}$ | 1.63±0.04$^{b}$ |
| 4. Food convertion ratio (FCR) | 1.52±0.03$^{c}$ | 1.47±0.17$^{b}$ | 1.25±0.02$^{a}$ | 1.34±0.09$^{b}$ |
| 5. Survival (%) | 70±10$^{c}$ | 76.67±5.77$^{b}$ | 93.33±5.77$^{a}$ | 83.33±5.77$^{b}$ |

Description :: The same sign of the superkrip shows no significantly difference
Treatment with the addition of protease papain enzyme in artificial feed with different dose ie
Treatment protease papain enzyme used dose A (0 mg / kg of feed), B (350 mg / kg of feed), C (700 mg / kg feed), D (1050 mg / kg feed ).

Table 4. Results of Water Quality Parameters on Red Tilapia During in the research.

| Treatment | temperate (°C) | pH | DO (mg/L) | $NH_3$ (mg/L) | Salinity (‰) |
|---|---|---|---|---|---|
| A | 25-31.5 | 7,5-7,8 | 5,97 6,75 | <0,001-0,006 | 18-20,5 |
| B | 25-31.5 | 7,4-7,5 | 5,87-6,39 | <0,001-0,005 | 18-20 |
| C | 25-31 | 7,5-7,9 | 5,89-6,59 | <0,001-0,002 | 18-9,37 |
| D | 25-32 | 7,5-7,7 | 5,83-6,49 | <0,001-0,005 | 19-20 |
| literature | 25 – 32* | 6,5 – 8,5* | ≥3* | <0,20* | 10-20** |

Description: *: SNI [25]
**: Setiawati and Suprayudi [24]

The result of measurement of water quality parameter showed that the water quality parameter value during the research was still in proper condition to be used as a medium of red tilapia saline fish culture in the medium of hybridity, it is based from the library about the optimum water quality condition for red tilapia saline (*O. niloticus*) in the medium of salinity.

**Discussion**

**Relative growth rate (RGR)**

Variation analysis results (Table 3) showed that the addition of protease papain enzymes with different doses had a very significant effect (P <0.01) on the relative growth rate of red salt tilapia (*O. niloticus*), it was suspected that protease papain enzymes could reduce anti-nutrients in the feed ingredients used. These results are in accordance with the opinion of [1,2,15] that papain protease enzymes added to artificial feeds can function well to reduce the phytic acid content present in plant-based foods and promote relative growth and feed efficiency.

The highest relative growth rate was found in C treatment with dose of protease papain enzyme 600 mg / kg of feed that is equal to 7.75 ± 0.10% / day, presumably the dose is the best dose to hydrolyze phytic acid which can inhibit the decomposition of nutrients and minerals in feed, nutrients can be well absorbed by fish. According to Chung [10], Hunter [19] papain protease enzymes function to increase the absorption of nutrients and regulate the excretion of nutrients (phosphorus, nitrogen and minerals) by hydrolyzing phytic acid found in artificial feed into inositol and phosphoric acid. The lowest relative growth result was obtained at treatment A of 4.09 ± 0.04% / day, suspected in the treatment of phytic acid contained in the feed still not decompose. Based on the research of Rachmawati and Hutabarat [23], the treatment which is not added by papain protease enzyme still contain phytic acid which still not hydrolyzed so protein and mineral complex that can not be utilized by body for growth.

The results of the research on the relative growth rate were rated higher when compared to Revindran (2000), Olusola and Nwanna [22], Rachmawati and Samidjan [9] studies, on freshwater tilapia *(O. niloticus)* of 2.50 ± 0.09% / day, Rachmawati and Hutabarat [23] research on fish tiger grouper (*Epinephelus fuscoguttatus)* of 3.39 ± 0.08% / day, Amin et al. (2010), 4.13% / day in catfish (Pangasius pangasius), Amin et al. (2011), in catfish (*Clarias* sp.) Of 4.18 ± 0.14% / day, and Hassaan et al. (2013), 2.11% / day in tilapia tilapia. The difference in value of the relative growth rate can be due to the age, type and size of the test fish used. The amount of minerals and nutrients needed for each type of fish to support growth is also different.

Growth on red salted tilapia fish can be seen that fish are able to utilize nutrient feed well. Utilization of nutrients in saline red tilapia was suspected because the bonds of anti-nutritional substances (phytic acid) with proteins and mineral complexes in the feed has been interrupted by the hydrolysis of protease papain enzymes. Hydrolyzed phytic acid will be inositol and phosphoric acid, so nutrients in the feed can be absorbed more leverage. According to Kim et al.[20], Rachmawati and Hutabarat (2006), Kumar et al. [7] the hydrolysis reaction is suspected to decrease anti physiological substance of phytic acid so that there will be termination of bond between phytic acid with protein and mineral complex become inositol and phosphoric acid. The breaking of the bonds according to Hossain and Jauncey (1993) will have a positive effect on the activity of trypsinogen into the enzyme trypsin solver protein to amino acid pengusunya. The complex bonds formed between phytate and multivalent cations can also decompose. The breakdown of these substances can cause the protein and minerals to not settle so that it can dissolve and can be utilized by the body [28, 13, 29].

**Efficiency of feed utilization (EPP)**

The results of the variance analysis (Table 3) showed that the addition of protease enzyme papain artificial feed was very significant ($P < 0.01$) to the efficiency of feed use of red tilapia.salin fish. High feed efficiency indicates that the feed can be well utilized by fish, it is suspected that protease papain enzymes can reduce phytic acid in feed, so that feed can be absorbed and digested properly by fish. The statement is in accordance with the opinion of Huet [18], Lovell [21], Watanabe [30], Olusola and Nwanna [22], improving the efficiency of feed utilization in fish can be attributed to the release of phytate feed by protease papain enzyme. The feed can be utilized by fish for better performance. High digestibility performance can also derive nutrient feed that is suspected due to the addition of protease papain enzymes.

The highest efficiency of feed utilization was obtained at the C treatment of the best dose of protease papain enzyme of 700 mg / kg (treatment C) of feed capable of producing EPP, $55.25 \pm 0.33^{a}$%, for red tilapia saline fish, it was suspected that the dose of the papain protease enzyme added to the C treatment of 700 mg / kg of feed was the best dose for the red salt tilapi. The dose was allegedly capable of over hauling phytic acid contained in the feed ingredients, thus improving the nutritional quality [29], that the protease papain enzyme added to the artificial feed works well to reduce the phytic acid content present in the plant-based diet as well as to increase relative growth and feed efficiency.

The efficiency of feed utilization is closely related to growth. The better the utilization efficiency of feed by the fish, then the growth of the fish is also getting better. High feed utilization shows that fish can utilize nutrient feed to be stored in the body and convert it into energy. the statement is in accordance with the opinion of Huet [18], that the high utilization efficiency of feed indicates that a small amount of food is overhauled to meet energy needs and the rest is used for growth.

The relative growth rate of the results of this study is higher, when compared with the results of Olusola and Nwanna [22], of 25.65 ± 0.01% in tilapia tilapia. The high efficiency of feed utilization in this study indicates that the amount of feed given, as well as the feed ingredients used are suitable to meet the needs of the fish, resulting in growth. According to Olusola Nwanna [22], feed efficiency indicates how much feed can be utilized by fish. Low feed efficiency values indicate that fish need more feed to increase their weight because only a fraction of the energy from the feed given is used by fish for growth. Bulbul et al., [1], adds that the addition of fitase in soy flour can improve the efficiency of feed compared with fish meal.

### The protein efficiency ratio (PER)

Variation analysis results (Table 3) showed that the addition of protease papain enzyme in the artificial feed had significant effect (P <0,05) to protein salivary protein efficiency ratio of red saline. According to Olusula and Nwanna [22], the value of high protein efficiency ratio is due to the protein can decompose into amino acids and its constituents, so that the absorption of protein in the fish body will be easier.

The results of this study indicate that the highest protein-efficiency ratio value was obtained in treatment C. The best dose of protease papain enzyme of 700 mg / kg (treatment of C) of feed was able to produce PER, respectively of 2.76 ± 0.01a%, for red tilapia (*O.niloticus*) while the lowest protein efficiency ratio value was obtained at treatment A of 0.95 ± 0.03%. These results suggest that the administration of papain protease enzymes in the diet was thought to be able to hydrolyze phytic acid and break the bond between phytic acid and protein and mineral complexes, resulting in more protein being absorbed by the body used for growth. The statement was in accordance with the opinion of Rachmawati and Samidjan [9], the high ratio of protein efficiency can be caused by protease papain enzyme contained in the feed that can reduce and decompose phytic acid and break the bond between phytic acid with protein and mineral complex, digestive enzymes, especially protein-breaking enzymes in breaking down proteins into their constituent amino acids. Winarno (1995) adds that perfect protein hydrolysis will produce high amino acids, so the more that the body absorbs.

This study shows the results of higher protein efficiency ratios from Rachmawati and Samidjan [9] studies, on freshwater tilapia of 0.64 ± 0.09%, Olusula and Nwanna [22], on tilapia tilapia (*O. niloticus*) of 0.17 ± 0.02% and Rachmawati and Hutabarat [23], in tiger grouper of 0.74 ± 0.04%. PER value is influenced by the quality of feed nutrition. The protease papain enzyme added in the diet is thought to break down proteins and mineral complexes bound to phytic acid. The decomposition of protein and mineral complexes will improve feed quality. Protein contained in pakanpun can be well absorbed for energy availability, and then will be used for growth. Kim et al. [20], adding that the protein absorbed from the feed will be used as energy and if there is excess protein then the protein will be used for growth. This ability is influenced by several factors namely the quality of the feed, where the higher the protein dimaanfaat by the body then the use of protein will be more efficient.

### Feed conversion rate (FCR)

The result of variance analysis (Table 3) showed that the addition of protease papain enzyme in feeding had significant effect (P <0,05) to conversion ratio of red tolapia saline *(O. niloticus*) feed. Optimal feed utilization will provide good feed conversion ratio value. The well-expended feed will produce energy for growth. Energy is produced from proteins that decomposed into amino acids that can be absorbed properly by the cultivation so that the nutrients in the feed will be utilized optimally. Maximum protein absorption due to phytic acid content in the feed can be hydrolyzed by protease papain enzyme [29].

The results of this study indicated that the value of feed conversion ratio the lowest value obtained at the C treatment of 1.25 ± 0.02[a]. for red tilapia fish. The results show that the dose of protease enzyme papain 700 mg / kg of feed on treatment C is the best dose, because it has the lowest FCR value FCR value is related to feed quality, the lowest FCR value is supposed that nutrient (2012), the value of feed conversion ratio is closely related to the quality of feed, so the lower the value the better the quality of feed and the more efficient the fish in using the feed consumed for growth.

The value of feed conversion ratio in this study is lower than Hassaan et al [4] , Tawwab [29] with the lowest value of feed conversion ratio of 3 and Rachmawati and Hutabarat [23] studies on *E. fuscoguttatus* of 2.09 ± 0.19, but higher than research Tawwab [29] of 1.1. These results indicate that the absorption of nutrients and minerals vary in each species, age and size of fish. According to Debnath et al. [15], the addition of papain protease enzyme is able to produce better feed conversion value compared to feeds that are not given protease papain enzymes in catfish (Pangasius pangasius).

Based on the results obtained in this study can be seen that the ratio of feed conversion on red tilapia fish less good. Feed conversion ratio ratio of 1.99 is still said high, because based on research Mirea et al [8], a good feed conversion ratio value for tilapia ranged from 1.43 to 1.7. The Olusola and Nwanna studies [22], on tilapia indigo fish also showed a low feed conversion value of 0.89 ± 0.01. Based on the results can be seen that the role of papain protease enzyme in the conversion ratio of red saline tilapia fish in this study is considered not maximal.

**Survival Rate**

The result of variance analysis (Table 3) showed that the addition of papain protease enzyme in artificial feed with different doses did not give significant effect (P> 0,05) to survival on saline red tilapia. The results of this study showed the highest survival value obtained by treatment C of 93.33 ± 5.77a%

The obtained survival value is high for tilapia maintained on the hygiene medium. Factors affecting the high survival rate of tilapia are abiotic and biotic factors, among others, competitors, population density, age and ability of organisms to adapt to the environment (Simanjutak, 2007).

The rate of survival of red tilapia in this hybrid medium showed the same value compared to Rachmawati and Samidjan [9], which was 93.33% and higher than Rachmawati and Hutabarat [23], on *E. fuscoguttatus* of 86.66%. Deaths that occur are suspected due to stress fish due to handling at the time of maintenance. The addition of papain protease enzyme did not significantly affect the survival of red saline tilapia, but the survival obtained in this study was high. The high of this life can also be affected by feeding. The amount of feed and the frequency of feeding according to need, will reduce mortality rate in fish. According to Tawwab [29], Hassaan et al.[4] reported the success of the cultivation business is intrinsically influenced by the growth rate and high livelihood. Growth rates and livelihoods are influenced by the availability of environmental quality feeds, pests, and diseases. The availability of feed is a very important factor, because it will determine the growth and survival rate of fish. Feeding will also suitably affect the survival that may reduce fish mortality.

## 4. Conclusions and recommendations

Conclusion

The results showed that the addition of papain protease enzyme in artificial feed gave a very highly significantly effect (P <0.01) to RGR, EPP, PER, FCR, SR red saline tilapia (*O. niloticus*). The best dose of protease papain enzyme of 700 mg / kg (treatment of C) of feed was able to produce EPP, RGR, PER, FCR, and SR respectively $55.25 \pm 0.33^{a}$%, $7.75 \pm 0.10$% / day, $2.76 \pm 0.01^{a}$% $1.25 \pm 0.02$a $1.25 \pm 0.02^{a}$ and $93.33 \pm 5.77^{a}$% for red salted tilapia. Water quality on maintenance media is within a reasonable range for red saline tilapia.

Suggestion

Suggestions that can be given from the as follows:

1. It is advisable to use a dose of papain protease enzyme greater than 700 mg / kg of feed on a red saline tilapia.
2. It is recommended to conduct further research with different species, weights, and stadia of fish.

**Acknowledgements**

Our thanks go to the Director of Directorate of Research and Community Service Directorate General for Research and Development of the Ministry of Technology and Higher Education Research In accordance with the Letter of Appointment Agreement of Community Service Program Implementation no: 103-24 /UN7P4.3/PM/2018 Prof. Dr. Ocky Karna Radjasa who has provided research funding on Hi-Link program, Marine and Fisheries Mrs Liliek office 0f Marine and Fisheries of Pekalongan city, Dean of Fisheries and Marine Faculty, Diponegoro University and Chairman of LPPM Undip and Mr. Miftahuddin partner of UKM who has provided pond facilities for this research.

**References**

[1] Bulbul, M., Md. A. Kader, M. A. Ambak, Md. S. Hossain, M. Ishikawa dan S. Koshio. 2015. Effects of Crystalline Amino Acids, Phytase and Fish Soluble Supplements in Improving Nutritive Values of High Plant Protein based Diets for Kuruma Shrimp, *Marsupenaeus japonicus*. Aquaculture Elsevier.,438:98-104.

[2] Baruah, K, Sahhu NP, Pal AK dan Debnath D. 2004. Dietary Phytase: an ideal Approach For a Cost Effective and Low-Polluting Aqua Feed. NAGA World Fsh Center Quarterly. 27(3&4): 15-19 p.

[3] Chervinski, J. 1982. Environmental physiology of tilapia. In R.S.V. Pulin, T. Bhukaswan, K. Thongtai & J.L. Mackan (Eds.). The Second International Symposium on Tilapia in Aquaculture. ICLARM. Conference Proceeding. Department of Fisheries. Bangkok,

Thailand and Int. Centre for Living Aquatic Resources Managment. Manila. Philipines. 119-128 pp.

[4] Hassan, M. S., M. A. Soltan, H. M. Agouz dan A. M. Badr. 2013. Influences of Calcium/Phosphorus Ratio on Supplemental Micerodedial Phytase Efficiency for Nile Tilapia (*Oreochromis niloticus*). Egyptian Journal of Aquatic Research., 39:205-213.

[5] Cao, L., W. Wang, C. Yang,Y. Yang, J. Diana, A. Yakupitiyage, Z. Luo dan D. Li. 2007. Application of Micerodedial Phytase in Fish Feed. Enzym and Micerodedial Technology., 40:497-507.

[6] Hossain. P, Jauncey.H. 1993. Effect of dietary Phytase supplementation on Phytate Phosphorus Utilization and Plasma Inorganic Phosphorus Concentration in Common carp ( *Cyprinus carpio* ). Hunter, B. (Ed) Phytase in Aplication in Aquaculture Asia Pasific. Bangkok. 8 pp.

[7] Kumar, V., A. K. Sinha, H. P. S. Makkar, G. De Boeck dan K. Becker. 2011. Phytate and Phytase in Fish Nutrition. Journal of Animal Physiology and Animal Nutrition., 96(2012):335-364.

[8] Mirea, C., V. Cristea, I. R. Grecu and L. Dediu. 2013. Influence of Different Water Temperature on Intensive Growth Perfomance of Nile Tilapia (*Oreochromis niloticus*, Linnaeus, 1758) in A Recirculating Aquaculture System. University of Agricultural Sciences and Veterinary Medicine lasi., 60:227-231.

[9] Rachmawati, D dan I. Samidjan. 2014. Penambahan Fitase dalam Pakan Buatan sebagai Upaya Peningkatan Kecernaan, Laju Pertumbuhan Spesifik dan Kelulushidupan Benih Ikan Nila (*O. niloticus*). Jurnal Saintek Perikanan., 10 (1) : 48-55.

[10] Chung, T.K. 2001. Sustaining Livestock Production and Environment. Food and Agriculture Asia Pacific Development. Singapore. 52-54 pp.

[11] Vandenberg, G.W., S.L. Scott., P.K. Sarker., V. Dallaire, dan J. de la Noüe. 2011. Encapsulation of micerodedial phytase: Effects on phosphorus bioavailability in rainbow trout (Oncorhynchus mykiss). Animal Feed Science and Technology 169: 230–243.

[12] Sajjadi, M. dan C. G. Carter. Dietary Phytase Supplementation and Utilisation of Phosorus by Atlantic Salmon (*Salmo salar* L.) Fed a Nacona-Meal-Based Diet. School of Aquaculture. Tasmanian Aquaculture and Fisheries Institute, University of Tasmania. Australia. 417-431 pp.

[13] Suprayudi., M. Agus, D. Harianto dan D. Jusadi. 2012. Kecernaan Pakan dan Pertmbuhan Udang Putih *Litopenaeus vannamei* Diberi Pakan Mengandung Enzim protease papain Berbeda. Jurnal Akuakultur Indonesia., 11 (2) : 103-108.

[14] De Silva, S.S. 1987. Finfish Nutrition Research in Asia. Proceeding of The Second Asian Fish Nutrition Network Meeting. Heinemann, Singapore. 128 pp.

[15] Debnath, D., Pal, A. K., Sahu, N. P., Jain, K. K., Yengkokpam, dan Mukherjee, S. C. 2005. Effect Dietary Micerodedial Phytase Supplementation on Growth and Nutrien Digestibility of Pangasius pangasius fingerling. Aquaculture Research. 36(2). Hal: 180-187.

[16] Effendi, M. I. 2002. Biologi Perikanan. Yayasan Dewi Sri. Bogor. 216 hlm.

[17] Fox, J. M., Addison L. L., Anthony J. S., D. Allen D., Denis R. M., Elizabeth C. D., and Tzachi M. S. 2006. Phytase Supplementation In Aquaculture Diets Improves Fish, Shrimp Growth Performance. Global Aquaculture Alliance. Hlm: 66.

[18] Huet, M. 1970. Textbook of Fish Culture Breeding and Cultivation of Fish. Fishing News (Book Ltd). London. 436 pp.

[19] Hunter B 2002 Phytase Aplications in Aquaculture (Roche Aquaculture Center Asia Pasific Bangkok) 425 p

[20] Kim, K., T.B. Kayes, and C.H. Amundson. 1991. Purified diet development and reevaluation of the dietary protein requirement of fingerling rainbow trout (Oncorhyncus mykiss). Aquaculture. 96: 57-67 pp.

[21] Lovell T. 1989. Nutrition ang Feeding of Fish. Van Nostrand Reinhold. New York. 260 pp.

[22] Olusola, S. E. dan L.C. Nwanna. 2014. Research Article Open Access Growth Performance of Nile Tilapia (*Oreochromis niloticus*) Fed Processed Soybean Meal Based Diets Supplemented With Phytase. International Journal of Aquaculture., 4 (8) : 48-54.

[23] Rachmawati, D dan J. Hutabarat. 2006. Efek Ronozyme P dalam Pakan Buatan terhadap Pemanfaatan Pakan dan Pertumbuhan Ikan Kerapu Macan (*Epinephelus fuscoguttatus*). Ilmu kelautan., 11 (4) : 193 – 200.
[24] Setiawati, M. dan M. A. Suprayudi. 2003. Pertumbuhan dan Efisiensi Pakan Ikan Nila Merah (*Oreochromis* sp.) Yang Dipelihara Pada Media Bersalinitas. Jurnal Akuakultur Indonesia., 2 (1) : 27-30.
[25] SNI. 1999. Benih Ikan Nila Hitam (*Oreochromis niloticus* Bleeker) Kelas Benih Sebar. 01-6140-1999.
[26] Srigandono, B. 1992. Rancangan Percobaan. Fakultas Peternakan, Universitas Diponegoro, Semarang, 178 hlm.
[27] Steffens, W. 1989. Principle of Fish Nutrition. Ellis Horwood Limited, West Sussex. England. 384 pp.
[28] Tacon, A. E. J. 1987. The nutrition and Feeding Formed Fish and Shrimp. A Training manual Food and Agriculture of United Nation Brazilling, Brazil. 108 p.
[29] Tawwab, M A. 2012. Effect of Dietary Protein Levels and Rearing Density on Growth Performance and Stress Response of Nile Tilapia, *Oreochromis niloticus* (L.). International Aquatic Research., 4(3):1-13
[30] Watanabe, T. 1989. Fish Nutrition and Marine Culture. JICA Text Book the General Aquaculture. Course Department of Aquatic Broscience. Tokyo University of Fisheries, 233 p.
[31] Wilson, R.P. 1982. Energy Relationships in Catfish Diets. *In*: R.R. Stickney and R.T. Lovell (Eds.). Nutrition and Feeding of Channel Catfish. Southern Cooperative Series. 54 p.
[32] Xue, Y. 2014. The Effect of Dietary Phytase Supplementation and Incubation in Soy Protein Concentrate Based Diet Fed to Nile Tilapia. Thesis. Norwegian University of Life Science. Faculty of Veterinarian Medicine and Bioscience. Norwegia. 44 p.

IOP Conf. Series: Earth and Environmental Science **406** (2019) 012031 doi:10.1088/1755-1315/406/1/012031

# THE ACCUMULATION OF HEAVY METALS KADMIUM (Cd) IN WATER, SEDIMENTS AND AQUACULTURE BIOTA WHICH CONTAMINATED BY BATIK WASTE IN MULYOREJO VILLAGE PEKALONGAN

**Siregar A.S[1], Prayogo N.A[1], and Harisam T[1]**
[1]Faculty of Marine and Fisheries University of Jendral Soedirman, Purwokerto
Corresponding Author: asrul_sir@yahoo.com

**ABSTRACT.** This research entitled "The Accumulation of Heavy Metals Kadmium (Cd) in water, sediments and aquaculture biota which contaminated by Batik waste in Mulyorejo village Pekalongan". Purposes to know the heavy metal concentration Cd in water, sediment and bandeng's flesh, relation between heavy metal concentration Cd in water and sediment with fish's flesh, the pollution status in this aquaculture, and also the consumption properness from the aquaculture station. This study conducted by survey method with stratified random sampling on four stations with four replications. The data then analyzed by descriptive comparative, F-test, regression and correlation, Contamination Factor index, Geoaccumulation index and bioaccumulation factor. The result revealed that the heavy metal concentration in the water 0,003-0,009 mg/L, in sediments 1,272-2,208 mg/kg and fish's flesh 0,087-0,168 mg/kg. The heavy metal concentration (Cd) in the water and sediments with fish's flesh are have positive correlation which strongly (R=0,888 and R=0,906). The pollution status based on the contamination factor index (CF) categorized as very high contaminated and geoaccumulation index is mild contaminated (station III and IV) and mild to severe contaminated (station I and II). The fishes in this station are improper to be consumed except in station IV (distance ±300m from Bremi's riveroutlet).

**Keywords** : metals Cd, water, sediment, bandeng fish's flesh

IOP Conf. Series: Earth and Environmental Science **406** (2019) 012031 doi:10.1088/1755-1315/406/1/012031

## INTRODUCTION

The bandeng (*Chanos chanos*)pond in Mulyorejo Village, Pekalongan Regency is one of the productive fishpond. Bandeng is a fish that is cultivated in fishponds in the village of Mulyorejo for consumption. Bremi River is one of the tributaries of the Sengkarang River with a length of ± 9 km located in Tirto District. Along the downstream of the Bremi River there are many activities including the batik, textile, screen printing, settlement and aquaculture industries [3]. These industries produce waste containing toxic and dangerous chemicals (B3), one of which is heavy metals.

Cd heavy metal is usually used as a coloring agent in industry, which is toxic and does not decompose in aquatic environments [5]. Heavy metals that are in water over time will undergo a process of sedimentation with mud and settles to the bottom of the water [6]. In addition, heavy metals are known to accumulate in the body of fish for a long time as poisons that have a negative impact on biota [5]. If the fish are consumed by humans, it will endanger human health. The impact of Cd on humans, among others, causes disruption of lung health, bones, liver, reproductive glands, kidneys, heart, which is neurotoxic which results in damage to the sense of smell [4].

Research to determine the Cd metal content that has accumulated in water, sediments, and biota in bandeng ponds in Mulyorejo Pekalongan Village has never been done before, with known Cd levels in water and pond sediments, then the feasibility of the pond as a bandeng aquaculture area. In addition, knowing the level of Cd in bandeng that is cultivated can be determined for its safety for consumption. Therefore, there is a need for research related to the accumulation of heavy metals Cd in these farms.

## MATERIAL AND METHODS

The method in this study is a survey method with sampling techniques using the Stratified Random Sampling method. Sampling was conducted at the Bandeng Fishpond in the Village of Mulyorejo, Pekalongan in May 2016. The research location was divided into four stations, sampling was carried out 4 times in a composite manner. Determination of sampling stations based on the distance from the source of freshwater input to the pond (Sungai Bremi). The sampling locations are presented in Table 1.

**Tabel 1.** Sampling Location

| Station | Coordinate | Location |
|---|---|---|
| I | E : 6°52'14.1"<br>S : 109°38'58.4" | Bandeng Fishpond I (close to the inflow Bremi River to the pond) |
| II | E : 6°52'09.8"<br>S : 109°38'58.5" | Bandeng Fishpond II (located ± 100m from station I) |
| III | E : 6°52'05.3"<br>S : 109°38'58,6" | Bandeng Fishpond III (located ± 100m from station II) |
| IV | E : 6°52'00.2"<br>S : 109°38'58.6" | Bandeng Fishpond IV (located ± 100m from station III) |

### Sampling and Preservation of Samples for Cd Metal Analysis

600 mL of water was taken into a sample bottle, then 0.75 mL concentrated HNO3 was added. ± 250 grams sediment was taken using Ekman Grab at a depth of 15-20 cm. Fish samples are taken using a net, the meat is taken and mashed, then put in a plastic bag. Furthermore, the sample is cooled in an ice box and analyzed in a laboratory (SNI, 2009).

### Sample Preparation for Cd Metal Analysis

100 ml sample water is taken, put in an erlenmeyer and shaked. 5ml concentrated

HNO3 was added, heat it slowly until the volume becomes 15-20 ml and changed in color, after that the sample is cooled and added with 50 ml of distilled water. The sample was put into a 100 ml measuring cup through Whatman No. filter paper. 40 and added distilled water until the volume becomes 100 ml, the sample is ready to be analyzed by AAS (SNI, 2009).

Sediment samples and crystals were first cleaned and dried, then refined in porcelain frying cups. Samples were taken ± 5 grams, embedded in porcelain cups, and then heated in a furnace muffle at 500 ° C for 5 hours. The resulting ash was dissolved in aqua regia of 5 mL and heated to a soluble ash, transfer to a 25 mL gauze and add 5 mL of HNO3 1 M, and then add aquadest to a volumetric flask incubated with a volume of 25 ml. The solution was filtered with Whatman No. paper. 41 (Muwartiningsih et al., 2015). Filtrates are ready for analysis with AAS (SNI, 2009).

**Analysis of Cd**

The Cd metal standard solution was made to find out the equation of the Cd metal standard curve by dissolving Cd (NO3) 2 in 0.1 N nitric acid. Then it was diluted and made in various concentrations of 0.02; 0.50; 1,00; 1.50; 2.00; 3.00; 4.00; and 5.00 mg / L, then absorbance was measured using AAS (SNI, 2009).

Measurement of Cd metal content by the Flame Atomic Absorption Spectrometry method using a set of AAS tools. The prepared filtrate is sucked with a 20 ml respirator tube and put into a Nabulyzer, then atomized and evaporated. The vapors that are formed are burned with a burner flame and followed by an atomization process, then irradiated with cathode rays at a wavelength of 228.8 nm and a strong current of 3.5 mA. The absorbance value of the sample and the Cd standard solution will appear on the AAS screen along with the line equation (SNI, 2009). Calculation of Cd metal content in the sample using the formula:

$$Cd = \frac{AxB}{C}\ mg/l$$

ex: A= ppm (mg/l)
B= the final volume of the extracted product
C= initial sample weight

Calculation of pollution levels based on Contamination factor (CF) index, Geoaccumulation index (Igeo), and Bioaccumulation Factor (BAF).
Contaminant Factor (CF) formula:

$$CF = \frac{[\ Cd\ heavy\ metal\ ]}{[\ Background\ ]}$$

Igeo Calculation Formulas (Geoaccumulation Index)

$$Igeo = Log^2 \frac{[\ Cd\ heavy\ metal]}{1{,}5 * Background}$$

Bioaccumulation Factor (BAF) calculation formula

$$BAF = \frac{[CB]}{[CW]}$$

ex:
CB = chemical concentration in organisms
CW = Concentrated chemistry in sediments (chemical mass / L)

## RESULTS AND DISCUSSION

### Content of Cadmium (Cd) Heavy Metals in Water

The heavy metal content of each station in the bandeng ponds in Mulyorejo Village is shown in Figure 1.

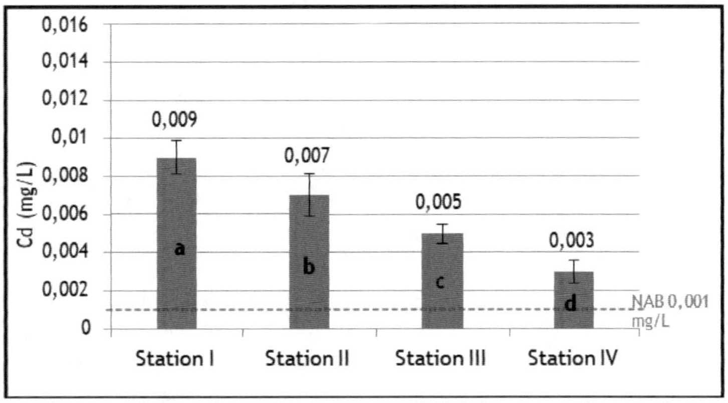


Figure 1. Average values and standard deviations of Cd in water for each station in bandeng fishponds in Mulyorejo Village, Pekalongan.

Figure 1 shows that the Cd heavy metal content between stations was significantly different ($P < 0.05$). This is caused by the distance of the station to the source of pollutants, sedimentation, and salinity. Station I is located closest to the Bremi River inflow to ponds and settlements, so that the highest heavy metal content in water. The source of fresh water in the ponds comes from the Bremi River, the heavy metal content of Cd in water at the source is 0.012 mg / L. The difference in heavy metal content is also caused by sedimentation. Water in ponds at station II, III, and IV is sourced from station I, so that heavy metals carried by the current from station I will experience a process of sedimentation continuously and decreasing. Metals that are difficult to undergo dilution in the water column will eventually descend to the bottom and settle in sediment, so that the metal content in water tends to be small.

At station I to station IV the salinity value increases. The salinity value at station I is the lowest compared to other stations which is 9.75 ppt, then the value has increased from station I to station II and so on up to station IV. Salinity values at stations II, III, and IV are 12.0, 13.2, and 16.2 ppt, respectively. This causes the heavy metal content from I to station IV to decrease. Low salinity will result in a decrease in complexing agent in waters ($Cl^-$), so that more heavy metals will be found in the form of free ions which are more easily entered into the body of biota, and vice versa the higher the salinity value then the complexing agent in waters ($Cl^-$) will increase and bind to heavy metals and settle to the bottom of the waters [2].

The high content of Cd in water affects the health of biota and humans who use the water. The Cd heavy metal content in all research stations is above the threshold value set by KMNLH No.51 of 2004, which is 0.001 mg / L (Figure 1).

### Content of Cadmium (Cd) Heavy Metals in Sediments

The heavy metal content of each sediment at the station in the bandeng fish pond of Mulyorejo Village is shown in Figure 2

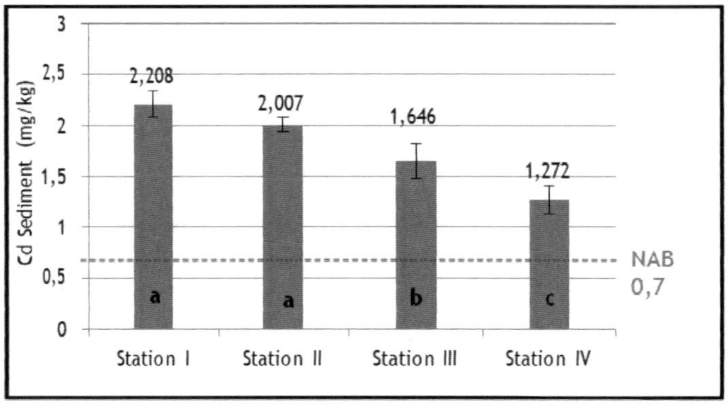


Figure 2. Average Cd values in the sediment of each station in bandeng fishponds in Mulyorejo Village, Pekalongan.

Figure 2 shows that the heavy metal content of Cd in sediments between stations was significantly different (P <0.05). Based on further tests of Cd heavy metal content in sediments at station I with stations III and IV were significantly different (P <0.05), while station I with station II were not significantly different (P≥0.05). The Cd metal content in sediments at stations I to IV has decreased. Heavy metals in sediments are sourced from heavy metals present in water, so decreasing heavy metal content in water will also cause a decrease in heavy metal sources which will settle in sediments (R = 0.893). The R value indicates that the Cd in water with Cd in sediments has a relationship that is directly proportional to the strong closeness. Each increase of 1 mg / L Cd in water was followed by an increase in Cd in sediment of 163,783 mg / kg (y = 0.783 + 163x). The relationship between Cd metals in sediments and Cd metals in water is shown in Figure 3.

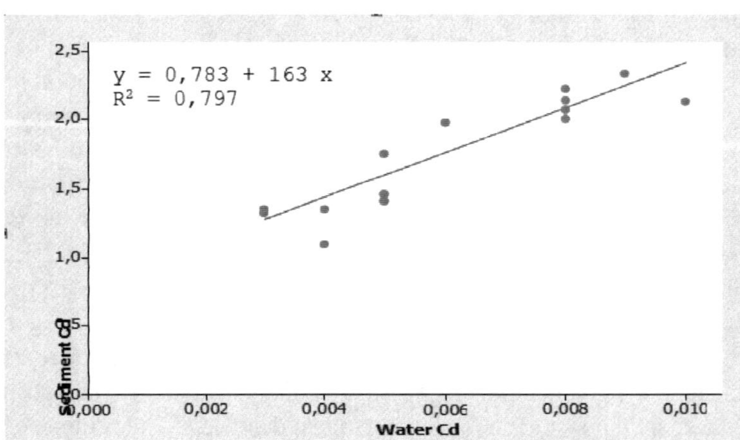


Figure 3. Relation between the Cd in sediment and Cd in water.

COD and TSS values at stations I to IV decreased, so that the content of heavy metals in sediments at stations I to IV also decreased. The average COD values at stations I through IV were 15.60, 14; 82, 9.75, and 7.22 ppm, while the mean TSS values were 1023.67, 862.33, 850.67, respectively. and 786.33 ppm. In addition, the salinity value at station I to station IV has increased, amounting to 9.75, 12.0, 13.2, and 16.2 ppt. The content of heavy metals in sediments between station I and station II is not significantly different because this is due to environmental conditions and the type of sediment that tends to have almost the same characteristics. The environmental conditions at stations I and II have relatively similar COD values which cause the metal content at the two stations is not significantly different. The COD content in the two stations is 15.60 and 14.82 mg / L.

IOP Conf. Series: Earth and Environmental Science **406** (2019) 012031 doi:10.1088/1755-1315/406/1/012031

The heavy metal content of Cd in sediments at all research stations was above the threshold value set by CCME [1], which is 0.7 mg / kg (Figure 2). This shows that the condition of aquaculture ponds in all stations has experienced heavy metal pollution and is not suitable as a habitat for aquatic biota, and is at high risk when used as a fishing business area, both for cultivation and capture activities. Cd **Heavy metal in Bandeng fish (Chanos chanos Forsskal)**

The heavy metal content of each station in bandeng fishponds in Mulyorejo Village is shown in **Figure4.**

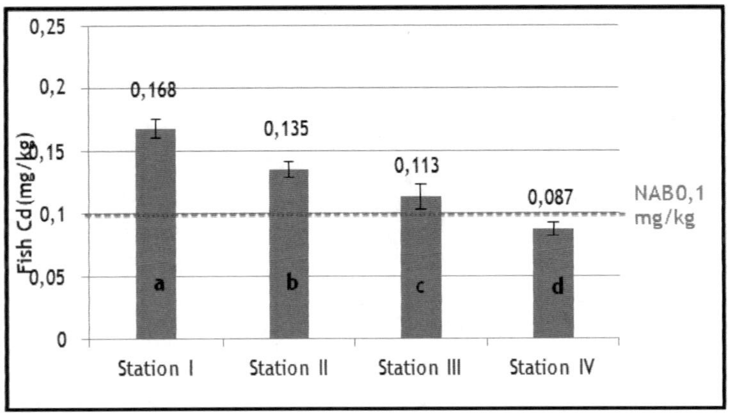


Figure 4. Average value of Cd in fish meat at each station in bandeng fishponds in Mulyorejo Village, Pekalongan.

Figure 4 shows that the heavy metal content of Cd in fish flesh between stations was significantly different ($P < 0.05$). This is caused by the distance of the station to the source of pollutants, Cd metal content in water, Cd metal content in sediments, and water quality (salinity). The source of fresh water in the ponds comes from the Bremi River, with heavy metal content of Cd in sediments and water at the source (Bremi River) in the amount of 2,380 mg / kg and 0.012 mg / L. Station I is closest to the Bremi River inflow, so the heavy metal content in water and sediment is highest, as is the heavy metal content in bandeng fish (Chanos chanos).

The decrease of heavy metal content in water from stations I to IV will cause the heavy metal content that accumulates in bandeng at stations I to IV to also decrease ($R = 0.888$). The correlation value (R) shows that between Cd in water and Cd in bandeng has a relation that is directly proportional to the strong closeness. Each increase of 1 mg / L Cd in water was followed by an increase in Cd in bandeng by 12,947 mg / kg ($y = 0.047 + 12.9x$). The relation between Cd metal in water and Cd metal in bandeng is shown in Figure 5.

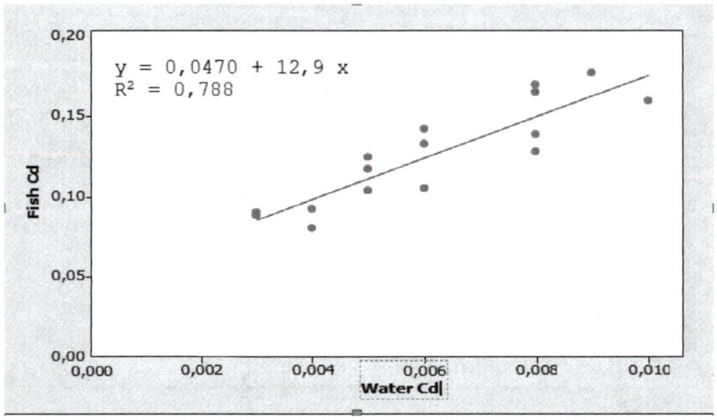


Figure 5. Relation between the content of Cd in water and Cd in fish meat.

The content of heavy metals in sediments at stations I to IV has decreased, so that the content of heavy metals in bandeng fish in stations I to IV has also decreased (R = 0.906). A positive correlation (R) value indicates that the Cd in sediments with Cd in bandeng fish has a relationship that is directly proportional to the strong closeness. Each increase of 1 mg / kg Cd in sediments was followed by an increase in Cd in bandeng fish by 0.0479 mg / kg (y = -0.024 + 0.0719x). The relationship between Cd metals in sediments and Cd metals in bandeng fish is shown in Figure 6.

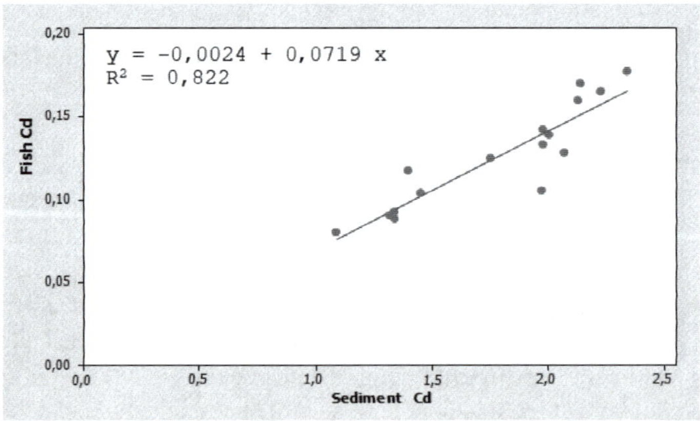


Figure 6. Relation between metal content of Cd in sediments and Cd in fish meat.

Water quality affects the content of heavymetals in the body of bandemg (Chanos chanos). Salinity from stations I to IV has increased, so that the content of heavy metal Cd in bandeng from station I to station IV has decreased. Salinity values at stations I through IV were 9.75, 12.0, 13.2, and 16.2 ppt.

The Cd heavymetal in bandeng (Chanos chanos) fish meat at stations I to III is above the threshold value, while at station IV it is below the threshold value set by SNI [8], which is 0.1 mg / kg (Figure 4). This shows that bandeng in fishponds at stations I to III are no longer suitable for consumption, while station IV is approaching the threshold value. The bandeng that is cultivated in the pond of Mulyorejo Village will have a dangerous health impact if consumed.

**Level of Heavy Metal Cadmium (Cd) Pollution**

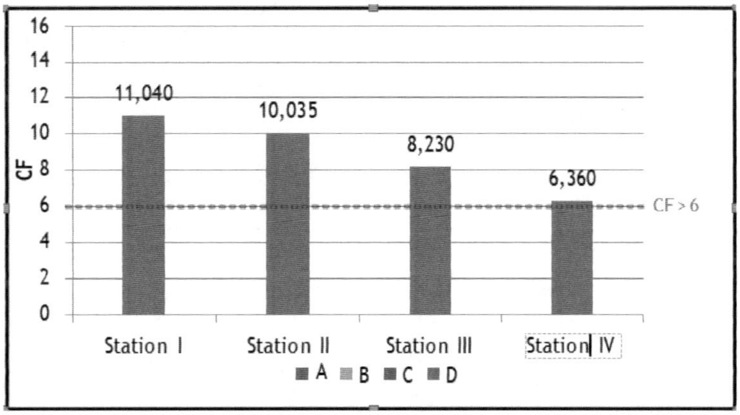


**Figure 7.** Contamination factor (CF) value of Cd heavy metals in the Bandeng pond in Mulyorejo Village.

Figure 7 shows the results of calculating the CF value at each station ranging from 6.360 to 11.040. CF values from station I to station IV belong to the very high contamination category with CF values> 6. This is due to the activity around the Bremi River which is used as a source of fresh water in the bandeng ponds in Mulyorejo Village, which is the waste that is mostly generated from the batik industry. CF values from station I to station IV are decreasing. This is caused by the distance of the station from the pollutant source in the fishpond. Station I is located not far from the existence of settlements, besides that it is located close to the Bremi River inflow to the pond, while in stations II to IV it is further away from the Bremi River inflow. The water from the Bremi River that enters the heavy metal containment ponds at station I is partly heavy metal deposited in sediments, then at station II freshwater is sourced from station I so that the heavy metal content in sediments is less.

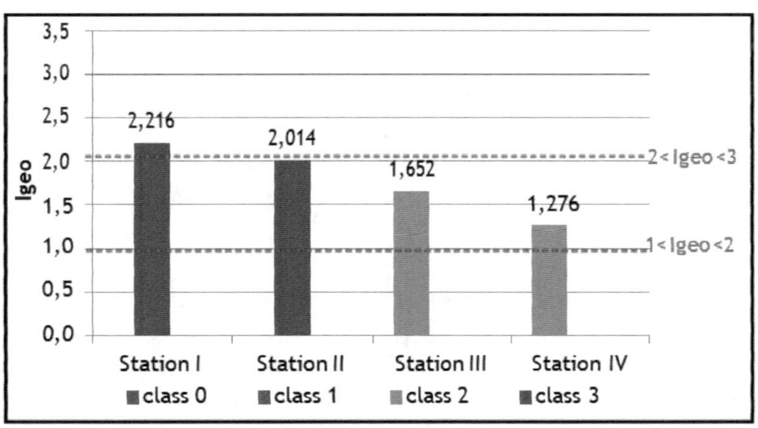


Figure 8. Geoaccumulation Index (Igeo) of heavy metal Cd in the bandeng pond in Mulyorejo Village.

Figure 8 shows the results of the calculation of the Igeo value at each station ranging from 1.276-2.216. The Igeo values from station I and station II are in the class 3 category, namely moderate to heavily polluted with igeo values of 2-3, while for station III and station IV they are classified in class 2 categories, which are moderately contaminated with igeo 1-2 values. This is due to the station I closest to the source of pollutants, namely the input of water from the Bremi River. Igeo values from station I to station IV are decreasing. Igeo values at stations III and IV are classified as moderately polluted. This is caused by the distance of the station from the pollutant source in the fishpond. Stations III and IV are very far from the Bremi River inflow, so that the flow will cause the distribution of heavy metal Cd in sediments to be uneven and this station is classified as far from industrial activities so that the accumulation of heavy metals in sediments is low.

Bioaccumulation of Cadmium (Cd) in Bandeng (Chanos chanos) Based on BAF Value

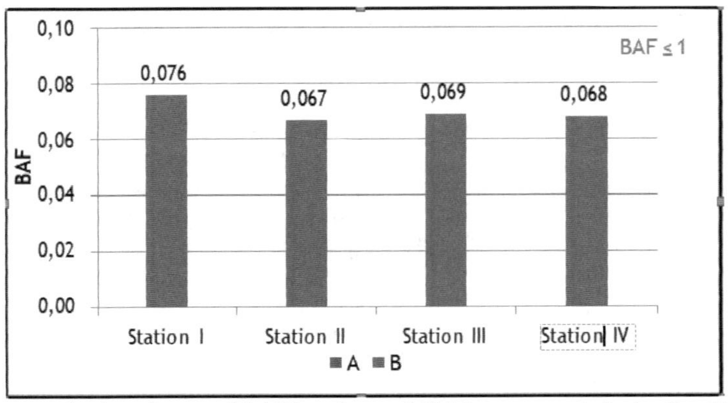


Figure 9. Bioaccumulation factor (BAF) value of Cd heavy metals in the Bandeng Farm in Mulyorejo Village.

Figure 9 shows the results of the calculation of the BAF value at each station ranging from 0.067 to 0.076. BAF values from station I to station IV belong to the category of organisms lacking in the ability to accumulate metals in their bodies with a BAF value ≤ 1. BAF values are obtained with a contaminant concentration ratio measured in biota in the field or under several exposure conditions with concentrations measured in sediments. This is related to the heavy metal content of Cd in water, the absorption process and the rate of accumulation of heavy metal Cd in bandeng. The low BAF value is influenced by the heavy metal content in the waters. The farther away from the source, the value of BAF in bandeng meat decreases, this is due to the heavy metal content in the media and the pond environment is also decreasing. The accumulation of Cd in bandeng occurs because the absorption of Cd from water is not from sediment. Bandeng is a pelagic fish that is in the water column so it is less able to accumulate heavy metal Cd in sediment and tends to accumulate Cd metal in the water column.

## CONCLUSION

Based on the results and discussion of the research it can be concluded that the content of cadmium (Cd) heavy metals in water, sediments, and fish meat ranges from 0.003 to 0.009 mg / L, 1,272-2,208 mg / kg, and 0.087-0,168 mg / kg. The higher the Cd metal content in water and sediments, the Cd metal content in bandeng increased, and vice versa (R = 0.888 and R = 0.906). The level of Cd metal pollution in bandeng ponds in the village of Mulyorejo is categorized as a contamination factor which is classified as very high contaminated, the value of the geoaccumulation index is classified as moderate to severe. The bandeng (Chanos chanos) at stations I to III are no longer suitable for consumption. Based on the value of bioaccumulation factor (BAF), bandeng are less able to accumulate heavy metals in their bodies.

## REFFERENCES

[1]. CCME. 2002. *Sediment Quality Guidelines*. Canadian Environmental Quality

IOP Conf. Series: Earth and Environmental Science **406** (2019) 012031 doi:10.1088/1755-1315/406/1/012031

Guidelines. 2 pp. Dwiono, S. A. P. 2003. Pengenalan Kerang Mangrove, *Geloina erosa* dan *Geloinaexpansa.Oseana*, **28(2)** : 31-38.

[2]. Handayani, U. S., Soegianto, A., dan Yuliani, R. 2001. Pengaruh Salinitas Terhadap Akumulasi Kadmium (Cd) pada Insang Udang *Macrobrachium sintangense* (de Man). *Jurnal Matematika dan Ilmu Pengetahuan Alam Universitas Airlangga*. **6(3)**: 159-167.

[3]. Mayagitha, K. A., Haeruddin, dan Siti, R. 2014. Status Kualitas Perairan Sungai Bremi Kabupaten Pekalongan Ditinjau Dari Konsentrasi TSS, BOD5, COD Dan Struktur Komunitas Fitoplankton. *Diponegoro Journal Of Maquares*, **3(1):** 177-185.

[4]. Palanisamy, P., Sasikala, G., Mallikaraj, D., and Natarajan, G.M. 2011. Study of Behavioural Response of Air Breathing Catfish Mystus cavasius (Hamilton) Exposed to Electroplanting Industrial Effluent Chromium. *IJPSR,* **2(7):** 1790-1792.

[5]. Palar, H. 2012. *Pencemaran dan Teksikologi Logam Berat*. Rineka Cipta. Jakarta

[6]. Rahman, A. 2006. Kandungan Logam Berat Timbal (Pb) dan Kadmium (Cd) pada Beberapa Jenis Krustacea di Pantai Batakan dan Takisung Kabupaten Tanah Laut Kalimantan Selatan. *Bioscientiae*. **3(3)** : 93-101.

[7]. Siregar, Y. I., dan Edward, J. 2010. Faktor Konsentrasi PB, Cd, Cu, Ni, Zn dalam Sedimen Perairan Pesisir Kota Dumai. *Maspari Journal*, **1(1)** : 1-10.

[8]. SNI 6989.16.2009, *Cara Uji Air dan Limbah cadmium (Cd) secara Spektofotometer Serapan Atom (SSA)*. Badan Standarisasi Nasional, Jakarta.

[9]. Wakida, D.F.T., Lara-Ruiz, E.J., dan Temorez, P. 2008. Heavy Metals in Sediments of the Tecate River, Mexico. *Environmental Geology*, **54** : 637-642.

[10]. Wardhana, W. A. 2004. *Dampak Pencemaran Lingkungan. Penerbit Andi.*Yogyakarta.

[11]. Wulandari, S. Y., Yuliyanto, B., dan Sukristyo. 2008. Pola Sebaran Logam Berat Pb dan Cd di Muara Sungai Babon dan Seringin, Semarang. *Ilmu Kelautan*, **13(4)**: 203-208.

[12]. Yudiati, E., Sedjati, S., Enggar, I., dan Hasibuan, I. 2009. Dampak Pencemaran Logam Berat Kadmium pada Salinitas yang Berbeda Terhadap Mortilitas dan Kerusakan Jaringan Insang Juvenile Udan Vename (*Litopeneus vanamei*). *Jurnal Ilmu Kelautan,* **14(4):** 29-35.

# THE EFFECT OF ENDOSULFAN (INSECTICIDE) ON EXPRESSION OF VITELOGENIN GENE IN FEMALE SILVER SHARKMINNOW (*Osteochilus hasseltii* C. V.)

Purnama Sukardi[1,2], Rita Eka Rosita[1,2], Asrul Sahri Siregar[2], Sonny Januar C[1,2], Taufan Harisam[1,2*], Farid Hidayat[1], Yasumasa Bessho[3], Norman Arie Prayogo[1,2]

[1]Research and Community Service, Jenderal Soedirman University, Indonesia
[2]Faculty of Fisheries and Marine Science, Jenderal Soedirman University, Indonesia
[3] Nara Institute of Science and Technology, Japan
*Coresponding author* : norman_s2biologi@yahoo.com

**Abstract.** Endosulfan is an organochlorine insecticide that has high toxicity, persistent and bioaccumulative. Endosulfan entering the waters can cause subletal toxicity to fish. Endosulfan residues that enter the body of the fish will inhibit their reproduction, such as the formation of vitelogenin. This study aimed to determine the effect of concentration and length of time endosulfan exposure which could reduce vitelogenin gene expression in female silver sharkminnow (Osteochilus hasseltii C. V.). Complete randomized design was applied using four treatments of endosulfan concentration i.e. 0 ppb, 0.88 ppb, 1.76 ppb and 2.64 ppb, respectively. The research was conducted for four weeks with sampling frequency every two weeks. The molecular study was carried out four steps include organ preparation, RNA isolation and DNAse treatment, measurement of RNA concentration and vitelogenin gene expression and RT-PCR was used in this study. Gene expression data analyzed by One Way ANOVA to each of the response variables to know whether differences existed among treatments. The results showed that the concentration and exposure time of endosulfan had not given different result on each treatment. This was presumably because the concentration of endosulfan given had not been able to influence changes in gene expression of vitelogenin in female silver sharkminnow.

**Keywords** : *Endosulfan, Sub Letal Toxicity, Vitelogenin Gene Expression*

## INTRODUCTION

Endosulfan is one of the cyclodien organochlorine insecticides whose use continues to increase. Since 1996, the use of endosulfan has been banned in Indonesia through Ministerial Decree. Agriculture No. 473 / KPTS / TP207 / 6/96. The use of endosulfan is restricted and prohibited due to its high toxicity, persistence, bioaccumulative nature and potential biomagnification. Biomagnification process occurs through the tropic level in living things [1], whereas bioaccumulative is characterized by absorption of endosulfan residues so that it accumulates in body tissues because it has high lipophility properties, which are easily bound in fat tissue [2].

Endosulfan that enters the waters can cause sub lethal toxicity to aquatic biota (such as fish), one of which is that it interferes with the reproductive system [3]. Endosulfan residues that enter the body of the fish will give a signal to the brain and passed on to the hypothalamus. The hypothalamus will respond to these signals by releasing GnRH which works in the pituitary gland to regulate gonadotrophic synthesis and secretion (GtHs). Endosulfan in chronic concentrations and long exposure times cause GtHR-Iα mutations. Fish that lose the function of GtHR-Iα will cause small and infertile ovaries due to inhibited vitelogenesis [4]. Endosulfan contamination also causes changes in the distribution of primordial germ cells that can affect the structure and function of the gonads. Vitelogenin is synthesized in hepatocytes in response to the activation of estrogen receptors by estrogenic compounds such as endosulfan [5]. Endosulfan is competitive against androgens and estrogens in binding to receptors, has estrogenic effects, stimulates estrogen receptor production and inhibits aromatase activity [6].

Female silver sharkminnow(*Osteochilus hasseltii* C. V.) is one of Indonesia's native species that is widely cultivated and found along the watershed in the Banyumas region and also widely cultivated by the community. The existence of agricultural activities around the river by using endosulfan will cause endosulfan residues into the waters to be used in the maintenance of cultured fish. This residue can cause fish reproduction to be disrupted. The disruption of reproduction in fish is also greatly influenced by the concentration and time of exposure of endosulfan that enters the waters, so it is necessary to conduct research on the effect of endosulvan ridid on the expression of the vitelogenin gene. This study aims to determine the concentration and duration of endosulfan exposure which can reduce the expression of the vitelogenin gene in female hard-lipped.

## MATERIALS AND METHODS

Completely randomized design research (CRD) was used in this experiment. First, this study carried out a preliminary test to get the LC value of 50-96 hours. LC value of 50-96 hours was obtained at 5.87 ppb. The study was conducted with 4 treatment series, ie 0 ppb endosulfan concentration (0% LC concentration 50-96 hours), 0.88 ppb (15% LC concentration 50-96 hours), 1.76 ppb (30% LC concentration 50-96 hour) and 2.64 ppb (45% LC concentration of 50-96 hours) with three replications of individual silver sharkminnowin each treatment each sample. The treatment is given for 4 weeks with sampling intervals every two weeks.

### Test Media Preparation and Manufacturing

The maintenance container is in the form of 4 round tubs of water fiber as much as 380 L. The fiber tub is made aeration and recirculation system to maintain water quality. After that, the four-month-old female silver sharkminnowwith a stocking density of 50 fish per tank was acclimatized in advance for three days. Preparation of a 10 ppm endosulfan stock solution was carried out by dissolving 2.86 mL of Akodan 35 EC in 10 mL aceton pro analyst (100%), then dissolving in 99,987.14 mL of water. The concentration of the treatment is obtained by the gradual dilution formula.

IOP Conf. Series: Earth and Environmental Science **406** (2019) 012032 doi:10.1088/1755-1315/406/1/012032

Female silver sharkminnowis maintained for 4 weeks. Every day the condition of the fish is checked, which is around 07.00, 12.00 and 18.00 WIB. Feed in the form of pellets (Prima Feed GP-2 brand with a protein content of 27-29%) is given as much as 3% of body weight, namely in the morning and evening.

**Tissue Preparation**

Female silver sharkminnowin each treatment were randomly taken with a total of 3 fish in each treatment tank. Fish brain and pituitary are taken by earpick, then the sample is weighed and put into a tube. Tube containing the sample is stored in a freezer of -80 ° C until the organ sample is isolated.

**RNA isolation**

Brain and pituitary samples were taken with tweezers and weighed ± 25 mg for each fish sample. Isolate RNA samples using products from Geneaid, namely the Total RNA Mini-Kit.

**DNAse *Treatment***

DNAse Treatment uses the DNA-free Thermo Scientific-Kit. DNAse treatment components consist of RNA *in RNase-free Water* (1 μL), 10X *reaction buffer with* $MgCl_2$ (1 μL), DNase I, RNase-*free* (1 U), DEPC-*treated Water* (#R0601) (8 μL). All DNAse components were put into a tube then incubated at 37 °C for 30 minutes. After that, 1 mL of 50 mM EDTA was added and incubated at 65 °C for 10 minutes.

**Measurement of Vitelogenin Gene Expression**

Vitelogenin encoding gene expression using KAPA ™ SYBR® FAST One-Step qRT-PCR Kit.

*Making the qPCR Mix Master*

Components in making a master mix consist of 2x Sensi FAST (5 μL), Forward Primer (0,4 μL), Reverse Primer (0,4 μL), RNA Tempelate (2 μL) and Nuklease is free water (up to 10 μL).

*Quantitative real-time analysis*

The One Step qPCR setting starts from synthesizing cDNA at 42 °C for 5 minutes, then deactivating RT (Reverse Transcript) at 95 °C for 2-5 minutes. Furthermore the denaturation process at a temperature of 95 °C for 3 seconds and aneling at a temperature of 60 °C for ≥20 seconds for as many as 40 cycles. The last step is the extension process for 5 minutes. The primers used in this study used specific real time primers amplifying the vitelogenin gene in female silver sharkminnow. Whereas house keeping gene uses primary beta actin, namely FA and RA (**Table 1**).

The results of the amplification using Real Time PCR are then used to compare the number of DNA molecules from the amplification of the vitelogenin coding gene with the results of the amplification of the β-Actin gene. The comparison values obtained were then compared again with the fish groups in various endosulfan treatments according to Forlenza et al. (2012) with the formula.

$\Delta\Delta CT$ = (CT vtg – CT actin) sampel – (CT vtg – CT actin) Kalibrator, Rvtg= $2^{-\Delta\Delta CT}$

Keterangan:

| | |
|---|---|
| $\Delta\Delta Ct$ | = Threshold Cycle |
| Ct vtg sample | = Vitogenogenin Ct value of the sample-i |
| Ct *β-actin* sample | = nilai Ct *β-actin sampel* ke-i |
| Ct vtg calibrator | = Vitogenogenin Ct values of samples with the lowest β-actin |
| Ct *β-actin* calibrator | = the lowest Ct β-actin value of the amplification results |
| Rvtg | = Vitelogenin gene expression level |

**Table 1.** The primary design used in Real Time PCR [16,17]

| No | Nama/kode primer | Sekuens komplemen DNA (primer) | Tm | Produk PCR |
|---|---|---|---|---|
| 1 | *Forward* Vitelogenin *Real Time* | CGTGGATCHYTGMARTACGA GTT | 62,81 | 1100 bp |
| 2 | *Reverse* Vitelogenin *RealTime* | ATGGTGGCRTCATTGAT | 62,83 | |
| 3 | *Forward Beta Actin* (FA) | GAG CTA TGA GCT CCC TGA CGG | 58,30 | 53 bp |
| 4 | *Reverse Beta Actin* (RA) | AAA CGC TCA TTG CCA ATG GT | 55,60 | |

**Data analysis**

Quantitative data on the level of vitelogenin gene expression were analyzed by One Way ANOVA to determine the concentration and duration of endosulfan exposure that could interfere with the expression of the vitelogenin gene in female silver sharkminnow (*Osteochilus hasseltii* C. V.).

**Results and Discussion**

The results of measurement of vitelogenin gene expression in female silver sharkminnow (*Osteochilus hasseltii* C.V.) exposed to endosulfan for 4 weeks with different concentrations can be seen in Figure 1.

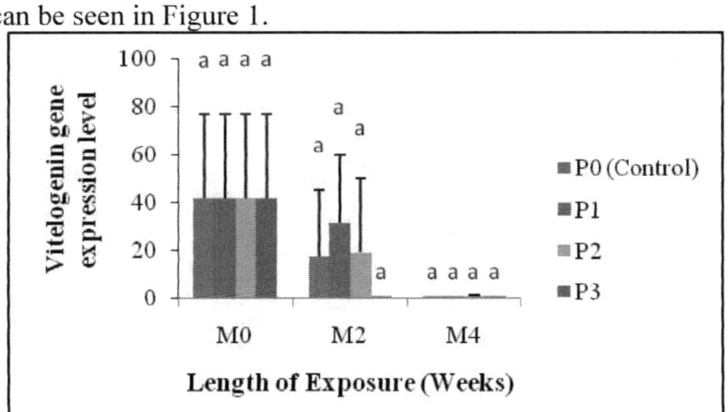


**Figure 1.** Vithelogenin Gene Expression Values of Endosulfan Fish Exposed for 4 Weeks

Based on Figure 1, it can be seen that the value of female nilema vithelogenin gene expression values at week 0 in each treatment concentration showed the same value, which was 41.8259. This is because at week 0 the female silver sharkminnow has not been given treatment so that there is no fluctuation in each concentration treatment.

At the second week the value of vitelogenin gene expression ranged from 0.9419 - 31.8288. The results of measurement of the silver sharkminnow vitelogenin gene expression at week 2 showed results that were not different in each treatment. This can be caused because the concentration of endosulfan given has not been able to affect changes in the expression of the vitelogenin gene in female nilem.

At the 4th week the expression value of the vitelogenin gene ranged from 0.9062 to 1.0667. The results of the measurement of vitelogenin gene expression at the 4th week showed that the inter-treatment did not differ. Apart from giving too low endosulfan concentrations, this can also be caused by the endosulfan concentration at the 4th week the level of toxicity has been reduced so that it has not so affected the expression of the silver sharkminnow vitelogenin gene. According to Reyes et al. [7], the half-life of endosulfan-α is 19 - 33 days, while endosulfan-β is 45 - 58 days.

The results of the measurement of vitelogenin gene expression at the 2nd and 4th week also showed no different results. This is allegedly due to exposure time that is too short. So that the exposure time of endosulfan which was carried out for 4 weeks has not been able to reduce the expression of the vitelogenin gene.

Measurement of vitelogenin gene expression in female silver sharkminnow that showed results that were not significantly different were estimated to be influenced by several factors, namely endosulfan concentrations given were still low and shorter exposure times. In the sublethal toxicity test, the concentration of treatment that is usually used is 0%, 20%, 40% and 60% of the LC-96 hour value [8,9,10,11,12,13]. Thus, endosulfan concentration and exposure time given have not been able to reduce the expression of the vitelogenin gene. Based on the results of research conducted by Hemmer et al. [14], endosulfan concentrations of 1.5 ppb which were exposed for 42 days on *Cyprinodon variegatus* have not been significantly different. Although endosulfan can bind to estrogen receptors at very high concentrations, pharmacokinetic considerations such as absorption, metabolism and distribution and concentration of end tissues can negate the potential estrogenic endosulfan so that it becomes weak[18]. In addition, fish used as test animals will also affect the level of endosulfan toxicity concentration.

The results of the analysis of the expression value of the vitelogenin gene in female silver sharkminnow exposed to endosulfan for 1 month with different concentrations can be supported by the results of the analysis of the value of the Somatic Gonado Index (IGS) because vitelogenin activity is closely related to the IGS value. The results of the analysis of IGS values showed no significant difference so that it showed that different endosulfan concentrations and exposure times did not have a significantly different effect on IGS values in female nilem.

The value of the Somatic Gonad Index (IGS) in silver sharkminnow (*Osteochilus hasseltii* C.V.) female used as a supporting factor in measuring the expression value of the vitelogenin gene can be seen in Figure 2.

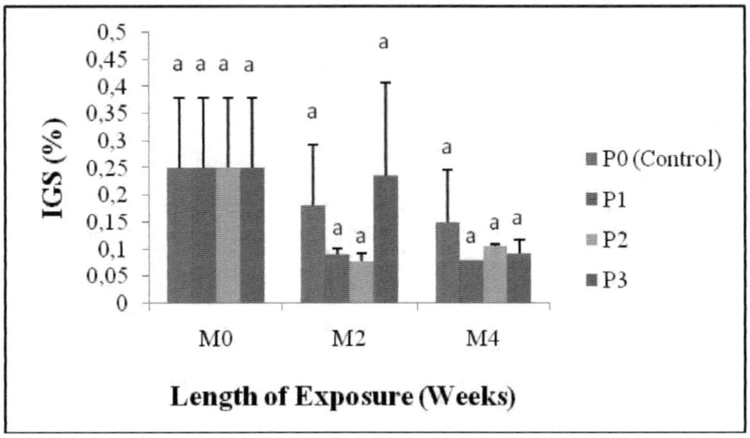


**Figure 6.** Index Gonad Somatic (IGS) in Female silver sharkminnow Exposed to Endosulfan for 4 Weeks

Based on Figure 2, the value of silver sharkminnow IGS percentage in each endosulfan treatment which was exposed for 4 weeks showed fluctuations. The IKG value of week 0 for each treatment is the same, that is 0,250%. While the second week the IKG values ranged from 0.077 to 0.235% and the fourth week ranged from 0.080 to 0.150%. IKG values in each treatment have a gonad maturity stage <4% which means the fish are not ready to spawn [15]. In the condition of fish that are not ready for spawning, gonad damage can be seen by observing gonad histology.

## CONCLUSION

Based on the results of the study, it can be concluded that the level of endosulfan concentration with various concentrations given (<2.64 ppb) and the length of time of endosulfan exposure for 4 weeks did not give different results on the decrease in the value of vitelogenin gene expression in silver sharkminnow (*Osteochilus hasseltii* C.V.).

## REFERENCES

[1] Kostoer, Y. 2006. Kimia dan Ekotoksikologi Pencemaran. UI-Press, Jakarta.

[2] Taufik, I. 2005. *Pengaruh Lanjut Bioakumulasi Insektisida Endosulfan Terhadap Pertumbuhan dan Kondisi Hematologis Ikan Mas* (*Cyprinus carpio*). Tesis. Institut Pertanian Bogor, Bogor.

[3] Putri, A. C., Razak, A., Sumarmin, R. 2015. *Pengaruh Insektisida Organoklorin Endosulfan Terhadap Daya Tetas Telur Ikan Nila* (*Oreochromis niloticus*). Universitas Negeri Padang, Padang.

[4] Murozumi, N., Nakashima, R., Hirai, T., Kamei, Y., Ishikawa-Fujiwara, T., Todo, T., Kitano, T. 2014. Loss of Follicle-Stimulating Hormone Receptor Function Causes Masculinization and Suppression of Ovarian Development in Genetically Female Medaka. *Endocrinology*, **155** (**8**) : 3136–3145.

[5] Chow, W. S., Chan, W. K. L., Chan, K. M. 2013. Toxicity Assessment and Vitellogenin Expression in Zebrafish (*Danio rerio*) Embryos and Larvae Acutely Exposed to Bisphenol A, Endosulfan, Heptachlor, Methoxychlor and Tetrabromobisphenol A. *Journal of Applied Toxicology*, **33** : 670–678

[6] Mnif, W., Hassine, A. I. H., Bouaziz, A., Bartegi, A., Thomas, O., Roig, B. 2011. Effect of Endocrine Disruptor Pesticides : A Review. *Int. J. Environ. Res. Public Health*, **8** : 2265-2303.

[7] Reyes, J. G. G., Rodríguez, G. R. G., Osuna, M. D. C. C., Jaward, F. M. 2014. Bioaccumulation and Evidence of Hormonal Disruptions in Tilapia Fish (*Oreochromis* ssp.) Exposed to Sub-Lethal Concentrations of Pesticides in Sinaloa, Mexico. *International Journal of Biochemistry Research*, **4** (**4**) : 333-343.

[8] Rudiyanti, S., Astri, D. E. 2009. Pertumbuhan dan Survival Rate Ikan Mas (*Cyprinus carpio* Linn) pada Berbagai Konsentrasi Pestisida Reagent 0,3 G. *Jurnal Saintek Perikanan*, **5** (**1**) : 39-47.

[9] Prayogo, N. A., Hidayati, A., Siregar, A. S., Yunasfi. 2016. Uji Toksisitas Letal dan Subletal Logam Berat Merkuri Terhadap Ikan Nilem (*Osteochilus hasseltii*). *Omni Akuatika*, **12** (**1**) : 86-94.

[10] Prayogo NA, Wijayanti GE, Murwantoko, Kawaichi M, Astuti P. 2012. Effect of photoperiods on melatonin levels, estradiols level and the expression of cGnRH-II and sGnRH genes, in hard-lipped barb (osteochilus hasselti c.v.). Journal of global veterinaria, 8(6) : 591-597.

[11] Prayogo NA, Wijayanti GE, Sulistyo I, Sukardi P. 2016a. Cloning and expression cgnrh-ii and sgnrh genes in hard-lipped barb (osteochilus hasselti c.v.). Biodiversitas, biodiversitas. 17(29) : 523-530

[12] Prayogo NA, Siregar AS, Sukardi P. 2016b. The disruptive effect mercurychloride (hgcl) on gene expression of cgnrh-ii sgnrh, and estradiol level in silver sharkminnow (osteochillus hasseltii c.v.). Turkish journal fishery and aquatic science. 16(2) : 1003-1009

[13] Prayogo NA, Siregar AS, Sukardi P, Bessho Y. 2018. Molecular Cloning Of The Vitellogenin Gene In The Hard-Lipped Barb (Osteochillus Hasseltii C.V) And Photoperiod's Effects On Gene Expression. Biotropia journal. 25(3) : 211-233.

[14] Hemmer, M. J., Hemmer, B. L., Bowman, C. J., Kroll, K. J., Folmar, L. C., Marcovich, D., Hoglund, M. D., Denslow, N. D. 2001. Effects Of *P*-Nonylphenol, Methoxychlor and Endosulfan on Vitellogenin Induction and Expression in Sheepshead Minnow (*Cyprinodon variegatus*). *Environmental Toxicology and Chemistry*, **20** (**2**) : 336–343.

[15] Sumantadinata, K. 1983. *Pengembangan Ikan-Ikan Peliharaan di Indonesia*. Satra Hudaya, Jakarta.

[16] Siregar, A. S., Prayogo, N. A. 2017. The Disruptive Effect of Mercury Chloride (HgCl) on Gene Expression of Gonadotrophin Hormones and Testosterone Level in Male Silver Sharkminnow (*Osteochilus hasseltii* C.V.) (Teleostei: Cyprinidae). *The European Zoological Journal*, **84** (**1**) : 436–443.

[17] Siregar, A.S., Prayogo, N.A. 2018. The disruptive effect of mercury chloride (HgCl) on gene expression of gonadotrophin hormones and testosterone level in male silver sharkminnow (Osteochilus hasseltii C.V.) (Teleostei: Cyprinidae). European Zoological Journal 84(1), pp. 436-443

**[18]** Sukardi P, Hana H, Prayogo NA, Sulistyo I, Soedibya PHT, Harisam T, WInanto T. 2018. A lipid-walled microcapsule diet as co-feed for early feeding the Osphronemus gourami (Lacepede) larvae | [Dieta micro encapsulada com lipídeos (Lipid-walled) como co- alimentação na alimentação precoce da larva de Osphronemus gourami Lacepede]. Acta Scientiarum. Animal Sciences, v. 40, e38335. Doi: 10.4025/actascianimsci.v40i1.38335.

IOP Conf. Series: Earth and Environmental Science **406** (2019) 012003 doi:10.1088/1755-1315/406/1/012003

# ANALYSIS OF BIOACTIVE COMPONENTS AND PHYTOCHEMICAL OF POWDERS STEM AND LEAVES OF KECOMBRANG (*Etlingera elatior*)

**Khairunnisa Nisrina Effendi, Nurahmahdiani Fauziah, Rumpoko Wicaksono, Erminawati, Poppy Arsil, and Rifda Naufalin*)**

Food Science and Technology Study Program, Faculty of Agriculture, Jenderal Soedirman University, Purwokerto 53122
*) Corresponding author, Email: rifda.naufalin@unsoed.ac.id

## ABSTRACT

Kecombrang (*Etlingera elatior*) has bioactive compounds such as phenols and flavonoids as antioxidants. This study aims to determine the effect of temperature and time drying on total phenols, total flavonoids, and phytochemical compounds from stem and leaves kecombrang powder using the cabinet dryer method. This study uses a completely randomized design (CRD). Factors tested in this study were kecombrang plant parts (stems and leaves) and drying temperature (50°C, 60°C, and 70°C). The variables observed in this study were total phenols, total flavonoids, water content and phytochemical compounds from stem and leaves kecombrang powder. Quantitative research results showed that kecombrang leaf powder with a drying temperature of 70°C had total phenols, total flavonoids and water content respectively of 35.89 mg TAE / g, 2.132 mg QE / g and 5.557%. Meanwhile, kecombrang stem powder with a drying temperature of 60°C has a total phenol, total flavonoid and water content respectively of 6.030 mg TAE / g, 0.160 mg QE / g and 7.873%. The results of phytochemical testing showed that the positive kecombrang leaf powder contained alkaloid, flavonoid, saponin, steroid, tannin, phenol and negative compounds containing glycoside compounds while the positive kecombrang stem powder contained alkaloid compounds, flavonoids, triterpenoids and phenols and negatively contained glycoside compounds, while kecombrang stem powder positively contained alkaloid compounds, flavonoids, triterpenoids and phenols while negatively containing tannins, saponins and glycosides.

Keywords: bioactive component, drying, kecombrang, stem, leaf

## 1. INTRODUCTION

Kecombrang plant (*Etlingera elatior*) is a spice plant that belongs to the Zingiberaceae group and has been used in making medicine as well as vegetables [1]. Kecombrang fruit and flowers are usually used as antioxidant and antimicobia. While kecombrang leaves and flower can be used as edible coating for sausage and fillet [2, 3, 4].

This plant is commonly found in mountainous areas or shady areas near water with a height of 800 meters above sea level. The chemical composition of kecombrang plants include polyphenols, alkaloids, flavonoids, essential oils, and saponins [1]. The biaoactive component can also be used as a food and beverage additive. However, the addition of the bioactive component of the kecombrang plant in fresh form is considered to be less effective both in terms of storage and use. Therefore, the use of kecombrang plants in fresh form needs to be converted into powder form to facilitate the use of bioactive components contained in kecombrang plants through the drying process.

Drying is the process of removing water internally from the material to produce a dry product. According to [5], drying is a fairly complex product preservation method mainly due to undesirable changes in the quality of the dry product. The basic objective in drying agricultural products is to reduce water in materials to a certain extent, where spoilage microbes and damage caused by chemical reactions can be minimized so that the quality of the dry product can be maintained.

However, during the drying process, temperature and time need to be considered. The process of drying fresh materials using temperatures that are too high with a long time can actually reduce the bioactive compounds in it because the bioactive compounds are damaged [6]. However, the use of temperatures that are too low is difficult to achieve water levels that are safe to store [7]. Therefore, it is needed the right temperature and time to produce dry simplicia which has a water content according to the standard and the bioactive compounds can be maintained.

The purpose of this study was to determine the effect of treatment interactions on quantitative and qualitative variables of kecombrang stem and leaf powder.

## 2. METHODOLOGY RESEARCH

### 2.1 Experimental design

The experimental design used in this study was a Completely Randomized Design. Factors observed were kecombrang plant parts consisting of two types namely kecombrang stem and leaves and drying temperature consisting of three levels, namely 50ºC, 60ºC and 70ºC.

### 2.2 Material preparation

The first step of drying the kecombrang powder stem and leaves begins with wet sorting carried out to separate impurities or other foreign materials. The next step is washing carried out to remove soil and other impurities that are attached to the leaves of kecombrang so as to obtain herbs that are suitable for use. The third stage is the chopping done to simplify the process of drying, packing and grinding.

### 2.3 Drying Material

Drying is done using a tool that is a cabinet dryer with a predetermined combination of treatments. The temperature used are 50ºC, 60ºC, and 70ºC

### 2.4 Simplisia Dry Powder

Powdering is done using a disk mill with 60 mesh sieve. It is done to simplify the subsequent analysis process.

### 2.5 Quantitative Analysis of Kecombrang Stem and Leaf Powder

#### 2.5.1 Analysis of Total Phenols

400 mL of supernatant was added with 1.5 mL of 10% Folin-Ciocalteu and allowed to stand for 5 minutes at room temperature. After that added 1.5 mL of sodium bicarbonate (NaHCO3) 0.556M is shaken and left in a dark room for 90 minutes, then absorbance is measured using a spectrophotometer at a wavelength of 725 nm [8].

#### 2.5.2 Analisis of Total Flavonoid

100 mL or 0.1 mL sample supernatant was added with 1 mL of 2% AlCl 3 (2 g AlCl 3 in 100 mL of 5% glalic acetic acid

solution) and 1 mL of 120 mM potassium acetate solution (1,176g of potassium acetate in 100 mL of distilled water). After that it was incubated for 1 hour at room temperature and measured its absorbance using a spectrophotometer at a wavelength of 435 nm [9].

## 2.6 Analysis of Qualitative Kecombrang Stem and Leaf Powder

### 2.6.1 Analysis of Alkaloid

A total of 0.05 grams of sample was added with 5 mL of 70% ethanol then shaked for 10 minutes at 200 rpm and filtered. As much as 2 mL of filtrate was added 10 mL of HCl 2% and heated on waterbath for 10 minutes at a temperature of 100ºC. Next 1 mL of filtrate was taken, put into a test tube, and added 2 drops of dragendrof reagent. The presence of red-orange deposits indicate the sample contains alkaloids [10].

### 2.6.2 Analysis of Flavonoid

Samples of 0.5 grams plus 10 mL of 70% ethanol are then shaken and filtered. A total of 2 mL of filtrate was taken and 0.1 gram of magnesium powder was added, 3 mL of amyl alcohol, and 1 mL of concentrated HCl. The formation of red, yellow, or orange in the amyl alcohol layer indicates the presence of flavonoids [11].

### 2.6.3 Analysis of Saponin

A sample of 0.5 gram was put into a test tube and then added with 10 mL of hot water. Then cooled and dropped 1 drop of 2N hydrochloric acid. Next, shake it continously for 10 seconds to form a stable white foam for about 10 minutes as high as 1-10 cm. Froth that does not disappear indicates that the simplicia contains saponins [12].

### 2.6.4 Analysis of Steroid /Triterpenoid

A sample of 0.5 grams was added with 10 mL of 70% ethanol. Next, shaked for 10 minutes at 200 rpm and filtered.

Filtrate of 2 mL plus 3 drops of concentrated HCl and 1 drop of concentrated H2SO4. A change in color to green indicates that a positive sample contains a steroid compound and a change in color to red or yellowish red indicates the presence of a triterpenoid [12]
.

### 2.6.5 Analysis of Tannin

A sample of 0.5 grams was added with 10 mL of distilled water, then boiled. After chilling, 5 mL FeCl3 1% (w / v) is added. if the color changes to dark blue, the sample contains tannin [12].

### 2.6.6 Analysis of Phenol

A total of 0.5 gram of sample was extracted with 10 mL of 70% ethanol. As much as 1 mL of the extract was added 3-4 drops of 5% FeCl3 solution. The formation of green or blue green indicates the presence of phenol compounds in the sample [12].

### 2.6.7 Analysis of Glycoside

1 gram of sample was added with 5 mL of anhydrous acetic acid. Then dropped with concentrated sulfuric acid. The color change to blue or green indicates the sample contains glycoside compounds [13].

### 2.7 Analysis of Statistic

Data were analyzed using the F test (analysis of variance) at a 95% confidence level with ANOVA. If the results of the analysis show a significant effect, then proceed with further tests using the DMRT test (Duncan Multiple Range Test) at 5% level. The software used for data analysis is SPSS 25.

## 3. RESULT AND DISCUSSION

### 3.1 Analysis of Quantitative

### 3.1.1 Analysis of Total Phenol

The average value of total phenol kecombrang powder of stem and leaves during drying can be seen in Figure 1.

IOP Conf. Series: Earth and Environmental Science **406** (2019) 012003 doi:10.1088/1755-1315/406/1/012003

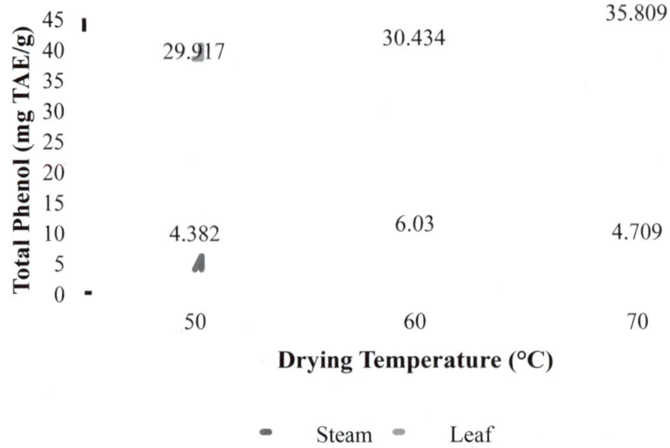


The average total phenol in kecombrang leaf powder is higher than that of kecombrang stem powder. The difference in total phenol levels can be caused by environmental factors for different kecombrang plants such as differences in plant growth, soil composition, temperature, rainfall and UV radiation [14].

According to [14], high levels of total phenols are also influenced by other compounds present in kecombrang leaves, namely tannins. The tannin content also influences the high level of total phenol produced from kecombrang leaves because tannins which are hydrophilic are extracted in a polar ethanol solution and are read at the time of measurement.

The highest total phenol content in the leaf plant section was obtained at a 70°C drying treatment at 35.809 mg TAE / g and the lowest total phenol content was obtained at a 50°C drying treatment at 29.917 mg TAE / g. This is in accordance with [15] which states that drying with high temperatures in a short time can reduce food damage more than a longer drying time with shorter temperatures. In addition, high levels of total phenols are also caused by the use of young leaves which have soft texture and higher water content compared to old leaves. During wilting and drying, young leaves provide less heat penetration so that the polyphenol enzyme is not much damaged [16].

On kecombrang stems the highest total phenol content was obtained at a 60°C drying treatment of 6.03 mg TAE / g and the lowest total phenol content was obtained at a 50°C drying treatment which was 4.382 mg TAE / g. These results are in accordance with the literature which states that oven drying at 60°C at a fast time provides high phenolic content and the most powerful antioxidant power [17]. Drying with a treatment temperature of 50°C has the lowest total phenol because drying using temperatures that are too low can result in the decomposition of phenolic compounds by the help of the enzyme polyphenol oxidase found in plants [10].

### 3.1.2 Analysis of Total Flavonoid

The average total flavonoid powder of the stem and leaves of the kecombrang during drying can be seen in Figure 2

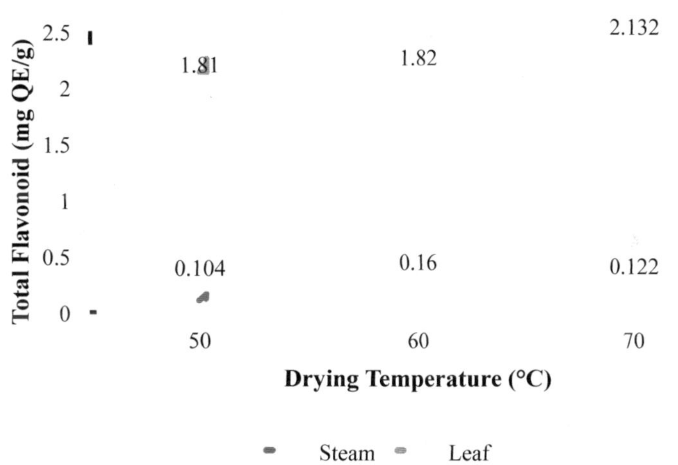


Average total flavonoids in kecombrang leaf powder is higher than kecombrang stem powder. Based on Figure 2, the highest total flavonoid content of the kecombrang leaf section was obtained at a 70ºC drying treatment temperature of 2.132 mg QE / g and the lowest total flavonoid content was obtained at a 50ºC treatment temperature which was 1.81 mg QE / g. This happens because the drying process on kecombrang leaves can destroy the waxy layer that is on the outer surface of the leaf and will further break down the cell wall so that it will facilitate the flavonoid compound to diffuse into the solvent. The higher the drying temperature causes the higher the damage to the cell wall. Carbohydrate and protein on cell wall will be broken because it, so the higher the amount of flavonoid compounds that come out with solvent [18].

On kecombrang stems the highest total flavonoid levels were obtained at 60ºC which is 0.160 mg QE / g and the lowest total flavonoid levels were obtained at 50ºC temperature treatment which was 0.104 mg QE / g. These results are in line with the literature which states that the optimal temperature used in drying simplicia is around 60oC [19]. Drying with a treatment temperature of 50°C has the lowest total phenol. This happens because flavonoids are polyphenol compounds that can be oxidized in the presence of the enzyme polyphenol oxidase in plants. Drying using temperatures that are too low can result in the decomposition of phenolic compounds by the help of the enzyme polyphenol oxidase found in plants [10].

### 3.1.3. Water Content

The average water content of kecombrang powder stem and leaves decreased with the increase in drying temperature which can be seen in Figure 3.

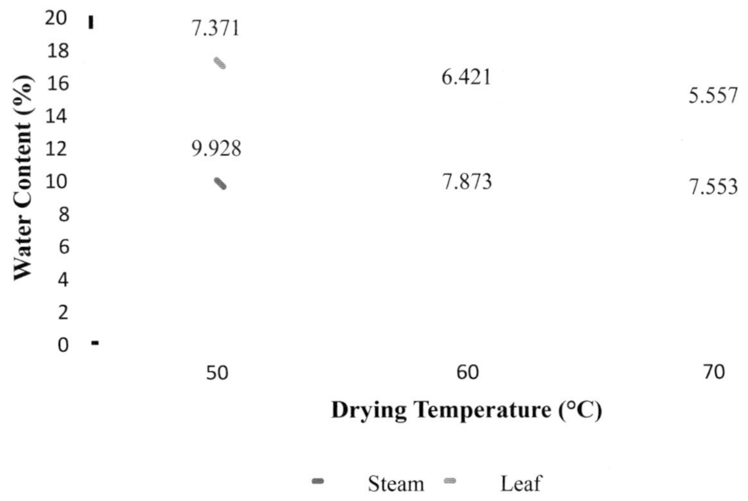


Based on Figure 3, the results of the water content of all treatments have the requirements as dry simplicia with a moisture content that does not exceed 10%. The lowest water content of the kecombrang leaves was obtained at a drying treatment temperature of 70°C which was 5.557% and the highest water content was obtained at a drying treatment temperature of 50°C which was 7.371% while in the kecombrang stem has the lowest water content was also obtained at a drying treatment temperature of 70°C which is equal to 7.553% and the highest water content is obtained at a treatment drying temperature of 50°C which is equal to 9.928%.

The highest drying temperature causes the evaporation of water contained in the kecombrang plant parts to be higher so that the water content is lower. This is supported by the statement of [20], the rate of evaporation besides being influenced by the humidity level is also influenced by the temperature around the dried material.

An increase in temperature on the surface of the material caused by the supply of heat energy from combustion. [21] revealed that slicing dried material will expand the surface of the

IOP Conf. Series: Earth and Environmental Science **406** (2019) 012003 doi:10.1088/1755-1315/406/1/012003

material and a broad surface can make it easier for water to come out. The composition of water in foodstuffs such as free water and bound water can affect the rate or duration of food drying.

### 3.2 Analysis of Qualitative

The qualitative variables observed in this study were alkaloids, flavonoids, saponins, steroids, tannins, phenols and glycosides. The results of phytochemical analysis of dried kecombrang stem and leaves can be seen in Table 1.

Table 1. Phytochemical analysis of dried kecombrang stem and leaf

| Treatment | Stem | Leaf |
|---|---|---|
| Alkaloid | ++ | +++ |
| Flavonoid | +++ | ++++ |
| Saponin | - | +++ |
| Steroid/Triterpenoid | + | +++ |
| Tanin | - | ++++ |
| Fenol | +++ | ++++ |
| glycoside | - | - |

Note: - (negative), + (positive of weak), ++ (Positive), +++ (positively strong), ++++ (positively very strong)

#### 3.2.1 Analysis of Alkaloid

Based on the results of qualitative analysis of alkaloid compounds in Table 1, it can be seen that the stem and leaf extracts of kecombrang with Dragendorff reagents contain the highest positive alkaloids of ++ for positive and positive stems +++ for leaves with strong positive marked by the formation of red brick deposits in large quantities [22].

According to [23] positive results of alkaloids in the Dragendorff test were also characterized by the formation of light brown to yellow (orange) deposits. These deposits are potassium alkaloids. The formation of orange deposits in the addition of dragendroff reagents occurs because nitrogen is used to form coordinate covalent bonds with $K^+$ which is a metal ion and yellowish white deposits are formed in the addition of major reagents because nitrogen in the alkaloids will react with $K^+$ metal ions from potassium tetraiodomerkurat (II) to form potassium complexes and form yellowish white deposits in the addition of major reagents because nitrogen in alkaloids will react with $K^+$ metal ions from potassium tetraiodomerkurat (II) to form potassium complexes. -alkaloid that settles [24].

### 3.2.2 Analysis of Flavonoid

Based on the phytochemical analysis of flavonoid compounds in Table 1, it can be seen that the dry simplicia of the stem and leaves of kecombrang which is extracted with ethanol solvent which is then reacted with concentrated HCl, magnesium powder and amyl alcohol produce orange-red color, so it can be concluded that the positive kecombrang stem extract contains compounds flavonoids with the highest flavonoid content of positive +++ which means strong positive and positively kecombrang leaf extract contain flavonoids with the highest flavonoid content of positive ++++ which means very positive positive. This is in line with the research of [25] which states that the positive kecombrang stem and leaf extract contains flavonoid compounds in the presence of orange-red formation.

### 3.2.3 Analysis of Saponin

Based on the results of phytochemical analysis of saponin compounds in Table 1, it can be seen that the stem powder extracted with distilled water solvent negatively containing saponin and kecombrang leaf powder extracted with distilled water solvent has the highest positive saponin compound content of 3 with strong positive. The chemical compound saponin in kecombrang leaf extract is characterized by the formation of froth for less than 10 minutes and is not lost on the addition of 2N hydrochloric acid. saponins are strong surface tension-reducing compounds that cause foam when shaken in water. Saponins work as antimicrobials by disrupting the stability of bacterial cell membranes, causing bacterial cells [25]

### 3.2.4 Analysis of Steroid/Triterpenoid

Based on the results of phytochemical analysis of steroid/triterpenoids compounds in Table 1 it is known that kecombrang stem extracts with Lieberman Burchard reagents produce pink formation which indicates the presence of triterpenoid compounds while kecombrang leaf extracts with Lieberman Burchard reagents produce green formation that indicates the presence of steroid compounds. This is in accordance with [26] which states that if there is a change in color to red, pink or violet, it is positive to contain triterpenoids and if the color changes to green, the sample is tested positive for steroids..

### 3.2.5 Analysis of Tannin

Based on the phytochemical analysis of tannin compounds in Table 1, it can be seen that the kecombrang stem extract with $FeCl_3$ reagents did not change color indicating that kecombrang stem extract negatively contained tannin compounds while the kecombrang leaf extract with $FeCl_3$ reagent formed a blackish green color with the highest tannin content of ++++ with a very strong positive statement. This is in accordance with [27] which states that kecombrang stem powder negatively contains tannin compounds and kecombrang leaf powder positively contains tannin. [25] also stated that the presence of tannins was marked by the formation of blue or blackish green.

### 3.2.6 Analysis of Phenol

Based on the phytochemical analysis of phenol compounds in Table 1 it can be seen that the kecombrang stems and kecombrang leaves extracted using ethanol solvent then added with $FeCl_3$ reagent turn dark green, this indicates that stem and leaf extracts of kecombrang positively contain phenols. Color changes indicate the presence of oxidized phenol compounds. The profile of phenolic compounds in sample extracts can change with differences in the degree of polarity of the solvents used [28]. This is in accordance with [29] which states that the phenol component contained in the fraction of kecombrang plant parts is thought to have a polarity close to ethanol polarity, so that the use of ethanol solvents is more effective to dissolve phenol compounds.

### 3.2.7 Analysis of Glycoside

Based on the results of phytochemical analysis of glycoside compounds in Table 1, it can be seen that the stem and leaves of kecombrang extracted using methanol solvent were then added to glacial acetic acid containing 1 drop of 1% $FeCl_3$ and also added concentrated sulfuric acid did not produce reddish brown rings. This shows that the stem and leaf samples of kecombrang do not contain glycoside compounds. The absence of glycosides in the stem and leaves of kecombrang is thought to be due to environmental factors of kecombrang growing. This is in accordance with [14] which states that differences in the content of compounds in the sample are caused by environmental factors for different kecombrang plants such as differences in plant growth, soil composition, temperature, rainfall and ultraviolet radiation..

## 4. CONCLUSION

Kecombrang stems and leaves were respectively dried using 60°C and 70°C temperatures to produce the best powder quality of the kecombrang plant parts as seen from the total phenols, flavonoids, water content and phytochemical analysis results. Kecombrang stem powder produces total phenols, total flavonoids, and water content respectively of 6.030 mg TAE / g, 0.160 mg QE / g, and 7.873% while kecombrang leaf powder produces total phenols, total flavonoids, and water content respectively of 35,809 mg TAE / g, 2,132 mg QE / g, and 5.557%.

**Acknowledgements**

Authors wish to thanks to the DRPM Kemenristek Republic of Indonesia which has provided research funding through Applied Research Grant Scheme 2019 also to Institute for Research and Community Service of Jenderal Soedirman University for the support.

**REFERENCES**

1 Naufalin, R., 2019. Natural preservation opportunities and challenges in improving food safety. In *AIP Conference Proceedings.* Vol. 2094, No. 1, p. 020032. AIP Publishing.
2 Latifasari, N., R. Naufalin, R. Wicaksono. 2019. Edible coating application of kecombrang leaves to reduce gourami sausage damage. IOP Conf. Series: Earth and Environmental Science 250 (2019) 012055: 1-7.
3 Hanifah, M.R., R. Naufalin, R. Wicaksono. 2019. The effect of edible coating contained Kecombrang leaves concentrate on gourami fish fillet quality. IOP Conf. Series: Earth and Environmental Science 250 (2019) 012057: 18.
4 Putri, F.A., R. Naufalin, R. Wicaksono. 2019. Antimicrobial edible coating application of Kecombrang flower concentrate to reduce microbial growth on gourami fish sausage. IOP Conf. Series: Earth and Environmental Science 250 (2019) 012056: 1-8.
5 Mujumdar, A.S., Menon, A.S.1995. Drying of solid : principle, classification and selection of dryers. *Handbook of industrial drying Marcel dekker.* Vol 1, New York.
6 Manalu, L.P., Adinegoro, H. 2016. Kondisi proses pengeringan untuk menghasilkan simplisia temuputih standar. *Jurnal Standardisasi,* 18 (1): 62-68.
7 Babalis, S.J., Belessiotis, V.G. 2004. Influence of the drying conditions on the drying constants and moisture diffusivity during the thinlayer drying of figs. *Journal of Food Engineering,* 65: 449–458.
8 Singleton,V.L. & J.A Rossi. 1965. Colorimetry of total phenolic with phosphomolybdic-phosphotungstic acid reagent. *American Journal Enology and Viticulture*. 16: 147
9 Azizah, D.N. 2014. Penetapan kadar flavanoid metode AlCl3 pada ekstrak metanol kulit buah kakao (*Theobroma cacao L.). Jurnal Ilm. Far.* [Online] Vol. 2(2): 45-49.
10 Rivai, H., Nurdin, H., Suyani, H., & Bakhtiar, A. 2010. Pengaruh cara pengeringan terhadap perolehan ekstraktif, kadar senyawa fenolat dan aktivitas antioksidan dari daun dewa (*Gynura pseudochina* (L.) DC.). *Majalah Obat Tradisional*, 15 (1): 26-33.

11 Noer, S. 2016. Uji kualitatif fitokimia daun *Ruta angustifolia*. *Jurnal Factor Exacta,* 9 (3): 200-206.
12 Harbone J. B., 1987. *Metode Fitokimia: Penentuan Cara Modern Menganalisis Tumbuhan* (Penerjemah Padmawinata, K dan I. Soediro). Institut Teknologi Bandung, Bandung.
13 Astarina, N.W.G., Astuti, K.W., & Warditiani, N.K. 2013. Skrining fitokimia ekstrak metanol rimpang bangle (*Zingiber purpureum* Roxb.). *Jurnal Farmasi Udayana*, 2 (4): 1-7.
14 Borges, L., S. Alves, B. Sampaio, E. Conceicao, M. Bara & J. Paula. 2013. Environmental factors affecting the concentration of phenolic compounds in *Myrcia tomentosa* leaves. *Brazilian Journal of Pharmacognosy*. Vol. 23(2): 230-238.
15 Estiasih, T., Ahmadi. 2009. *Teknologi Pengolahan Pangan*. Jakarta: Bumi Aksara.
16 Rauf, A., Pato, U., D. F. Ayu. 2017. Aktivitas antioksidan dan penerimaan panelis teh bubuk daun alpukat (*Persea Americana Mill.*) berdasarkan letak daun pada ranting. *Jurnal Faperta*. Vol. 4(2): 5.
17 Rivai, H., Hasnah, Syarif, M. 2009. Pengaruh cara pengeringan terhadap perolehan kadar senyawa fenolat dan aktivitas antioksidan dari daun jambu biji (*Psidium guajava* Linn.). *Jurnal Farmasi Higea,* 1 (1): 46-51.
18 Susanti, D.F. 2008. Efek suhu pengeringan terhadap kandungan fenolik dan kandungan katekin ekstrak daun kering gambir: 1-13. Makalah disampaikan dalam *Prosiding Seminar Nasional Teknik Pertanian Yogyakarta, 18-19 November 2008.*
19 Supriningrum, R., Fatimah, N., & Wahyuni, S.N. 2018. Penetapan kadar flavonoid ekstrak etanol daun pacar kuku (*Lawsonia inermis* L.) berdasarkan perbedaan cara pengeringan. *Jurnal Ilmiah Manuntung,* 4 (2): 156-161.
20 Widjanarko, A. 2012." Hubungan anemia dengan pengetahuan gizi, konsumsi Fe, Protein, Vitamin C, dan pola haid pada mahasiswa putri" dalam media gizi mikro Indonesia. *Jurnal Balai Penelitian dan Pengembangan kesehatan (Balitbangkes)*. Vol. 4(2): 5158.
21 Dimas, T. P. 2008. *Teh dan Pengolahannya*. Universitas Brawijaya Press. Malang.
22 Titis, M.B.M., Fachriyah, E., D. Kusrini. 2013. Isolasi, identifikasi dan uji aktifitas senyawa alkaloid daun binahong (*Anredera cordifolia (Tenore) Steenis*). *Chem. Info*. Vol. 1(1): 196 – 201.
23 Ergina, S. N., Pursitasari, I.D. 2014. Uji kualitatif senyawa metabolit sekunder pada daun palado (*Agave angustifolia)* yang diekstraksi dengan pelarut air dan etanol. *J. Akad. Kim.* Vol. 3 (3): 165-172. ISSN 2302-6030.
24 Marliana, S.D., Suryanti, V., Suyono. 2005. Skrining fitokimia dan analisis kromatografi lapis tipis komponen kimia buah labu siam (*Sechium edule Jacq. Swartz.)* dalam ekstrak etanol. *Jurnal Biofarmasi*. Vol. 3(1): 26-31.
25 Kusumawati, E., Supriningrum, R., Rozadi, R. 2015. Uji aktivitas antibakteri ekstrak metanol daun kecombrang *Etlingera elatior* (Jack)

R.M.Sm terhadap *Salmonella typhi*. *Jurnal Ilmiah Manuntung*. Vol. 1(1):1-7.

26 Santoso, J., S. Anwariah., R.O. Rumiantin., A.P. Putri., N. Ukhty, Y. Yoshie-Stark. 2012. Phenol conten, antioxidan activity and fibers profile of four tropical seagrasses from Indonesia. *Journal Of Medical Plants*. 10 (37): 73-79.

27 Naufalin, R. 2017. *Antimikroba dan Pemanfaatannya Sebagai Pengawet Pangan*. Universitas Jenderal Soedirman Purwokerto, Purwokerto.

28 Kahkonen M.P., A.L. Hopia., H.J. Vourela., J.P. Rauha., Pihlajak., T.S Kujala, M. Heinonen. 2001. Antioxidant activity of extract containing phenolic compounds. *J Agric Food Chem*. (47): 3954-62.

29 Naufalin, R, Rukmini, H.S. 2010. Potensi antioksidan hasil ekstraksi tanaman kecombrang (*Nicolaia speciosa Horan*) selama penyimpanan: 1-13. Makalah disampaikan dalam *Seminar Nasional*, Surakarta, 8 Juni 2010.

IOP Conf. Series: Earth and Environmental Science **406** (2019) 012004 doi:10.1088/1755-1315/406/1/012004

# COLOR-BASED ANALYSIS FOR NON-DESTRUCTIVE QUALITY EVALUATION OF SIAMESE ORANGE (*Citrus nobilis*) DURING STORAGE IN ROOM AND COLD TEMPERATURE

**Susanto B. Sulistyo*, Agus Margiwiyatno, Krissandi Wijaya, Poppy Arsil, Furqon, Arief Sudarmaji, and Purwoko H. Kuncoro**

Department of Agricultural Engineering, Faculty of Agriculture, Jenderal Soedirman University, Purwokerto, Indonesia

susanto.sulistyo@unsoed.ac.id; agmargi@yahoo.co.id; krissandi.wijaya@unsoed.ac.id; poppy.arsil@unsoed.ac.id; furqon.8@gmail.com; arief.unsoed@gmail.com; kuncoro.hari@yahoo.com

**Abstract.** This research aimed to develop image processing program to extract several features of Siamese orange images and to determine the appropriate key features of the orange images which were highly correlated with physical and chemical characteristics of Siamese orange. A sample size of 210 oranges was stored in a room temperature (25-27 °C) and cold temperature (8-10 °C) for 30 days. All oranges were captured their images using a webcam and were subsequently measured their physical and chemical characteristics, i.e. weight, hardness, and total soluble solids (TSS). The visual parameters of images measured were binary area in pixels, RGB color, HSV color, CIE-Lab color and gray value. The results showed that there was a high correlation between fruit weight and binary area in both room and cold temperature storage with $r^2 = 0.8197$ and $r^2 = 0.8291$, respectively. On the other hand, color values cannot be used to estimate fruit hardness since the correlation coefficient was too small. The highest correlation coefficient between them was $r = 0.114$ which was achieved from the correlation between fruit hardness and hue value. The $r$ value of TSS and some color components was nevertheless relatively strong. Based on the experiments, hue value of HSV color model and a* component of CIE-Lab color model have fairly strong correlations with TSS of oranges stored in room temperature which were indicated from the $r^2$ value of 0.7473 and 0.7029, respectively.

**Keywords:** image processing, physical characteristics, chemical characteristics, total soluble solids, RGB

## 1. Introduction

Citrus is one of the most common commercialized horticulture products which may easily deteriorate its quality during storage. The quality deterioration can be investigated through the changes of its physical as well as mechanical, chemical and optical properties. The information of quality properties of citrus are important to be understood since they have a significant role in machinery design as well as in harvesting process and other postharvest activities, such as sorting, grading, storage, packaging, transportation, and fruit processing (e.g. drying, cooling, heating, etc.).

A lot of studies have been established to identify quality properties of citrus. Rafiee et al. (2007) did a research to figure out physical properties of Iranian fruit, i.e. Bergamot (*Citrus medica*). Sharifi et al. (2007) performed a study to identify physical properties of Tompson citrus. Abdullah et el. (2012) studied the physical properties of musk lime (*Citrus microcarpa*). Moreover, Pawar et al. (2015) identified physical and mechanical properties of Kagzi-lime. Joshi and Awate (2016) studied physical and mechanical characteristics of citrus for sorting and grading. Olabinjo et al. (2017) conducted a research to identify physical and chemical of sweet orange (*Citrus sinensis*) peel.

Quality evaluation of citrus is commonly conducted destructively. For instance, fruit hardness is usually measured by penetrating a plunger tip into citrus peel using a fruit penetrometer. In addition, total soluble solid (TSS) which is one of the most important chemical properties of fruits is commonly measured by placing a couple of drops of citrus juice onto the prism of a refractometer. To extract citrus juice, indeed, some samples of citrus must be squeezed and such practice is obviously destructive.

Recent developments of computation technology has led to enhancement of agricultural practices, both on-farm and off-farm activities. One of the renowned methods that has been widely used to examine horticultural products non-destructively is image processing. Therefore, this research aimed to analyze the changes of citrus quality during storage in room and cold temperature as well as to develop image processing program to extract several features of Siamese orange images and to determine the appropriate key features of the orange images which were highly correlated with physical and chemical characteristics of Siamese orange during storage in room and cold temperature.

## 2. Materials and methods

### *2.1. Raw materials preparation*

Harvested siamese oranges were obtained from a local citrus farm in Kalikajar Village, Purbalingga Regency, Indonesia. All samples of the oranges were picked from trees at relatively homogeneous period, i.e. around 200 days after blossoms. A sample size of 210 oranges was prepared prior to data acquisition in the Laboratory of Food and Agricultural Product Processing, Department of Agricultural Technology, Jenderal Soedirman University. These fruits were firstly rubbed using a clean damp cloth to remove dirt and foreign materials from the epicarp.

After cleaning the oranges peels, the fruits were then split into two storage treatment, i.e. in room temperature (25-27 °C) and in cold temperature (8-10 °C). All samples were stored under the two conditions for 30 days. Several tools and devices were utilized to measure fruit quality parameters, i.e. electronic balance, fruit penetrometer, hand refractometer, lightmeter, and a closed box with a webcam and controlled lighting system for image acquisition.

### *2.2. Data acquisition*

There were four types of fruit's properties measured in this research, i.e. physical properties (weight), mechanical properties (hardness), chemical properties (total soluble solid) and optical properties (color). During storage, every five days, five oranges were randomly chosen for measurement of their properties.

To determine fruit weights, five orange samples were randomly selected from each treatment and were carefully weighed using an electronic balance to an accuracy of 0.01 g. The weighing procedure is replicated three times, and the average value was taken and recorded. Fruit hardness was measured using fruit penetrometer by manually penetrating a plunger tip into orange peel. This procedure was also replicated three times and the average value was calculated. To measure total soluble solid, citrus juice must firstly be extracted by squeezing some orange samples. A couple of drops of citrus juice was then placed onto the prism of a refractometer. Refractometer is an instrument to quantify substances dissolved in water and certain oils using the principle of light refraction through liquids.

To determine color values of Siamese orange, an image processing approach was applied. Firstly, oranges were put in the middle of the closed box base and exactly under the webcam attached. The light intensity in the closed box was maintained relatively uniform with a range of 280-300 lux. The images of oranges were then captured by means of a webcam and subsequently stored as JPG files in a computer.

IOP Conf. Series: Earth and Environmental Science **406** (2019) 012004 doi:10.1088/1755-1315/406/1/012004

The recorded images were then processed to extract their visual features using a developed image processing technique. Some measurement activities can be seen in Figure 1.

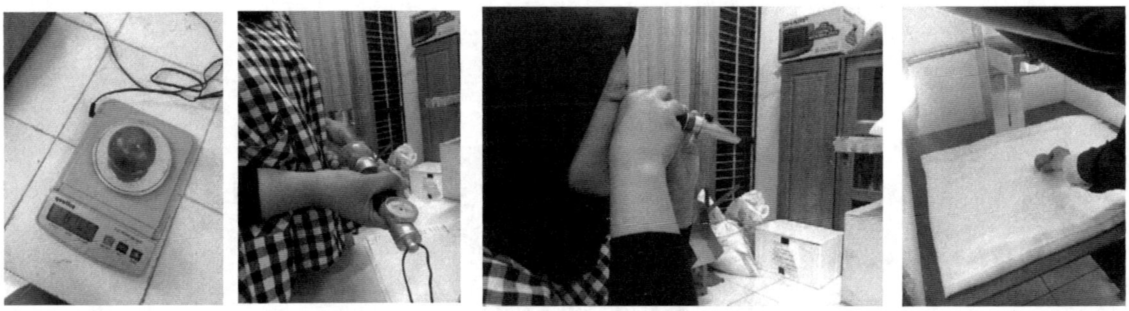

(a) (b) (c) (d)

Figure 1. Measurements of orange quality parameters: (a) weight, (b) hardness, (c) total soluble solid, and (d) color measurement using image processing technique.

### *2.3. Image processing technique*

Orange images captured using a webcam were then processed through several steps of image processing technique. All image processing program in this research was developed using Matlab R2009b software. After image acquisition, the next step was image segmentation. In this step, an orange image was distinguished from its background through a thresholding method. In order to ease and speed up advanced processing without significantly losing important information of image data, we selected a smaller region of each segmented image which was located in the middle of image area. This region selection was conducted by determine center of gravity (COG) of each orange image then cropped a rectangular area of 101×101 pixels with the COG as the center of the rectangular. An example of the image segmentation process is shown in Figure 2.

Once orange images were segmented, a number of image features were subsequently extracted. In this research, the main features extracted included binary area, mean RGB, mean HSV and mean L*a*b*. RGB color values were converted to HSV and L*a*b* color by using the following formulae:

- RGB to HSV color conversion:

$$\mathrm{H} = \tan\left(\frac{3(G-B)}{(R-G)+(R-B)}\right)$$

$$S = 1 - \frac{\min(\mathrm{R,G,B})}{\mathrm{V}}$$

$$\mathrm{V} = \frac{\mathrm{R+G+B}\,B}{3}$$

- RGB to L*a*b* color conversion:

$$\begin{bmatrix} X \\ Y \\ Z \end{bmatrix} = \begin{bmatrix} 0{,}4125 & 0{,}3576 & 0{,}1804 \\ 0{,}2127 & 0{,}7152 & 0{,}0722 \\ 0{,}0913 & 0{,}1192 & 0{,}9502 \end{bmatrix} \begin{bmatrix} R \\ G \\ B \end{bmatrix}$$

$$L^* = 116 f\left(\frac{Y}{Y}\right) - 16$$

$$a^* = 500[f(\frac{X}{X}) - f(\frac{Y}{Y})]$$
$$b^* = 200[f(\frac{Y}{Y}) - f(\frac{Z}{Z})]$$

$$a^* = 500[f(\frac{X}{X}) - f(\frac{Y}{Y})]$$

$$b^* = 200[f(\frac{Y}{Y}) - f(\frac{Z}{Z})]$$

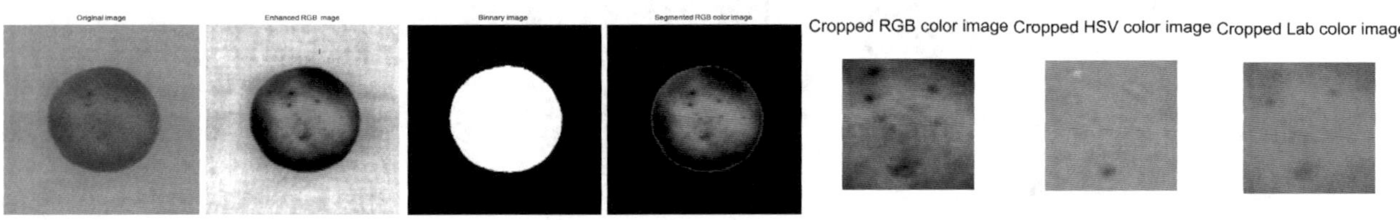


(a) (b) (c) (d) (e) (f) (g)

Figure 2. An example of image segmentation and color conversion processes: (a) original image, (b) contrast enhancement, (c) binary image, (d) color segmented image, (e) cropped RGB color image, (f) cropped HSV color image, and (g) cropped L*a*b* color image.

### *2.4. Data analysis*
Data analysis employed in this research was correlation analysis between visual features of orange images and fruit quality parameters (weight, hardness and TSS). Results of the correlation analysis were then utilized to develop mathematical model to estimate changes of orange quality ($Q$) according to storage time ($t$) and temperature ($T$):

$$Q = f(t,T)$$

## 3. Results and Discussions
The changes of orange weight against storage time at room and cold temperature followed order-4 polynomial graph with mean squared errors of room and cold temperature treatment were 1.08 and 1.04, respectively. In addition, the changes of orange hardness against storage time at room and cold temperature followed order-3 polynomial graph with mean squared errors of room and cold temperature treatment were 0.0018 and 0.0086, respectively. However, the changes of orange TSS against storage time at room and cold temperature followed order-2 polynomial graph with mean squared errors of room and cold temperature treatment were 0.2682 and 0.0521, respectively. The actual changes of orange quality parameters against storage time are shown in Figure 3.

Orange weights stored in both room and cold temperature had a strong relationship with image area, which was indicated by the high value of $r^2 = 0.8197$ and $r^2 = 0.8291$, respectively (Figure 4). The visual parameter of image area, therefore, can be used as a parameter to estimate orange weight during storage. On the other hand, fruit hardness did not have a strong correlation with the visual parameters of the image as indicated by small correlation coefficients of various visual parameters of the image with the hardness of the orange. The biggest correlation coefficient only reached $r = 0.114$ from the relationship between the average hue value and hardness. TSS of oranges stored in room temperature had a fairly strong relationship with the average hue value of the HSV color model and the average chroma a* of the CIE-Lab color model with coefficients determination of $r^2 = 0.7473$ and $r^2 = 0.7029$, respectively.

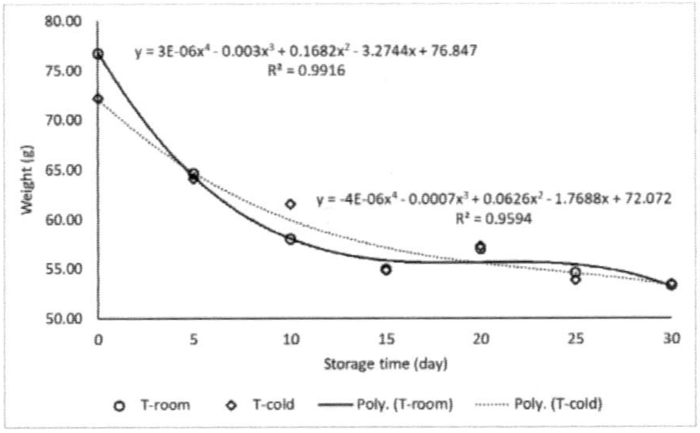


(a)

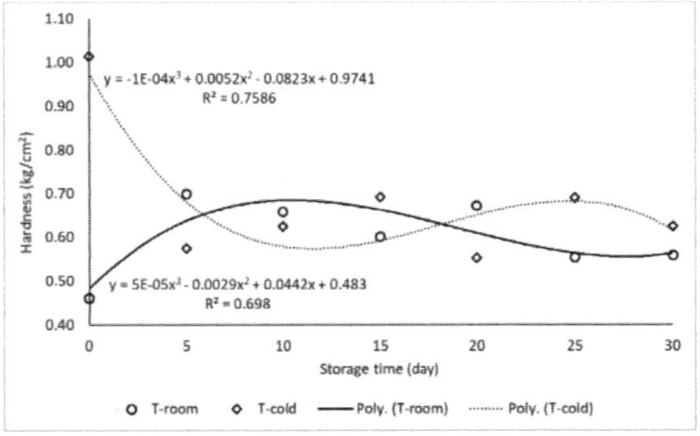


(b)

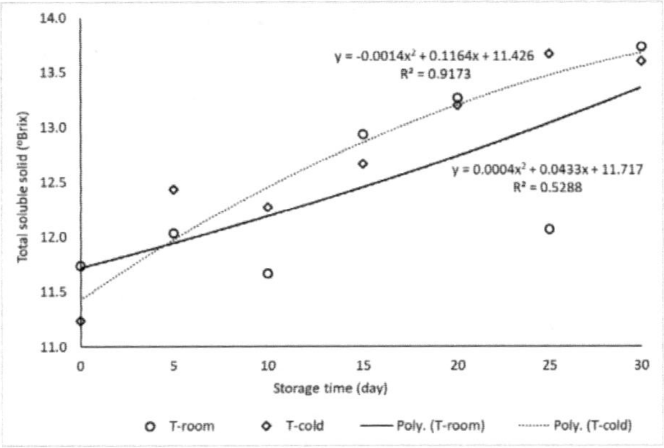


(c)

Figure 3. The changes of orange quality against time during storage in room and cold temperature: (a) weight, (b) hardness, and (c) total soluble solid.

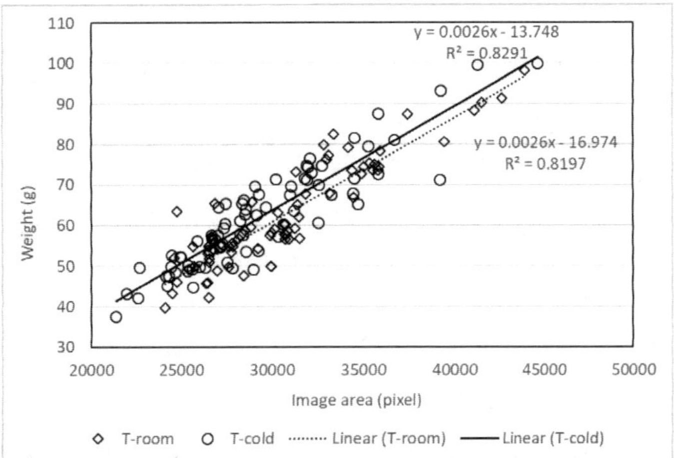


Figure 4. Correlation between orange weight and binary image area.

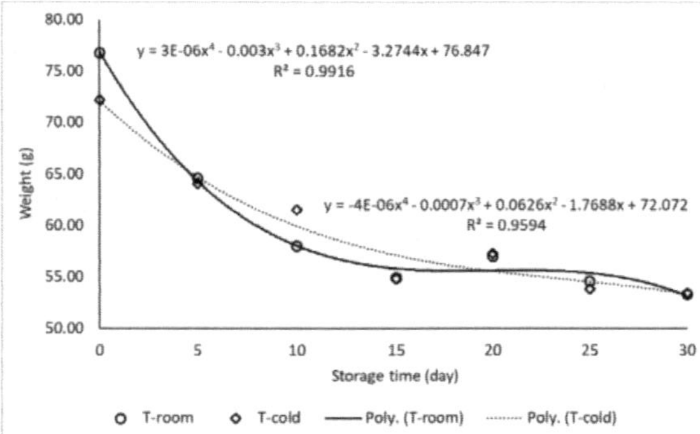


(a)

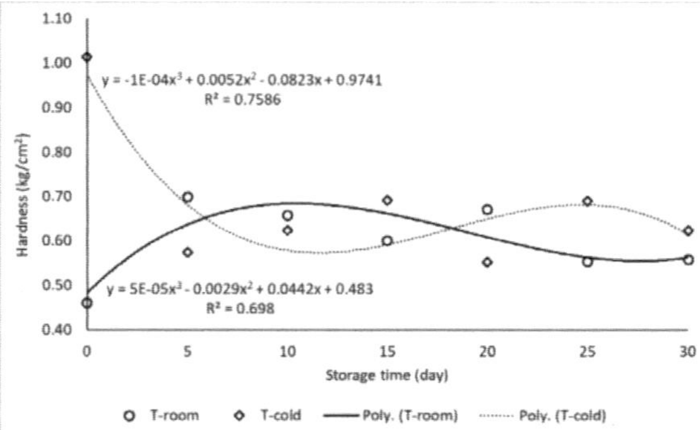


(b)

Figure 5. Correlation between orange total soluble solid and: (a) mean hue and (b) mean a*.

## 4. Conclusion

a. Binary area of orange image can be used to estimate changes in weight of orange fruits during storage.
b. None of visual parameters of orange images can be used to estimate the hardness of oranges because the correlation between them is very small.
c. TSS has a fairly strong relationship with the average hue value and the average chroma a* in room temperature storage.

## Acknowledgment

This research was fully funded by Research and Community Service Institution, Jenderal Soedirman University, Purwokerto, Indonesia through a scheme of Riset Institusi Unsoed (RIU) with Contract No P/484/UN23/14/PN/2019.

## References

[1]. Aghistni F. 2008. Model Perpindahan Massa pada Mikrofiltrasi untuk Penghilangan Limonin dan Naringin dari Jus Jeruk Siam (*Citrus nobilis* var microcarpa) [disertasi]. Sekolah Pasca Sarjana. Institut Pertanian Bogor. Bogor.

[2]. Handoko, D. D., Napitupulu, B., dan Sembiring, H. 2000. Penanganan Pasca Panen Buah Jeruk. Balai Pengkajian Teknologi Pertanian Sumatera Utara, Medan.

[3]. Hulme, A.C. 1971. The Biochemistry of Fruit and their product. Vol. I. Academic Press, London.

[4]. Kamran, G., Youcef, G., and Ebrahimzadeh, M. A. 2009. Antioxidant activity, phenol and flavonoid contents of 13 citrus species peels and tissues. Pakistan Journal of Pharmaceutical Sciences. 22 (3): 277-281.

[5]. Kefford, J.F. Chandler, B.V. 1970. The Chemical Constituents of Fruits. Academic Press, New York and London.

[6]. Olabinjo, O.O., Ogunlowo, A.S., Ajayi, O.O., and Olalusi, A.P. 2017. Analysis of Physical and Chemical Composition of Sweet Orange (Citrus sinensis) Peels. International Journal of Environment, Agriculture and Biotechnology (IJEAB). 2(4): 2201-2206.

[7]. Pantastico, E. R. B, 1993. Fisiologi Pasca Panen. UGM-Press, Yogyakarta.

[8]. Pracaya, 2000. Jeruk Manis, Varietas, Budidaya dan Pascapanen. Penebar Swadaya, Jakarta.

[9]. Rafiee, S., Jahromi, M.K., Jafari, A., Sharifi, M., Mirasheh, R., and Mobli, H. 2007. Determining Some Physical Properties of Berganot (citrus medica). International Agrophysics. 21: 293-297.

[10].Saputera, Sutrisno., S. Susanti dan I. W. Budiastra, 2000. Pengkajian Penyimpanan Jeruk Besar (Citrus grandis L) Pengolahan Minimal dengan Kemasan Atmosfer Termodifikasi. Buletin Keteknikan Pertanian.

[11].Simbolon, F. P. M., 2008. Pengaruh Konsentrasi Emulsi dan Lama Penyimpanan Terhadap Mutu Buah Jeruk Manis (Citrus Sinensis, Linn). USU-Press, Medan.

[12].Ting, S.V. dan J.A. Attaway. 1971. Citrus Fruit. in Hulme, A.C. The Chemical Constituents of Fruits.
[13].Academic Press, New York and London.

IOP Conf. Series: Earth and Environmental Science **406** (2019) 012005 doi:10.1088/1755-1315/406/1/012005

# The antioxidant activity of Roselle (Hibiscus *sabdariffa Linn*) phenolic compounds in different variations microwave-assisted extraction time and power

**Ike Sitoresmi M Purbowati[1*], Ali Maksum[1]**
[1]Departement of Agricultural Technology, Agricultural Faculty, University of Jenderal Soedirman, Indonesia
*Corresponding Author: Ikesitoresmi@yahoo.co.id

**Abstract.** The roselle flower bioactive compounds must be extracted out of the cell to facilitate their use. The use of microwave-assisted extraction methods is considered more effective and efficient, but improper use of microwave power and extraction time will actually cause damage to the bioactive compound. *The focus of these study was to evaluate the effect of different* microwave power and extraction time on chemical characteristics and antioxidant activity of roselle flower extract. The research design used in this study was randomized block design (RBD) trial with 2 (two) factors. The first factor is microwave power (100, 175, 250, 325 and 400 watts). Second one is extraction time (1, 3, 5, 7 and 9 minutes). The effect showed that the use of microwave power which give the best results is 325 watts with total anthocyanin 13.8591 mg/100g, vitamin C 110.88 mg/100g,. The best extraction time was 5 minutes with total anthocyanin 13.8381 mg/100g, vitamin C 121.44 mg/100g. The bioactive content is directly proportional to the antioxidant activity produced, namely $IC_{50}$ of 0.515 µg/ml.
*Keywords : Microwave Assisted Extraction (MAE; roselle; antioxidant*

## 1. Introduction

Pollution of the environment in where we live and unhealthy lifestyles such as smoking, drinking alcoholic beverages and physical activity that can stimulate the free radicals cause damage in the body. Increasing the potential of free radicals causes the body must get additional antioxidants from the outside. However, synthetic antioxidant ingredients are not recommended for consumption because it can cause cancer effects in the long term. There are fears of the possibility of side effects from synthetic antioxidants causing natural antioxidants to become alternatives that are needed [5].

Roselle, one alternative plant that can be used as a source of natural antioxidants. Research conducted by [11] states that roselle extract contains a total anthocyanin bioactive compound of 14.80 ± 0.08 mg / g, with a total phenol of 23.77 ± 0.25 mg / g, vitamin C 10.74 ± 0.14 mg / g, IC50202.47 µL / mL, and microbial activity on *Staphilococcus aureus* 11.6 ± 0.3 and *Escherichia coli* 12.6 ± 0.6 mm. The content of these bioactive materials makes roselle has the ability as antioxidant and antibacterial agent.

The content of antioxidant compounds in roselle needs to be extracted out of the cell to facilitate its utilization. Several extraction techniques have been used to extract bioactive components. Extraction methods that have long been developed are conventional extraction methods such as maceration and soxhlet method. These techniques usually require more extraction time. so that it runs the risk of severe thermal degradation for most bioactive components such as antioxidants [7]. According to [1,12]. Conventional extraction methods

IOP Conf. Series: Earth and Environmental Science **406** (2019) 012005 doi:10.1088/1755-1315/406/1/012005

can spend 8, 16, 24 hours or more in an extraction process. In addition, conventional extraction techniques also require sufficient organic solvents and thermal energy to be considered less efficient. More innovative extraction methods are currently being developed such as ultrasound-assisted extraction, enzyme-assisted extraction and microwave-assisted extraction (MAE). Of the several methods, MAE is a method that is considered more environmentally friendly and effective. In addition, when compared with conventional methods, MAE has several advantages, namely requiring shorter extraction time and lower temperature compared to conventional extraction, thereby reducing the decrease in thermolabile components (Li et al., 2017). MAE is also suitable for extraction using water as a solvent because water has a dielectric coefficient of 78.3 so that it can absorb microwaves well. In addition, the advantage of water as a solvent for extraction is its abundance is large, non-toxic and will minimize the use of organic solvents [10,15,16].

The purpose of this study was to determine: 1) the effect of variations in microwave power on the chemical and physical characteristics of the extracts of roselle flower petals produced; 2) the effect of variations in the extraction duration of roselle calyx on chemical characteristics and physical extracts produced; 3) the interaction of microwave power variations and extraction time to the antioxidant activity of the extract produced.

## 2. Materials and Methods

This research was conducted in January-May 2018 in the Laboratory of Agricultural Technology, Study Program of Food Science and Technology, Department of Agricultural Technology, Faculty of Agriculture, Jenderal Soedirman University.

The materials used in this study were red roselle calyx (Hibiscus sabdariffa) by drying for 7 hours using a cabinet dryer obtained from Kediri and Bumi Herbal Dago (Bandung), aquadest, ethanol pa, acidic ethanol, 1% starch, iodine solution 0.01 N. The tools used in this study are microwave (electron), erlemeyer (pyrex), measuring cup (pyrex), digital scales (AND), analytical scales, blenders, sieves, UV-VIS spectrophotometers (Shimadzu), cuvettes, test tubes (pyrex), spatulas, pp plastic, filter paper, funnels, aluminum foil, fillers, 10 mL and 100 mL volumetric flasks, brown glass bottles, and measuring pipettes.

**Extraction:** The dried roselle calyxes were ground for 1 minute using grinder and sieved in 60 mesh. The extraction with various concentrations of ethanol (50%, 60%, 70%, 80% and 90%), microwave power extraction (100, 175, 250, 325, and 400 W) and time of extraction (1, 3,5,7, 9, and 11 minutes) were conducted for each parameter using non factorial design. The extraction was done in the ratio dried roselle and solvent 1:10 w/v (wet weight), 10 g dried roselle powder, in 100 mL solvent. The slurry was radiated in microwave oven at regular intervals (one minute radiation and two minutes off) to keep the temperature not rise above the boiling point [16]. Roselle extract was filtered and concentrated with vacuum evaporator at 70°C, 44 cmHg and blowing with $N_2$ to ensure the solvent totally evaporate. The main response parameter that analyzed was total phenols.

### Anthocyanin (Fuleki et al., 1968)

A total of 1 mL sample was added with 1 mL of acidic ethanol (95% ethanol +1.5 N HCl (85:15 v / v)) Stored overnight at 40 ° C. Then diluted to 10 mL and centrifuged 5 minutes 1000 rpm. Measured absorbance at 535 nm.

### Antioxidant activity (Al Hashimi, 2012)

0,6 mL of sample was dissolved in 0.12 mL of 98% ethanol and 2.88 mL of a 2.51% linoleic acid and 9 mL of a 40 mM phosphate buffer (pH 7.0) were added. The mixture was incubating at 40°C in a test tube in the dark for 3 days (72 hours). After incubation, a 0.1 mL was taken from the mixture and diluted with 9.7 mL of 75% ethanol, followed by the addition of 0.1 mL of 30% ammonium Thiocyanate. Precisely three minutes after adding the 0.1 mL of 20 mM Ferrous chloride in 3.5% hydrocloride acid, the absorbance of the red color was measured at 500 nm. The level of lipid peroxidation inhibition by each fraction was calculated from the absorbance ratio to that of a blank without any sample. A half lipid peroxidation inhibition expressed with $IC_{50}$.

The design used was a randomized block design (RCBD) with two factors. The factors used were microwave power (100, 175, 250, 325 and 400 watts) and extraction time (1, 3, 5, 7 and 9 minutes). The factors were arranged factorial so that 25 treatment combinations were obtained. Each treatment was repeated 2 times to obtain 50 experimental units. Total anthocyanin data and vitamin C of dried roselle flower extracts were analyzed using the F test (variance test) at a confidence level of 95% ($\alpha$ = 5%) with ANOVA and if there was a significant effect, then continued with the DMRT test (Duncan Multiple Range Test) at a 95% confidence level. Data on antioxidant activity were analyzed using linear regression.

## 3. Results and Discussion

### Effect of variations in microwave power used on the total anthocyanin produced

The results showed that variations in microwave power (P) and extraction time (T) significantly affected the total content of anthocyanin.

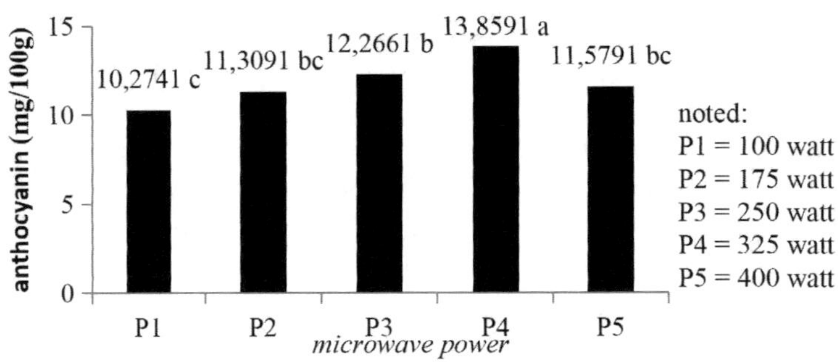


Figure 1. Total anthocyanin in microwave power variations.

Figure 1 shows that extracted anthocyanin levels tend to increase from 100 watts of power and then reach the maximum point at 325 watts and decrease in subsequent microwave power variations. [2] state that when the microwave power increases, the extraction yield will increase. According to Chen et al (2008), the higher microwave power resulted the higher effect of microwave energy on biomolecules by ionic conduction and dipole rotation which results in power dissipated inside the solvent and plant material. This condition can generate molecular movement and heating. Linear to some extent before it begins to decrease or become insignificant. This is caused by the degradation or decomposition of heat-sensitive compounds (thermolabile). It is suspected that at 325 watts of power it is able to damage the cell walls of the roselle calyx so that the addition of power does not increase the amount of anthocyanin extracted , it will reduce the amount of anthocyanin compounds extracted because the higher exposure to microwaves can accelerate the rise in temperature which can damage the anthocyanin compounds. In addition, high exposure to microwaves causes reduced water as a solvent. The reduction of water as a solvent causes the bioactive compounds will not be maximally taken up in the solvent [4]. This is similar to the study of

[9] who reported that it requires 300 watts of power to produce Rosa-damascena flower extracts with the highest total anthocyanin.

**Effect of variation of extraction time used on total anthocyanin produced**

Figure 2 is a graph that shows the results of various analysis of the variation of extraction time variations on total anthocyanin.

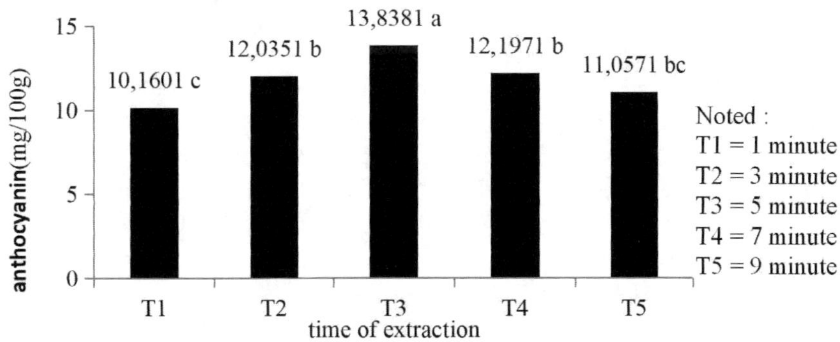


Figure 2. Total anthocyanin in the extraction time variation.

Figure 2 shows that the total anthocyanin extract increased from the extraction time of 1 minute to 5 minutes. This is presumably because the longer the material is exposed to microwave radiation, it causes the rupture of the tissue of the material so that it will release the solute into the solvent [17]. This result also similar with P[15] that the yield of phenols is increasing until 5 minutes of extraction time. However, the use of time over 5 minutes will reduce the total anthocyanin produced. This is due to the longer extraction causes the resulting temperature to be higher, causing damage to anthocyanin compounds. In addition, the extraction time that exceeds the maximum time will cause the solution to become saturated and the anthocyanin decreases, so that the change in time will not give a real change in concentration [17].

[6] reported that it took 10 minutes to extract anthocyanin compounds from mangosteen peel waste. This could be due to [6] research using only 160 watts of power, allegedly the difference in power usage was also quite large. It will make a big difference in the extraction time. The results of research conducted by [9] showed that it only takes 5 minutes to extract the anthocyanin compound extract of Rosa-Damascena flower to the maximum with a power of 300 watts, the use of extraction time of more than 15 minutes will cause a decrease and a stable amount stable of the compound extracted anthocyanin.

**The effect of variations in microwave power used on vitamin C produced**

Figure 3 showed the results of the analysis of the influence of microwave power on vitamin C produced.

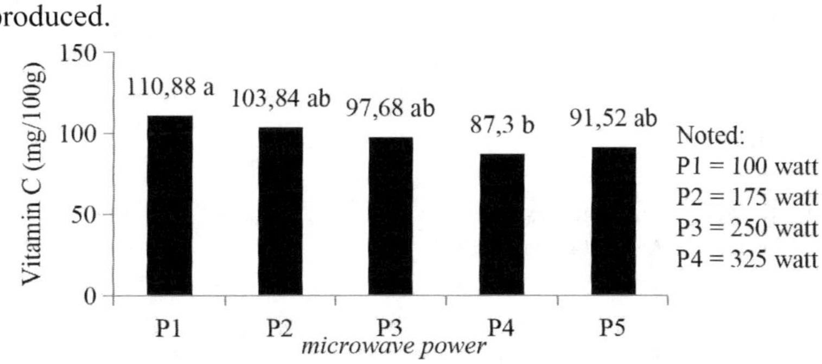


Figure 3. Vitamin C on microwave power variations.

Figure 3 vitamin C content produced in microwave power of 100 watts produces the highest vitamin C when compared to other microwave power variations of 110, 88 mg / 100g. This can be caused because vitamin C has thermolabile characteristic especially in a state of dissolution so that the greater the power used, vitamin C will be more degraded due to rising temperatures. According to [18] the use of microwave power that exceeds 160 watts will cause the bioactive component to decrease especially vitamin C in the extraction of Clinachantus nutans plants. This is supported by the statement of [18] who extracted vitamin C from guava fruit by comparing the Nitrogen-protected microwave-assisted extraction (NPMAE) method to the usual MAE method, stated that in addition to rising temperatures and light factors, a decrease in vitamin C could also occur due to oxidation reactions due to extraction process that does not use nitrogen protection (closed system) so that the addition of power will accelerate the rate of oxidation. This is caused because vitamin C is very sensitive to changes in temperature, light, oxygen and pH.

**Effect of variations in extraction time on vitamin C produced**

Figure 4 is the result of the analysis of the effect of extraction time various on vitamin C produced, showing that the variation of extraction time produced the highest vitamin C content of 121.44 mg / 100 g using extraction variations for 1 minute.

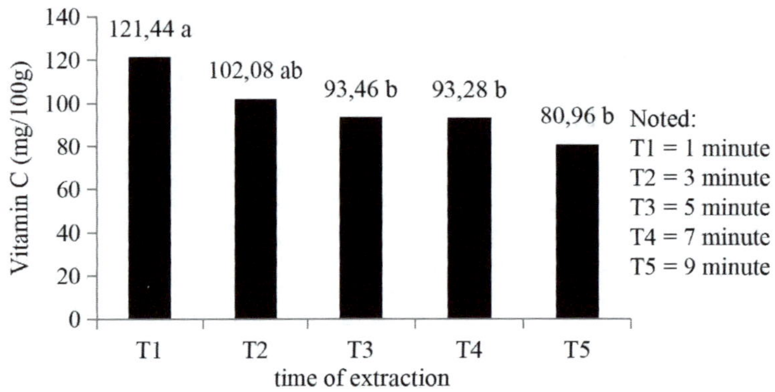


Figure 4. Vitamin C on the extraction time variation.

[18] reported that extracting the bioactive component from the Clinachantus nutans plant using an optimal power of 160 watts requires a maximum extraction time of 60 seconds (1 minute). The use of extraction time that exceeds 1 minute will reduce the yield of extracted bioactive components. This is due to the longer extraction time will increase the microwave energy which causes high evaporation of the substrate so that vitamin C which dissolves easily in water will also evaporate. Different bioactive substances show different tendencies because the properties of substances are determined by their structure. The length of time of microwave radiation will damage the molecular structure of the substance thereby reducing the extraction yield.

**Antioxidant Activity**

Antioxidant activity expressed with $IC_{50}$ value. $IC_{50}$ is the concentration can inhibit 50% of free radicals. It means that the lower IC50 value, related to higher antioxidant activity. Figure 4 shows the results of data analysis of antioxidant activity using linear regression in each combination factors. These showed that the treatment of variations in power of 325 watts with 5 minutes extraction time has the highest antioxidant activity is along with the low $IC_{50}$ value, while the lowest antioxidant activity is the extraction result using 400 watts of power with 9 minutes extraction time.

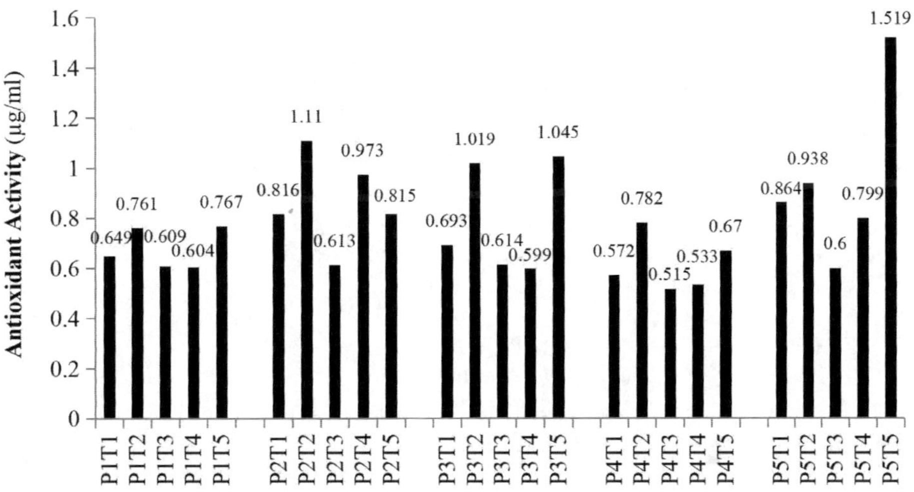


Noted:
P1T1: 100 watts of power for 1 minute
P1T2: 100 watts of power for 3 minutes
P1T3: 100 watts of power for 5 minutes
P1T4: 100 watts of power for 7 minutes
P1T5: 100 watts of power for 9 minutes
P2T1: 175 watts of power for 1 minute
P2T2: 175 watts of power for 3 minutes
P2T3: 175 watts of power for 5 minutes
P2T4: 175 watts of power for 7 minutes
P2T5: 175 watts of power for 9 minutes
P3T1: 250 watts of power for 1 minute
P3T2: 250 watts of power for 3 minutes
P3T3: 250 watts of power for 5 minutes
P3T4: 250 watts of power for 7 minutes
P3T5: 250 watts of power for 9 minutes
P4T1: 325 watts of power for 1 minute
P4T2: 325 watts of power for 3 minutes
P4T3: 325 watts of power for 5 minutes
P4T4: 325 watts of power for 7 minutes
P4T5: 325 watts of power for 9 minutes
P5T1: 400 watts of power for 1 minute
P5T2: 400 watts of power for 3 minutes
P5T3: 400 watts of power for 5 minutes
P5T4: 400 watts of power for 7 minutes
P5T5: 400 watts of power for 9 minutes

Figure 5. Comparison of IC50 microwave power and extraction time of roselle calyx.

Figure 5 indicates that the antioxidant activity is higher when the antioxidant compounds contained in roselle calyx are extracted using a variety of microwave power and the appropriate extraction time. It is suspected that the antioxidant activity of each extract correlates with the content of vitamin C and anthocyanin. Vitamin C has a group free hydroxy which acts as a free radical scavenger and if it has a polyhydroxy group will increase antioxidant activity [13]. Anthocyanin compounds also greatly affect the antioxidant activity of extracts. This is supported by the statement of [3] which states that in several studies mentioned that the antioxidant activity of anthocyanins is 2-6 times greater than other common antioxidants such as ascorbic acid or better known as vitamin C. [8] reported in his research that pure extracts of the anthocyanin compound roselle have excellent antioxidant activity. Anthocyanin in the form of alcohol will reduce antioxidant activity while in the form of diglucoside will increase antioxidant activity. This is consistent with research conducted by [14] that high bioactive compounds such as total phenols, anthocyanins and vitamin C result in high antioxidant activity.

**4. Conclusion**

1. Variations in microwave power have a significant influence on the chemical and physical characteristics of roselle calyx extract. The best microwave power is at 325 watts of power, producing a total anthocyanin of 13.8591mg / 100g, vitamin C 110.88 mg / 100g.
2. Variation of extraction time has a significant influence on the chemical and physical characteristics of roselle calyx extract. The best extraction time for 5 minutes produced a total anthocyanin of 13.8381 mg / 100g, vitamin C of 121.44 mg / 100g.
3. Variations in microwave power and extraction time affect the antioxidant activity of roselle calyx extract. The best antioxidant activity is found in the combination of microwave power of 325 watts with extraction time of 5 minutes, which produces an IC50 of 0.515 μg / ml.

**REFERENCE**

[1]. Afoakwah, A.N., J. Owusu, C. Adomako, and E. Teye. 2012. Microwaved Assisted Extraction (MAE) of Antioxidant Constituents in Plant Materials. *Global Journal of Bio-Science & Biotechnology*, 1(2): 132-140.

[2]. Alias, N.H and Z. Abbas. 2017. Microwave-assisted Extraction of Phenolic Compound from Pineapple Skins: The Optimum Operating Condition and Comparison with Soxhlet Extraction. *Malaysian Journal of Analytical Sciences*, 21(3): 690-699.

[3]. Anggriani, R., N. Ain, and S. Adnan.2017. Identifikasi Fitokimia dan Karakterisasi Antosianin dari Sabut Kelapa Hijau (*Cocos nucifera* L var varidis).*Jurnal Teknologi Pertanian*, 18(3): 163-172.

[4]. Bintari Y.R. and H. Elyani.2017. Ekstraksi Senyawa Bioaktif dari *Cladophora sp.* Dengan Metode Solvent Free Microwave Assisted Extraction (SFMAE).*Journal of Islamic Medicine Research*, 1(1):1-11.

[5]. Erawati. 2012. Uji Aktivitas Antioksidan Ekstrak Daun *Garciniadaedalanthera* Pierre dengan Metode DPPH (1,1-Difenil Pikrilhidrasil) dan Identifikasi Golongan Senyawa Kimia dari Fraksi Paling Aktif. *Skripsi.*Fakultas Matematika dan Ilmu Pengetahuan Alam, Universitas Indonesia, Depok.

[6]. Farida, R. and F.C. Nisa. 2015. Ekstraksi Antosianin Limbah Kulit Manggis Metode Microwave Assisted Extraction (Lama Ekstraksi dan Rasio Bahan : Pelarut. *Jurnal Pangan dan Argoindustri*, 3(2): 362-373.

[7]. Gharekhani, M., M. Ghorbani, and N. Rasoulnejad. 2012. Microwaved-Assisted Extraction of Phenolic and Flavonoid Compounds from *Eucalyptus camaldulensis Dehn* Leaves as Compared with Ultrasound-Assisted Extraction. *Latin American Applied Research*, 42: 305-310.

[8]. Kouakou, T.H., N.G. Konkon, A.P. Obouayeba, Z.H. Abeda, and M. Kone. 2015. Anthocyanin Production in Calyxand Callus of Roselle *(Hibiscus sabdariffa Linn)* and Its Impact on Antioxidant Activity. *Journal of Pharmacognosy and Phytochemistry*, 4(3): 9-15.

[9]. Lasunon, C., P. Ponpesh, and C. Satirapipathkul. 2018. Pigment Extraction from Flower using Microwave Assisted Extraction. *International Journal of Science, Engineering and Management* (IJSEM), 3(5): 37-39.

[10]. Li, Y., S. Li, S.J. Li, J.J. Zhang, C.N. Zhao and H.B. Li. 2017. Microwaved-Assisted Extraction of Natural Antioxidants from the Exotic Gordonia axillaris Fruit: Optimization and Identification of Phenolic Compounds. *Molecules Article,* 22(1481): 1-16. (*On-line*).http://www.mdpi.com/1420-3049/22/9/1481/html diakses 1 Mei 2018.

[11]. Maksum, A. and I.S.M. Purbowati.2017. Optimasi Ekstraksi Senyawa Fenolik dari Kelopak Bunga Roselle (*Hibiscus sabdariffa*) Berbantu Gelombang Mikro. Agrin, 21(2): 91-104.

[12]. Mandal, V., Y. Mohan and S. Hemalatha. 2007. Microwave Assisted Extraction – An Innovative and Promising Extraction Tool for Medicinal Plant Research. *Pharmacognosy Reviews*, 1(1): 7-18.

[13]. Nadia, S., Riyanti, and R. Nirmala. 2016. Uji Aktivitas Antioksidan Kombinasi dari Kulit Buah Naga (*Hylocereus costaricensis*) dan Bunga Roselle (*Hibiscus sabdariffa* L.) dengan Metode DPPH (*1,1-Dyphenil 2-pycrylhidrazyl*) beserta Bentuk Tunggalnya. *Jurnal Kesmadaska*, 94-99.

[14]. Purbowati, I.S.M. 2014. Nanoenkapsulasi ekstrak kelopak bunga roselle (*Hibiscus sabdariffa* L) hasil optimasi ekstraksi berbantu gelombang sebagai bahanantibakteri dan antioksidan. *Disertasi.*Sekolah Pascasarjana. IPB. Bogor.

[15]. Purbowati, I. S. M., K. Syamsu, E. Warsiki, and H. Sri. 2016. Stabilitas Senyawa Fenolik dalam Ekstrak dan Nanokapsul Kelopak Bunga Rosella pada Berbagai Variasi pH, Suhu dan Waktu. *Agrointek*. Vol. 10, No. 1: 31-40.

[16]. Purbowati, I.S.M., Sujiman and Maksum, A. 2018. Antibakterial activity of roselle (Hibiscus sabdariffa) extract phenolics compound produced with variying drying methods and duration. *Jurnal Teknologi Industri* 28 (1):19-27

[17]. Putri, A.R.W. and F.C. Nisa.2015. Ekstraksi Antosianin dari Bunga Mawar Merah (Rosa damascene Mill) Sortiran Metode Microwave Assisted Extraction. *Jurnal Pangan dan Agroindustri*, 3(2): 701-732.

[18]. Yu, Y., B. Chen, Y. Chen, M. Xie, H. Duan, Y. Li, and G. Duan. 2015. Nitrogen-protected Microwave-assisted Extraction of Ascorbic Acid from Fruit and Vegetables. *Journal Sep. Sciences*, 32: 4227-4233.

IOP Conf. Series: Earth and Environmental Science **406** (2019) 012006 doi:10.1088/1755-1315/406/1/012006

# The Effect Of Kelor Leaves (Moringa oleifera) Ethanol Extract On Serum Uric Acid And Tumor Necrosis Factor-α Of Hyperuricemic White Rats (Rattus norvegicus)

**Fajar Wahyu Pribadi[1], Catharina Widiartini[2]**

[1,2]Faculty of Medicine, Jenderal Soedirman University, Purwokerto, Central Java, Indonesia

fajarwahyupribadi@yahoo.com, widiartini.catharina@gmail.com

**Abstract.** Chronic hyperuricemia leads to various complications due to the inflammation and oxidative processes. However, the current chemical medications have caused serious side effects. Ethanol extract of Kelor (Moringga oleifera) leaves contains various active substances potential for decreasing serum uric acid and inflammatory cytokine el. The aim of this experiment is to study the effect of Kelor (Moringga oleifera) leaves ethanol extract to serum uric acid and TNF-α level. The Method used is experimental study with post test only with control group design. Twenty-five white rats were randomly assigned into 5 groups: A (healthy control), B (diseased control; hyperuricaemic), C,D,E (hyperuricemic, given ethanol extract 300,600,1200 mg/KgBW/day). Hyperuricemia inductioan was conducted by administration of 20 g raw cow brain/rat/day for 10 days and was continued until day 17. The result is the average of uric acid (mg/dL) in group A: 4,18±0,55, B: 15,02±0,71, C: 8,46±0,12, D: 6,55±0,14, E: 6,45±0,13. The average of TNF-α (mg/dL) in group A: 29,59±10,67, B: 122,58±40,45, C: 108,80±19,89, D: 60,54±8,40, E: 36,12±8,07. Data analysis used One Way ANOVA, followed by post hoc LSD for uric acid and Kruskall-Wallis, followed by Mann-Whitney tests for TNF-α reveal significant differences among groups. The conclucion of this study is the administration of Kelor (Moringga oleifera) leaves ethanol extract in various doses significantly decrease serum uric acid and TNF-α level of hyperuricemic model white rats.

**Keywords:** moringa oleifera; leaves, uric acid; TNF-α; hyperuricaemic

## 1. INTRODUCTION

Hyperuricemia is a condition when there is high level of serum uric acid, i.e more than 7,0 mg/dL in male and more than 5,7 mg/dL in female (Soeroso, 2011). One of the causes of hyperuricemia is daily protein intake found in meat which is high in purine. High purine intake causes an increased serum uric acid level as its end metabolic result (Pribadi & Ernawati, 2010, Manampiring, 2011). Chronic hyperuricemia may lead to the development of serious health problems such as gout arthritis and cardiovascular diseases (Liu *et al*., 2011). As reported in Basic Health Research 2012, the prevalence of hyperuricemia in Indonesia was 11,9 % (Depkes RI, 2013), whereas in Central Java was 24,3% in male and 11,7% in female aged 15-45 year old (Purwaningsih, 2010). Hyperuricemia prevalence is variable based on age and starts to increase from age 30 y.o. in male and 50 y.o. in female. The higher prevalence (17-28%) were reported in studies conducted in hospitals due to the underlying diseases and side effects of medications used (Liu *et al*., 2011).

Uric acid is one of inflammation agents that can stimulate macrophage to releases pro-inflammatory cytokines, such as tumor necrosis factor-α (TNF-α) (Putra, 2009). Billiet *et al.* (2014) reported that in male and female, serum uric acid level positively correlates with CRP, IL-6 dan TNF-α level. Inflammation process releases reactive oxygen species (ROS) that causes oxidative stress (Zhou *et al.*, 2018).

Anti-inflammatory medications and antioxidants are potential for decreasing TNF-α level. However, the current medications may cause serious side effects (Katzung *et al.*, 2012). Kelor plant (*Moringa oleifera*) has been widely known as one of vegetables and herbal medicine plants in many places in Indonesia. Rahmawati (2015) proved that daily administration of 3,75 g/kgBW kelor leaves brewed in 3,6 ml warm water for 14 days decreased serum uric acid level from 2,463 mg/dL to 1,788 mg/dL.However, the effective dose of Kelor leaves ethanol extract to decrease serum uric acid level in hyperuricemia has not been established yet.

Kelor leaves contain various active substances that are potential as antioxidant (i.e flavonoid, vitamin C and vitamin E), as hepatoprotector, immunomudulator, and as anti inflammation agent (Witt, 2013). The actve substances may decrease serum uric acid and TNF-α level through inhibition of uric acid synthesis, oxidative stress and inflammation (Giancarlo *et al.*, 2010). It was hypothesized that the administration of Kelor leaves ethanol extract decrease serum uric acid and TNF-α level of hyperuricaemic white rats (*Rattus norvegicus*).

## 2. METHOD

The experimental method used completely randomized design with post test only with control group approach. Twenty five male wistar strain white rats (*Rattus norvegicus*) aged 2-3 months, weighted 150-200 g, healthy, active and have no observable physical anomaly, gained from MS. Bram Mouse Farm, Purbalingga, Central Java. The rats were randomly assigned into 5 groups: A (healthy control), B (diseased control, hyperuricaemic), C,D,E (hyperuricaemic, given ethanol extract 300,600,1200 mg/KgBW/day). Hyperuricemia induction was conducted by administration of 20 g raw cow brain/rat/day for 10 days and was continued until day 17.

The independent variable was various ethanol extract dosage (300,500, 1200 mg/kgBW/day). The dependent variable was serum uric acid and TNF-α level. Kelor leaves powder was bought from *Pusat Informasi dan Pengembangan Tanaman Kelor Indonesia* (The Indonesian Center for Information and Development of Kelor Plant), non governmental organization: Media Peduli Lingkungan, Blora, Central Java. Masseration technique with 96% ethanol to gain ethanol extract was conducted in Faculty of Pharmacy, Jenderal Soedirman University. Uric acid level (in mg/dL) was measured using spectrophotometer from 5 mL blood sample taken from the orbital plexus. TNF-α level was measured using ELISA. The absorbance read on 492 nm wave length (Kusumastuty *et al.*, 2015). Data analysis used One Way ANOVA, followed by post hoc LSD for uric acid and Kruskall-Wallis, followed by Mann-Whitney tests for TNF-α. Differences were considered as significant when $p<0.05$.

## 3. RESULT AND DISCUSSION

The data of uric acid and TNF-α level (presented as mean ± standard deviation) are as shown in Figure 1 and 2. As seen, the pattern of uric acid and TNF-α level result were almost similar. The least level was found in normal group (A), significantly lower from other groups. This showed the success of induction process through daily intake of cow brain which is high in purine (100-1000 mg purine /100 gram food) (Lestari *et al.*, 2014)

Previous study by Pribadi & Ernawati (2010) reported that the administration of 20g/rat/day of cow brain for 8 days caused significant increase in serum uric acid level until reached hyperuricemic state. The result also showed that the induction of hyperuricemia have caused increase in serum TNF-α level. This was in accordance with Zhou *et al.*'s (2012) finding that there was an increase in TNF-α level in hyperuricemia white rats.

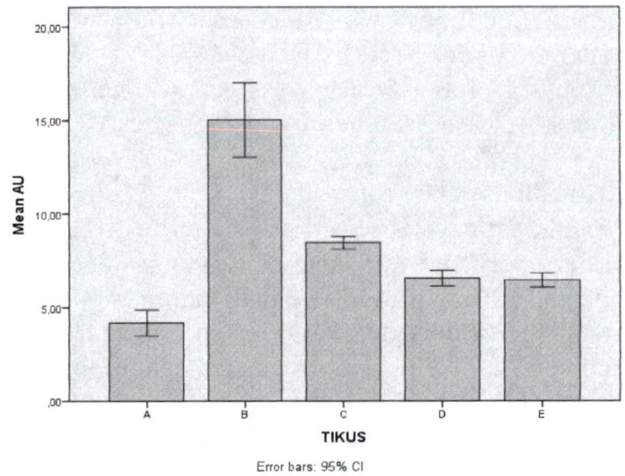


**Figure 1.** Serum Uric acid level average
Note: groups: A=4,18±0,55 (healthy control), B=15,02±1,59 (diseased control, hyperuricaemic), C=8,46±0,12, D=6,55±0,14, E=6,45±0,13 (hyperuricaemic, given ethanol extract 300,600,1200 mg/KgBW/day).

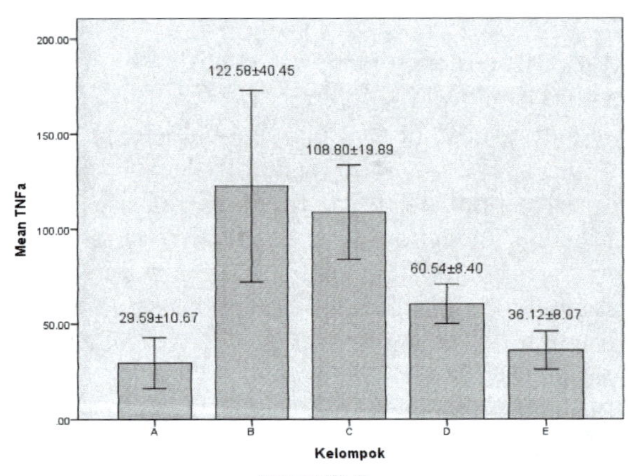


**Figure 2.** Serum TNF-α level average
Note: groups: A (healthy control), B (diseased control, hyperuricaemic), C,D,E (hyperuricaemic, given ethanol extract 300,600,1200 mg/KgBW/day).

**Tabel 1.** *One Way* ANOVA Result of Serum Uric Acid Level

| | N | Mean±Std. Deviasi | p *value* |
|---|---|---|---|
| Group A | 5 | 4,18 ±0,55 mg/dL | |
| Group B | 5 | 15,02 ± 0,71 mg/dL | |
| Group C | 5 | 8,46 ±0,12 mg/dL | 0,000 |
| Group D | 5 | 6,55 ±0,14 mg/dL | |
| Group E | 5 | 6,45 ±0,13 mg/dL | |

Note: groups: A (healthy control), B (diseased control, hyperuricaemic), C,D,E (hyperuricaemic, given ethanol extract 300,600,1200 mg/KgBW/day).

**Tabel 2.** *Post hoc LSD* Result of Serum Uric Acid Level

| Group | | *P* |
|---|---|---|
| Group A | Group B | .000 |
| | Group C | .000 |
| | Group D | .000 |
| | Group E | .000 |
| Group B | Group A | .000 |
| | Group C | .000 |
| | Group D | .000 |
| | Group E | .000 |
| Group C | Group A | .000 |
| | Group B | .000 |
| | Group D | .000 |
| | Group E | .000 |
| Group D | Group A | .000 |
| | Group B | .000 |
| | Group C | .000 |
| | Group E | .777 |
| Group E | Group A | .000 |
| | Group B | .000 |
| | Group C | .000 |
| | Group D | .777 |

*significancy at p<0,05

Note: groups: A (healthy control), B (diseased control, hyperuricaemic), C,D,E (hyperuricaemic, given ethanol extract 300,600,1200 mg/KgBW/day).

**Tabel 3**. *Kruskal-Wallis* Result of Serum TNF-α Level

| | **N** | **Median (min-max)** | **Mean±Std. Deviasi** | **p *value*** |
|---|---|---|---|---|
| **Group A** | 5 | 30,35 (13,58-43,64) | 29,59±10,67 mg/dL | |
| **Group B** | 5 | 146,35 (58,60-151,89) | 122,58±40,45 mg/dL | |
| **Group C** | 5 | 115,83 (85,50-127,69) | 108,80±19,89 mg/dL | 0,001 |
| **Group D** | 5 | 62,21 (47,03-68,04) | 60,54±8,40 mg/dL | |
| **Group E** | 5 | 33,43 (29,88-50,05) | 36,12±8,07 mg/dL | |

Note: groups: A (healthy control), B (diseased control, hyperuricaemic), C,D,E (hyperuricaemic, given ethanol extract 300,600,1200 mg/KgBW/day).

**Tabel 4** *Mann-Whitney* **Result of Serum** TNF-α **Level**

| *Group* | | *P* |
|---|---|---|
| **Group A** | **Group B** | **.009** |
| | **Group C** | **.009** |
| | **Group D** | **.009** |
| | Group E | .175 |
| **Group B** | **Group A** | **.009** |
| | Group C | .347 |
| | Group D | .076 |
| | **Group E** | **.009** |
| **Group C** | **Group A** | **.009** |
| | Group B | .347 |
| | **Group D** | **.009** |
| | **Group E** | **.009** |
| **Group D** | **Group A** | **.009** |
| | Group B | .076 |
| | **Group C** | **.009** |
| | Group E | .016 |
| **Group E** | Group A | .175 |
| | **Group B** | **.009** |
| | **Group C** | **.009** |
| | Group D | .016 |

*Significancy at p<0,05

Note: groups: A (healthy control), B (diseased control, hyperuricaemic), C,D,E (hyperuricaemic, given ethanol extract 300,600,1200 mg/KgBW/day).

As shown in Figure 1, the significant decrease in group C and D was not followed by group E. The result of One Way ANOVA (Table 1), followed by Post hoc LSD (Table 2) found significant differences (p=0,000) among groups, except between group D and E (p=0,777). This could happen because dose-response of active substance is dependent on the dosage.The increase in dosage causes receptor saturation as seen as the decrease in the effect observed (Katzung, *et al.*, 2012).

Nevertheless, as shown in Figure 2, the decrease in TNF-α level is directly proportional to the increase in ethanol extract dosage, i.e the higher the dosage of ethanol extract, the greater the decrease in TNF-α level. The same as the result of Kruskal-Wallis (Table 3),followed by Mann-Whitney test (Table 4) showed the significant difference (p=0,009) between group A and group B/C/D, except between group A and E (p=0,175).

### *3.1. OXIDATIVE STRESS AND INFLAMMATION IN HYPERURICEMIA*

Uric acid is one of inflammatory agents. Increase in serum uric acid over 6,8 g/dL leads to the formation and deposition of uric acid crystal called monosodium urate monohidrate. The crystals then fagosited by macrophage, thus triggers inflammation response. The stimulated macrophage releases pro-inflammatory cytokines such as interleukin 1β (IL-1β), interleukin-6 (IL-6), interleukin 8 (IL-8) and tumor necrosis factor-α (TNF-α) (Putra, 2009). Billiet *et al.* (2014) found that in male and female, serum uric acid level positively correlates with CRP, IL-6 dan TNF-α level.

TNF-α is a pleiotropic proinflammation cytokine for host defense against viral, bacterial and parasitic infection. It is released by macrophage, activated by lymphosite T, NK cell, antigen and Mast cell, thus it is frequently found in serum examination in inflammation and infection (Abbas *et al.*, 2012). TNF-α altogether with IL-1 work after released by peripheral serum monosite and trigger the expression of endothelial E-selectin, intercelluler adhesion molecule 1 (ICAM-1) dan vascular cell adhesion molecule 1 (VCAM-1). The group of immunoglobuline cause leukocyte attraction to the urate crystal deposition region (Ngestiningsih *et al.*, 2012). The progression of inflammation process also produces reactive oxygen species (ROS) that builds up and causes oxidative stress (Zhou *et al.*, 2018).

### *3.2. THE ROLE OF KELOR LEAVES*

Kelor (*Moringa oleifera*) leaves contain various active substances that potential for reducing the level of serum uric acid through several mechanisms. Inhibited processes are the synthesis of uric acid, stress oxidative and inflammation.

#### *3.2.1. Inhibition of Uric Acid Synthesis*

Generally, flavonoid substanes (in Kelor leaves consist of quercetine and kaempherol), alcaloid, tannine and saponin decrease serum uric acid level through inhibition of xanthine oxidase (Nijveldt, *et al.*, 2001; Anandagiri *et al.*, 2014). According to Sutrisna *et al.*, (2010), the 50% inhibition of xanthine oxidase causes 50% decrease in serum uric acid. One of common inhibition pathway is through n-hexane. When the n-hexane pathway is disrupted, xanthine oxidase cannot transform xanthine and hypoxanthine into uric acid. Therefore the level of serum uric acid is decreased (Yumita, *et al.*, 2013). Moreover, quercetine inhibit xanthine oxidase by its role as competitive inhibitor that occupy active site of the enzyme because of its shape similarity with the substrate. The transformation from xanthine and hypoxanthine is not happened, therefore the level of serum uric acid is decreased (Kusmiyati, 2008; Song, *et al.*, 2018)

#### *3.2.2. Inhibition of Oxidative Stress Progression*

Inhibition of oxidative stress progression is the result of characteristic molecular structure in flavonoid (quercetine and kaempherol) that maintain structural stability when scavanging free radicals and chelating transition metals. The free radicals scavenging protects cell from damage or inflammation process, thus prevent the increase in serum uric acid level (Hatano, *et al.*, 1990; Tion, *et al.*, 2013). Other uric substances in Kelor leaves with antioxidant potential are alcaloid, tannine, saponine, vitamin C and Vitamin E (Shankar, *et al.*, 2006, Anandagiri *et al.*, 2014)

#### *3.2.3. Inhibition of Inflammation Reaction*

Active substance of Kelor leaves with potential as antiinflammation is quercetine, saponine and vitamin C (Shankar, *et al.*, 2006, Anandagiri *et al.*, 2014). The inhibition of inflammation reaction conducted by transforming the expression and activity of various uric acid transporter. Verzola, *et al.* (2014) showed that in the regulation of uric acid transporter, URAT1 is an important transporter for uric acid to get into the renal proximal tubule cell and induce intracellular oxidative stress. Lipid peroxidation disrupts cell membrane and induces further inflammatory response by the release of pro-inflammatory cytokines such as IL-1, TNF-α, IL-6, IL-8 dan Cox-2. Damaged renal tubule cells cause decrease in uric acid excretion (Praditya *et al.*, 2018).

Nieman (2007, in Choo *et al.*, 2008) showed that the water extract of Kelor leaves give invitro antiinflammatory activity by decreasing TNF-α through the inhibition of Nuclear Factor Kappa B (NF-κB). Quercetine and vitamin C work by decreasing phosphorilation of Iκβ and reducing the activity of NF-κB, which later decreasing the endogenic production of TNF-α (Nair *et al.*, 2006; Carcamo *et al.*, 2002; Choo *et al.*, 2008)

Statistical analysis in this research showed that serum uric acid and TNF-α level in groups given ethanol extract of Kelor leaves (group C,D,E) were significantly lower than group B (diseased control). This proves the effect of various dosage (300, 600, 1200 mg/KgBW) Kelor leaves ethanol extract administration in reducing serum uric acid and TNF-α level. The most significant decrease, nearly reach (insignificantly different from) the normal level was in group E, which was given 1200 mg/kgBW Kelor leaves ethanol extract. Therefore, the dosage was the most effective in reducing uric acid level.

Kelor leaves' flavonoids (quercetine and kaempherol) are called nefroprotector due to their inhibitory ability to oxidative stress and inflammatory reaction process in renal tubule cells (Verzola, *et al.*, 2014). Protection to renal function to excrete uric acid maintains the normal level of serum uric acid (Wang, *et al.*, 2015). Moreover, Vitamin C in Kelor leaves increases the renal excretion of uric acid (Shankar, *et al.*, 2006, Anandagiri *et al.*, 2014)

## 4. CONCLUSION

The administration of Kelor (*Moringa oleifera*) leaves in various dosage were able to significantly decrease the serum uric acid and TNF-α in hypercaemic model white rats. The most significant decrease was foun in group given dosage of 1200 mg/KgBW/day for 7 days.

## 5. SUGGESTION FOR FURTHER STUDY

This study can be replicated with method improvements, such as using more variable dosage and using linear regression in the data analysis. More frequent assessment of uric acid and TNF-α level should be done, i.e: after acclimatization to measure the pre-induction uric acid and TNF-α level, after induction period to make sure the quantity of pre-post test difference in uric acid and TNF-α level. It would be interesting to investigate the specific flavonoid with potential for reducing uric acid and TNF-α level. Finally, it is also needed to study the toxic dose of Kelor (*Rattus oleifera*) leaves.

**Acknowledgement**

The researcher present his sincere appreciation goes to the Dean, the Head of Pharmacology Department Faculty of Medicine and the Head of LPPM Jenderal Soedirman University.

**References**

[1] Abbas, A.K., Lichtmann A.H., Shiv P. 2012. Cellular and Molecular Immunilogy $7^{th}$ed. Philadelphia : WB Saunders Co.

[2] Anandagiri, D A., Manuaba I.B., Suastuti N. G. 2014. Pemanfaatan Teh Kombucha Sebagai Obat Hiperurisemia Melalui Penghambatan Aktifitas Xantin Oksidase pada *Rattus norvegicus*. *Jurnal Kimia*. 8(2): 220-225.

[3] Billiet, L., Doaty S, Katz J.D, Velazquez M.T. 2014. Review of Hyperuricemia as New Marker for Metabolic Syndrome. *ISRN Rheumatology*. 1(1): 1-7

[4] Carcamo, J.M., Alicia P., Oriana B.O., David W.G. 2002. Vitamin C Suppresses TNF-α-Induced NFκB Activation by Inhibiting IκBα Phosphorylation. *American Chemical Society*. 41(43): 12995-13002.

[5] Choo, M.K., Hiroaki S., Dong-Hyun K., Ikuo S. 2008. A Ginseng Saponin Metabolite Suppresses Tumor Necrosis Factor-α Promoted Metastasis by Suppressing Nuclear Factor-κB Signaling in Murine Colone Cancer Cells. *Oncology Reports*. 19(3):595-600.

[6] Depkes RI. 2013. Riset Kesehatan Dasar. Jakarta: Badan Penelitian dan Pengembangan

Kesehatan Kementrian Kesehatan RI.

[7] Giancarlo, A., Kyung-Jin Y, Etsuo N, Robert MR. 2010. eds. *Biomarkers for Antioxidant Defense and Oxidative Damage: Principles and Practical Applications*. 1st ed. USA: Blackwell:10-12.

[8] Hatano T, Yasuhara T, Yoshihara R, Agata I. 1990.Effect of Interaction of Tannina with Co-existing Substance. VII Inhibitory Effect of Tannins and Related Polyphenol of Xanthine Oxidase. *Chemical Pharmaceutical Bulletin;* 38(5): 1224- 1229

[9] Katzung, B.G., Masters S.B., Trevor A.J. 2012. *Farmakologi Dasar dan Klinik edisi 12*. Jakarta : EGC

[10] Kusmiyati, A. 2008. Kadar Asam Urat Serum dan Urin Tikus Putih Hiperurikemia Setelah Pemberian Jus Kentang (Solanum tuberosum L.). *[Skripsi].* Jurusan Biologi Fakultas MIPA UNS, Surakarta.

[11] Kusumastuty, I., Adi P., Harti L.B., Nugroho F.A. 2015. Pengaruh Pemberian Daun Ubi Jalar Ungu (*Ipomoea Batatas Lam*) Terhadap Kadar TNF-α, Il-6 Dan Nf-Kb Pada Tikus Yang Dipapar Asap Rokok. *Jurnal Kedokteran Brawijaya.* 28(3): 228-232

[12] Lestari, E., Maryanto S., Paundrinagari, M.D. 2014. Hubungan Konsumsi Makanan Sumber Purin dengan Kadar Asam Urat pada Wanita Usia 35-59 Tahun di Desa Sanggrahan Kecamatan Kranggan Kabupaten Temanggung. *Skripsi*. Stikes Ngudi Waluyo, Semarang.

[13] Liu, B., Wang T, Zhao HN, Yue WW, Yu HP, Liu CX, *et al.* 2011. The Prevalence of hyperuricemia in China: a Meta-Analysis. *BMC Public Health.* 11:832.

[14] Manampiring, A.E. 2011. Hiperurisemia dan Respon Imun. *Jurnal Biomedik.* 3(2): 102-110

[15] Nair, M.P, Supriya M., Jessica L.R., Ravikumar A., Harikrishnan N., Stanley A.S. *et al.* 2006. The Flavonoid Quercetin Inhibits Proinfllamatory Cytokine (Tumor Necrosis Factor Alpha) Gene Expression in Normal Peripheral Blood Mononuclear Cells via Modulation of the NF-κB System. *Clinical and Vaccine Immunology*. 13(3): 319-328

[16] Ngestiningsih, D., Widiastuti I, Wahyu S.T, Hadi S, Suntoko B. 2012. Perbedaan Pemberian Ekstrak Herbal (Daun Salam, Jintan Hitam dan Daun Seledri) dengan Allopurinol terhadap Kadar IL-6 dan TNF-α Serum Penderita Hiperurisemia. *Medical Hospital.* 1(1): 20-24

[17] Nijveldt RJ, Nood E, Hoorn DEC, Boelens PG, Norren K, Leeuwen PAM. 2001. Flavonoids : a review of probable mechanisms of action and. *American Journal Clinical Nutrition*; 74: 418-425.

[18] Pribadi, F.W., Ernawati D. A. 2010. Efek Catechin Terhadap Kadar Asam Urat, C–Reactive Protein (CRP) Dan Malondialdehid Darah Tikus Putih (*Rattus Norvegicus*) Hiperurisemia. *Mandala of Health.* 4(1): 39

[19] Purwaningsih, T. 2010. Faktor-Faktor Risiko Hiperurisemia pada Studi Kasus di Rumah Sakit Umum Kardinah Kota Tegal. *Tesis*, Universitas Diponegoro, Semarang.

[20] Putra, T. R. 2009. Hiperurisemia. Dalam : *Buku Ajar Penyakit Dalam Jilid 3*. Edisi 5. Jakarta : Interna Publishing 2550-2555.

[21] Rahmawati. 2015. Pengaruh Pemberian Seduhan Daun Kelor (*Moringa oleifera Lamk*) Terhadap Kadar Asam Urat Tikus Putih (*Rattus norvegicus*). *Skripsi*, Universitas Diponegoro, Semarang

[22] Shankar A, Klein R, Klein B.E.K, Nieto F.J. 2006. The association between serum uric acid level and longterm incidence of hypertension : *population-based cohort study*: Vol 20:(12): 937-945.

[23] Soeroso, J., Algristian. H. 2011. Asam Urat. *Penebar Plus*, Jakarta.

[24] Song, S.H., Park, D.H., Bae, M.S., Choi, C.Y., Shim, J.H., Yoon, G., *et al.* 2018. Ethanol Extract of Cudrania tricuspidata Leaf Ameliorates Hyperuricemia in Mice via Inhibition of Hepatic and Serum Xanthine Oxidase Activity. *Evidence-Based Complementary and Alternative Medicine*. Vol : 2018 8037925

[25] Sutrisna, E., Azmy, U. & Wahyuni, A.S. 2010. Efek Ekstrak Etanol Daging Buah Mahkota Dewa *(Phaleria macrocarpa (Scheff.) Boerl.)* Terhadap Penurunan Kadar Asam Urat Pada

Mencit Putih Jantan Yang Diinduksi Potassium Oxonate. *Pharmacon*, 11(2), pp.62-9.

[26] Tion S.H, Looi CY, Hazni H., Arya, A., Paydar, M., Wong.W.F., *et al.* 2013.Antidiabetic and antioxidant properties of alkaloids from Catharanthus roseus (L.) G. Don. *Molecules* (Basel, Switzerland); 18(8): 9770-84.

[27] Verzola D, Ratto E, Villaggio B. 2014. Uric acid promotes apoptosis in human proximal tubule cells by oxidative stress and the activation of NADPH oxidase NOX 4. *PLoS One*. 9:12.

[28] Wang, Y.,Zhang g., Pan J., Gong D. 2015. Novel Insights Into The Inhibitory Mechanism of Kaempferol On Xanthine Oxidase. *Journal of Agricultural and Food Chemistry*. 63(2) : 526–534.

[29] Yumita, A., Asep, G. S., Elin, Y. S. 2013. Xantine Oxidase Inhibitor Activity of some Indonesian Medicinal Plants and Active Fraction of Selected Plants. *International Journal of Pharmacy and Pharmaceutical Science:* Vol. 5 (2).

[30] Witt, K. A. 2013. *The Nutrient Content of Moringa oleifera Leaves*. Messiah College Department of Nutrition and Dietetics.

[31] Zhou, Y., Li F., Lei J., Ping W., Hongdi C., Weichun H., *et al.*. 2012. Uric Acid Induces Renal Inflammation via Activating Tubular NF-κB Signaling Pathway. *Plos One*. 7(6): 1-10.

[32] Zhou, Y., Mingcai Z., Zheyan P., Guoqiang X., Xiangkun L.. 2018. Relation Between Oxidative Stress and Inflammation in Hyperuricemia. *Medicine*. 97(49): 1-8.

# Uncovering Potential of Neem and Pyrethrum Extract as Biolarvacide of Aedes aegypti for Dengue Control

**Firda Yanuar Pradani[1], Mutiara Widawati[1], Siwi Pramatama Mars Wijayanti [2]**

1) Loka Penelitian dan Pengembangan Kesehatan Pangandaran, National Institute Health of Research and Development

2) Public Health Department, Faculty of Health Sciences, Jenderal Soedirman University

Correspondence to: fhierda@gmail.com

**Abstract**. Currently, several problems raised from the frequent application of chemical insecticide for dengue control such as insecticide resistance and environmental health issues. This fact supporting the need for the alternative active ingredients which are natural and could not develop insecticide resistance. This study aims to find out the potential of Pyrethrum (Chrysanthemum cinerariifolium (Trevir.) Vi) and Neem (Azadirachta indica A. Juss) as biolarvacide against Aedes aegypti. This study used a quasi-experimental study using several combinations of Pyrethrum and Neem extracts. The larvae test used 25 larvae for every combination. Three repetitions applied for each treatment. Larval deaths were observed every hour until 24 hours, and several dead larvae were calculated. Results of this study showed that almost all formulations caused death of larvae in the first hour except formulation 5 (2 ml Pyrethrum + 1 ml DMSO + 24 ml liquid paraffin + 24 ml chitin), formulation 8 (49 ml neem + 1 ml Tween 20) and formulation 11 (49 ml neem + 1 ml DMSO). These results emphasised the potential of a formulation containing Pyrethrum and neem as biolarvacide against Aedes aegypti.... (Introduction, methods, results, discussion, conclusion)

**Keywords: dengue, pyrethrum, neem, larvacide**

IOP Conf. Series: Earth and Environmental Science **406** (2019) 012007 doi:10.1088/1755-1315/406/1/012007

## 1. INTRODUCTION

Dengue is a mosquito-borne tropical disease caused by dengue virus (DENV) infection. There are commonly four antigenically distinct serotypes of DENV which are serotype-1, 2, 3, 4[1]. In October 2013, the fifth serotype (DENV-5) was found and caused new challenges in dengue control [2]. Dengue usually spread in tropical areas, due to the breeding site of its vector which are Aedes aegypti as primary and Aedes albopictus as secondary vector. This disease is transmitted by the female Aedes sp. mosquito bites which containing DENV [3]. Currently, when dengue vaccine is available, yet dengue prevention and control are still relying on vector control and avoidance from mosquito bites. Several problems raised from frequent application of chemical insecticide for dengue control, such as insecticide resistance and environmental health issues [4][5]. Several previous studies showed the insecticide resistance occurred around the world to several types of mosquitoes insecticide. [6][7] This fact supporting the need for natural alternative active ingredients which could not develop insecticide resistance. Several plants are known to have active ingredients to be used as larvacide, such as Pyrethrum (Chrysanthemum cineraraefolium (Trevir.) Vi) and Neem (Azadirachta indica A. Juss). Both plants contain active ingredients of nerve poisons such as pyrethrin and azadirachtin and have the potential to kill insects [8].

Indonesia has biodiversity sources, including plants that have active ingredients as larvacides. Flowers of Pyrethrum (Chrysanthemum cinerariaefolium (Trevir.) Vis) contain pyrethrin active compounds which could act as insecticides. Several active compounds contained in C. cinerariaefolium are pyrethroids, sesquiterpenoids, flavonoids, coumarins, triterpenoids, steroids, phenolics, purines, lipids, aliphatic compounds and monoterpenoid [9]. Pyrethrum extract is one of the essential natural insecticides extracted from Pyrethrum plants. The biologically active constituents of Pyrethrum extract are the pyrethrins, also known as 'rethrins'. Previous research confirmed that pyrethrin and its derivatives are extensively used for insecticidal activity [10][11].

Another plant that has the potential to be a natural insecticide is Neem (Azadirachta indica A. Juss). A indica is a fast-growing, long live tree, 12 meters in height with unpleasant smelling wood. It has evergreen pinnate leaves and small fragrant yellow-white flowers, followed by green-yellow berries. Azadirachta indica also known as neem is a tree in the mahogany family Meliaceae. This plant is known as a plant that could be utilised responsibly for the pesticidal, larvicidal, antifeedant or repellant action on various insects[12]. The active compounds of Neem are Nimbinen, Nimbidine, Meliantriol, Azadirachtin and Salanin. Neem seeds extract can be used to control various types of pests such as Helopeltis sp., Jengkat caterpillar, Aphis sp., Nilarvata sp., and Sitophilus sp. The active ingredients of neem seeds are useful to repel disturbing insects, prevent plant-eating pests, dispel adult larvae and insects, prevent the change of larvae skin, reduce egg production in female insects, and prevent female insects from placing eggs. [12]

Neem seeds contain 60% oil or fat from stearic, palmitic, oleic, linoleic, lauric, uteric and small amounts of essential oils [13]. Other active compounds from neem seeds are phenols, quinones, alkaloids, triterpenoids, and flavonoids. Residues from neem seeds are easily broken down into non-toxic compounds, making them friendly and safe for the environment. [13] A preliminary test was conducted to determine the biological activity of neem seed ethanol extract against mosquito larvae of Aedes aegypti as well as phytochemical preliminary tests. The preliminary test results showed that the neem seed ethanol extract was active as a larvicidal agent with an LC50 value of 282.29 ppm. The phytochemical test results stated that the ethanol extract of neem seeds contained secondary metabolites of the triterpenoid group, flavonoids and alkaloids. [14]

As a single substance, pyrethrin and azadirachtin are known to be very effective in killing mosquito larvae. However, to obtain a single active substance, the cost needed is very expensive, and its effectiveness still needs to be studied again. Various formulations were made to increase the efficacy of extracts against larvae, including by combining active substances with other substances such as chitin. Besides, it showed the effect of the emulsifier used in the effectiveness and extract's effectiveness in killing larvae. The present study aims to examine the potential of various formulations of extracts of Pyrethrum and Neem plants as Aedes aegypti larvacides.

## 2. MATERIAL AND METHODS

### Research design

This study used a quasi-experimental design. Neem (Azadirachta indica) and Pyrethrum (Chrysanthemum cinerariaefolium) were exposed to larvae of Aedes sp.

**Extraction of Pyrethrum and Neem**

Fresh Neem leaves washed with water, then cut into small pieces. The leaf pieces dried at room temperature, then smoothed to powder. One hundred grams of dry powder macerated using 70% ethanol for 1 x 24 hours. The maceration process repeated for 3 (three) times by replacing ethanol every 24 hours. The extract was evaporated using a vacuum so that the thick extract was obtained. The concentrated extract dissolved with water (7:3) then ethanol was evaporated so that the remaining water extract is left. The water extract is partitioned using n-hexane solvent and evaporated again. The water layer was extracted with chloroform then chloroform evaporated. The remaining water extract was added with ethyl acetate and vaporised again using a vacuum evaporator so that Neem leaf extract was obtained. Formulations were made using solvents and surfactants [15].

**Pyrethrum Flower Extract**

Pyrethrum flowers were cut into small pieces, then 20 grams of it was wrapped using filter paper and inserted into soxhlet. The flowers were extracted using 96% ethanol for 100 mL at 780C until the solvent returned to its original colour and obtained from the pyretrial flower oil filtrate. The filtrate was purified by extraction using an extractor at a temperature of 780C until pure pyretrial flower oil was obtained. Formulations are made using solvents and surfactants [16].

**Larvacidal test**

The test was conducted at the Loka LitbangKes Pangandaran, National Institute of Health Research and Development, while Pyrethrum and Neem essential oil were obtained from the Balai Tanaman Obat and Aromatic Bogor (Balitro). The Formulations of Pyrethrum and Neem extracts 100% described in table 1. This study uses three content-based formulations of active ingredients (1.25% pyrethrin, 0,36% azadirachtin, and 0.6% azadirachtin).

**Table 1. Formulation of Pyrethrum and Neem Extract**

| No | Formulation | Content |
|---|---|---|
| 1 | 1 ml Pyrethrum + 1 ml Tween 20 + 24 ml palm oil + 24ml chitin | 1,25% pyrethrin |
| 2 | 1 ml Pyrethrum + 1 ml Tween 20 + 48 ml palm oil | 1,25% pyrethrin |

IOP Conf. Series: Earth and Environmental Science **406** (2019) 012007 doi:10.1088/1755-1315/406/1/012007

| | | |
|---|---|---|
| 3 | 1 ml Pyrethrum + 1 ml Tween 80 + 24 ml palm oil + 24ml chitin | 1,25% pyrethrin |
| 4 | 1 ml Pyrethrum + 1 ml Tween 80 + 48 ml palm oil | 1,25% pyrethrin |
| 5 | 2 ml Pyrethrum + 1 ml DMSO + 24 ml liquid paraffin + 24 ml chitin | 1,25% pyrethrin |
| 6 | 2 ml Pyrethrum + 1 ml DMSO + 24 ml palm oil + 24 ml chitin | 1,25% pyrethrin |
| 7 | 30 ml neem + 1 ml Tween 20 + 20 ml palm oil | 0,36% azadirachtin |
| 8 | 49 ml neem + 1 ml Tween 20 | 0,6% azadirachtin |
| 9 | 30 ml neem + 1 ml Tween 80 + 20 ml palm oil | 0,36% azadirachtin |
| 10 | 30 ml neem + 1 ml DMSO + 20 ml palm oil | 0,36% azadirachtin |
| 11 | 49 ml neem + 1 ml DMSO | 0,6% azadirachtin |

For the larvicidal test, instar 3 of Aedes aegypti larvae prepared in containers containing 25 larvae per cup. 10 mL (concentration of 10%) of formulations added to each cup. Three repetitions applied for each treatment. Larval deaths were observed every hour to 24 hours, and the number of dead larvae was calculated.

**Data Analysis**

The data were collected and grouped based on their lethal time. The data were categorized into a table to be discussed descriptively.

## 3. RESULTS

This study tested several formulations (Table 1) to Aedes sp. Larvae, then the larval mortality was observed up to 24 hours. The results can be seen in Table 2.

**Table 2. Test results of 11 formulations against *Aedes aegypti* for 24 hours**

| No | Formulation | Number of tested larvae | LT90 (hour) | The average number of active larvae (24h) | The average number of dead larvae (24 h) |
|---|---|---|---|---|---|
| 1 | Control | 25 | - | 25 | 0 |
| 2 | Formulation 1 | 25 | 1 | 0 | 25 |
| 3 | Formulation 2 | 25 | 1 | 0 | 25 |
| 4 | Formulation 3 | 25 | 1 | 0 | 25 |
| 5 | Formulation 4 | 25 | 1 | 0 | 25 |
| 6 | Formulation 5 | 25 | 2 | 0 | 25 |
| 7 | Formulation 6 | 25 | 1 | 0 | 25 |
| 8 | Formulation 7 | 25 | 1 | 0 | 25 |
| 9 | Formulation 8 | 25 | >24 | 3 | 22 |
| 10 | Formulation 9 | 25 | 1 | 0 | 25 |
| 11 | Formulation 10 | 25 | 1 | 0 | 25 |
| 12 | Formulation 11 | 25 | >24 | 11 | 14 |

Table 2 showed that almost all formulations caused mortality in the first hour except formulation 5, formulation 8 and formulation 11.

## 4. DISCUSSION

The result of tests carried out on 25 Aedes aegypti larvae showed that the formulation with the main ingredient of Pyrethrin flower extract could cause larval mortality effectively (with most of the lethal time of 90 (LT90) were one hour). The longer LT90 was indicated by formulation five consisted of Pyrethrum flower extract, DMSO, liquid paraffin and chitin (1.25% pyrethrin). Compared to other formulations (formulations 1, 2, 3, 4, and 6), extract number 5 uses additional liquid paraffin in its formulation. Liquid paraffin is a mixture of hydrocarbons obtained from mineral oil, which is a thick liquid, transparent, non-fluorescent, colourless, almost odourless and almost tasteless liquid. Paraffin is not soluble either in water or ethanol.[17]

Liquid paraffin also can maintain microbial viability by preventing drying of microbial culture storage media [18]. The nature of paraffin is predicted as a factor that makes the decrease in the effectiveness of the formulation. The thick properties of liquid paraffin cause the formulation to spread more slowly throughout the surface of the water.

Chrysanthemum extract has been known to have toxic effects on insects. The high pyrethrin content of this plant is widely used in various insect control activities. The effectiveness test of herbal mosquito repellent named 'Morizena', which has the composition of leaves of Passiflora putida (40%), Chrysantemum cinerariaefolium (40%), and essential oils from Cymbopogon nardus (20%), is known to be more effective in killing adult Aedes aegypti than commercial insecticide at the same concentration (LC90:2977 ppm).[19]

Pyrethrin has long been known as a natural insecticide that has a high economic value. Demand for pyrethrin is quite high, yet it is not balanced with the production level of the pyrethrin. Several studies were conducted to improve the productivity of pyrethrin by developing tissue culture or biosynthetic methods with help from microbes or precursors such as ethrel, Chlormequat chloride, Paclobutrazol, and glucose. [20][21][22].

In addition to pyrethrin content, Chrysanthemum cinerariaefolium is also known to have a high flavonoid content. Flavonoids are known as anti-virus, anti-asthma, and have the potential as anti-cancer. [23]. Repellent test results on beetles also show that Chrysanthemum cinerariaefolium flowers are useful to repel adult pulse beetle (Callosobruchus chinensis Linn). [24].

Formulations with Neem ingredients (Formulations 7-12) have diverse effectiveness as well. Formulations 7, 9, and 10 have almost the same efficacy (LT90= 1 hour) while two other formulations (formulations 8 and 12) have LT90 longer than 24 hours. Neem leaves are known to have an antimicrobial activity that can kill pathogenic bacteria. Neem is also widely known as an anti-parasitic and often used for animal food. It has proven to be effective in neutralising snake poisons and insects in arid/desert areas.[25] Neem can also be used for anti-inflammatory as an alternative to dexamethasone.[26]

In addition to leaves and seeds, the neem tree is known to have Limonoid Insect Growth Regulatory (IGRs) which functions to inhibit the growth and development of insects. It can be used as an alternative since azadirachtin that usually has a larvicidal activity, is not present in its stem. [27] Neem secondary metabolites inhibit the biological activity of adult Anopheles stephensi. The Azadirachtin test in the form of a commercial product has an effect in delaying oviposition, and the analysis shows a slightly toxic effect, it also affects both the follicular cell and the oxytocytes of An. Stephensi.[28] Another advantage of the neem is that it does not affect non-target insects so that the natural predators of the larvae will not be affected even though neem extracts were given in the form of granules. [29] The results of in vivo and in vitro test from the crude extracts of neem leaves are known to have the ability to inhibit dengue virus type two (DENV 2) replication. Minimum doses can inhibit the development of the virus so that it does not cause cytopathic effects. A maximum dose (120-130 mg/ml) can inhibit viral replication. It is confirmed from the absence of clinical symptoms, and RT-PCR examination confirmed the absence of virus-specific amplicons.[30]

Formulations 8 and 11 have lower effectiveness since it is suspected that there are effects of solvents or surfactants used in this formulation. Although the concentration of azadirachtin is 0.6% higher, the mixture of tween 20 and DMSO is considered to reduce toxic effects on larvae. Other formulations added to palm oil showed higher effectiveness for larval mortality even though the azadirachtin concentration in formulations was smaller (0.36%).

Palm oil is known as an antioxidant. An antioxidant is suitable for human well-being because it has carotenoid compounds. The human body can quickly absorb antioxidant [31]. The study predicts that in formulation containing a mixture of palm oil, the larvae rapidly consume the extract into its body. Besides, it is suspected that pyrethrin content is more optimal in oil because of its polarity properties. It was reported that the total pyrethrin content was higher in hexane extract, which happened to be a non-polar substance, compared to pyrethrin content in semi-polar methanol.[32]

This study's formulation might be useful for a researcher or practitioner to be developed further to achieve optimal use from essential oil. Further research is needed to improve the formulation and also to produce larvacide which could be widely implemented in community. Hopefully in the future, essential oil formulation will be the first choice for mosquito control.

The results of this study emphasised the potential of a formulation containing pyrethrum and neem as biolarvacide against Aedes aegypti. We concluded that almost all formulations contain mix of Pyrethrum, tween, DMSO, and Neem, tween, DMSO caused 100% mortality with the lethal time of less than an hour. Except for formulation contains Pyrethrum, DMSO liquid paraffin, and chitin; neem and tween; and neem and DMSO.

## 5.ACKNOWLEDGMENTS

We thank Balai Tanaman Obat Aromatik Bogor for the help in making essential oils, and to Laboratory of Loka Litbang Kesehatan Pangandaran for the assistance in conducting the larvae test.

## 6. REFERENCES

1. Hasan S., Jamdar S.F., Alalowi M. AAABSM. Dengue Virus : A Global Human Threat: Review of Literature. J Int Soc Prev Dent. 2016;6:1–6.
2. Mustafa M.S., Rasotgi V., Jain S. G V. Discovery of Fifth Serotype of Dengue Virus (DENV-5): A New Public Health Dilemma in Dengue Control. Med J Armed Forces India. 2015;71:67–70.
3. Y H. Dengue Vectors and Their Spatial Distribution. Trop Med Health. 2011;39:17–27.
4. Marcombe s., Mathieu r.B., Pocquet N, Riaz M.A. PR et. al. Insecticide Resistance in The Dengue Vector Aedes aegypti from Martinique: Distribution, Mechanism and Relations With Environmental Factors. PLoS One. 2012;7.
5. Amelia-Yap, Chee hang Chen, Mohd Sofian Azirun VLL. Pyrethroid Resistance in The Dengue Vector Ae. aegypti in Southeast Asia: Present Situation And Prospects For Management. Parasites and Vectors. 2018;11:332.
6. Vontas J, Kioulos E, Pavlidi N, Morou E, della Torre A, Ranson H. Insecticide resistance in the major dengue vectors Aedes albopictus and Aedes aegypti. Pestic Biochem Physiol [Internet]. Elsevier Inc.; 2012;104(2):126–31. Available from: http://dx.doi.org/10.1016/j.pestbp.2012.05.008
7. Lima E.P., Paiva M.H.S., de Araujo A.P., da Silva U.M. E al. Insecticide Resistance in Aedes aegypti Population From Ceara, Brazil. Parasit Vectors. 2011;4:5.
8. Khater HF. Prospects of Botanical Biopesticides in Insect Pest Management. Pharmacologia. 2012;3(12):641–56.
9. Bhakuni R.S., Kahol A.P., Singh S.P. KA. Composition of North Indian Pyrethrum (Chrysantemum cinerariaefolium) Flower Oil. 2013. p. 31–5.
10. Bowman N.M., Akialis K, Cave G., Barrera R., Apperson C.S. E al. Pyrethroid Insecticides Maintain Repellent Effect on Knock-Down Resistant Populations of Aedes aegypti Mosquitoes. PLoS One. 2018;13.

11. Johnson G.D., Goosey H.B., Rolston M.G., Miller W.L., Hokit D.G. E al. Evaluation of Mosquito Responses to Pyrethroid Insecticides Topically Applied to Sheep. J Am Mosquitoe Control Assoc. 2013;29:146–53.
12. Dapar M., Okwori V. E al. Mosquito Repellent Property of Azzadirachta Indica Extract (Fruit Bark and Seed Kernel). In 2015. p. 1–4.
13. Wiryowidagdo. Kimia dan Farmakologi Bahan Alam. Jakarta: UI Press; 2002.
14. Sikka A. Uji Efektivitas Larvasida Ekstrak Etanol Daun Mimba (Azadirachta indica) terhadap Larva Aedes aegypti. Semarang: Fakultas Kedokteran Univ. Diponegoro; 2009.
15. Sri Wahjuni., Ni Made Puspawati. NPREA. Isolasi dan Identifikasi Senyawa Aktif Anti Jamur dari daun Mimba (Azadirachta indica A, Juss) Sebagai Pengendali Jamur Fusarium Sp. Pada Tanaman Buah Naga (Hylocereus Sp.). J Chem. 2016;10(2):197–203.
16. Ikawati AL. Ekstraksi Minyak Atsiri Bunya Krisan (Chrysantemum cinerariaefolium) Dengan Pelarut Ethanol dan n-Heksana. Semarang; 2015.
17. Triayu SI. Formula Krim Obat Jerawat Munyak Atsiri Dau Jeruk Nipis (Citrus aurantifolia Swingle) dan Uji Daya Antibakteri Secara Invitro. Surakarta; 2009.
18. Machmud M. Teknik Penyimpanan dan Pemeliharaan Mikroba. Agrobio. 2001;4 (1):24–32.
19. Susilowati RP, Darmanto W, Aminah NS. The Effectiveness Of Herbal Mosquito Coils "Morizena" Against Aedes Aegypti Death. Indones J Trop Infect Dis. 2018;
20. Adnane Hilmi, Alain Coudret CB. The Production of Pyrethrins by Plant Cell and Tissue Cultures of Chrysantemum cinerariaefolium and Tagetes Species. Crit Rev Plant Species. 2010;19(1):69–89.
21. S Haque., A.H. A. Farooqi., M.M. Gupta., R.S. Sangwan. AK. Effect of Ethrel, Chlormequat chloride and Paclobutrazol on Growth and Pyrethrins Accumulation in Chrysantemum cinerariaefolium Vis. Plant Growth Regul. 2007;15(3):262–9.
22. Kazuhiko Matsuda, Yukio Kikuta, Atsushi Haba, Koji Nakayama, Yoshio Katsuda, Akikazu Hatanaka KK. Biosynthesis of Pyrethrin I in Seedling of Chrysantemum cinerariaefolium. Phytochemistry. 2005;66(13):1529–35.
23. Widi Purwianingsih, Santika Febri K. Formation Flavonoid Secondary Metabolites in

Callus Culture of Chrysantemum cinerariaefolium as Alternative Provision Medicine. In: AIP Conference Proceedings. 2016.

24. Chandel B.S. G. Evaluation of Repellent Potential of Cichorium intybus, Inula racemosa, Tagetes minuta and Chrysantemum cinerariaefolium Against Pulse Bettle, Callosobruchus chinensis Linn (Coleoptera: Bruchidae). 2nd ed. Gupta Publisher; 2017. 2455–4758 p.
25. Kumar D, Rahal A, Malik JK. Neem Extract. In: Nutraceuticals: Efficacy, Safety and Toxicity. 2016.
26. Maragathavalli, S., Brindha, S., Kaviyarasi, N. S., Annadurai, B. B., Gangwar SK. Antimicrobial Activity in Leaf Extract of Neem (Azadirachta Indica. Int J Sci Nat. 2012;
27. Howard AF V., Adongo EA, Hassanali A, Omlin FX, Wanjoya A, Zhou G, et al. Laboratory Evaluation of the Aqueous Extract of Azadirachta indica (Neem) Wood Chippings on Anopheles gambiae s.s. (Diptera: Culicidae) Mosquitoes . J Med Entomol. 2009;
28. Lucantoni L, Giusti F, Cristofaro M, Pasqualini L, Esposito F, Lupetti P, et al. Effects of a Neem Extract on Blood Feeding, Oviposition and Oocyte Ultrastructure in Anopheles stephensi Liston (Diptera: Culicidae). Tissue Cell. 2006;
29. Chandramohan B, Murugan K, Panneerselvam C, Madhiyazhagan P, Chandirasekar R, Dinesh D, et al. Characterization and Mosquitocidal Potential of Neem Cake-Synthesized Silver Nanoparticles: Genotoxicity and Impact on Predation Efficiency of Mosquito Natural Enemies. Parasitol Res. 2016;
30. Parida MM, Upadhyay C, Pandya G, Jana AM. Inhibitory Potential of Neem (Azadirachta indica Juss) Leaves on Dengue Virus Type-2 Replication. J Ethnopharmacol. 2002;
31. Ahmad Ghazali SS. Potensi Minyak Sawit Merah dan Karotenoid Sebagai Suplemen Antioksidan Dalam Pengujian Toleransi Glukosa Pada Tikus Putih (Preliminary Study). Trop Med Health. 2018;1:251–6.
32. Nagar A, Chatterjee A, Rehman LU, Ahmad A, Tandon S. Comparative extraction and enrichment techniques for pyrethrins from flowers of Chrysanthemum cinerariaefolium. Ind Crops Prod. 2015;76:955–60.

IOP Conf. Series: Earth and Environmental Science **406** (2019) 012008 doi:10.1088/1755-1315/406/1/012008

# QUALITY OF SIMPLICIAN BIOACTIVE COMPONENTS AND LIQUID EXTRACT OF KECOMBRANG FLOWER POWDER FROM TEMPERATURE AND TIME OPTIMIZATION RESULTS

**Dwiana Intan Pertiwi, Rifda Naufalin*), Poppy Arsil, Erminawati, Rumpoko Wicaksono and Taslimatul Auliya**

Food Science and Technology Study Program, Faculty of Agriculture, Jenderal Soedirman University, Purwokerto 53122
Corresponding author : rifda.naufalin@unsoed.ac.id

**Abstract.** Kecombrang flower has benefits as an antioxidant and antimicrobial. This capability can be maximized by processing through drying and liquid extraction. In the drying and extraction of liquid, the temperature and time that is not right can affect the content of bioactive components of the product. This study aims to determine the quality of bioactive components of powder and liquid extract of kecombrang flowers quantitatively and qualitatively based on temperature and time. The experimental design used in kecombrang flower powder is a Completely Randomized Design (CRD) with 9 treatment and 3 replication as long as the liquid extract of kecombrang flowers using the Responsse Surface Method (RSM) method with Expert Design 10 software. Central Composit Design (CCD) produced 13 optimization formulas. Factors that were tested on drying are 50°C, 60°C, 70°C, and time is 4, 5 and 6 hours. Liquid extraction temperature is 40oC, 50oC, 60oC and time is 3, 4 and 5 hours. Based on the results of the study the highest total phenol simplisia and liquid extract of kecombrang flowers namely 27.93 TAE / 100 g sample and 64.458 mg TAE / 100 g. The highest total of flavonoids is simplicia and liquid extract of kecombrang flower which is 2.073 QE / 100 g sample and 35.120 mg QE / 100 g and pH value (3.9). The compounds in the optimum formula product are phenolic, flavonoids, alkaloids, terpenoids and tannins.

*Keywords: kecombrang flower, liquid extraction, temperature and time drying.*

## 1. INTRODUCTION

One of the herbs and medicinal plants that have potential as functional foods in antioxidants and antibacterial is kecombrang (*Etlingera elatior*) [1, 2]. Kecombrang is one type of spice plant that has long been known and used by humans as food ingredients. Plant parts that are commonly used are flowers and stems. Compounds contained in kecombrang flowers are alkaloids, flavonoids, polyphenols, terpenoids, steroids, saponins, and essential oils [3, 4, 5].

The ability of kecombrang flowers can be maximized by further processing by drying. Drying aims to reduce the water content of the kecombrang flowers to the limits of the development of microorganisms and enzyme activities that can cause decay is inhibited or even stopped altogether. Thus, the dried material has a longer shelf life [6]. During the drying process, several things must be considered such as drying temperature and drying time. The drying process plays a very important role. If the drying temperature is too high it will cause a decrease in the active compound and change the color of the dried product. Meanwhile, if the temperature used is too low it is difficult to reach the desired water content.

The use of kecombrang flower powder can still be maximized by processing it in liquid extract form. Extraction is the activity of withdrawing soluble chemicals so that they are separated from insoluble substances by using a liquid solvent [7]. The method used in making liquid extract of kecombrang flower powder is by extraction using a water solvent.

Generally the solubility of the extracted active substance will increase in size with increasing temperature. However, the increase in extraction temperature also needs to be considered. This is because temperatures that are too high can cause damage to the material being processed [8]. Besides temperature, the use of extraction time also needs to be considered. This is in accordance with [9], which states that the right extraction time will produce optimal compounds. Extraction time that is too long will cause hydrolyzed extracts, while extraction time that is too short causes not all active compounds to be extracted from the ingredients.

Therefore, this study aims to determine the quality of the quantitative bioactive components of powder and liquid extract of kecombrang flowers, namely total phenols, total flavonoids and pH values as well as qualitative, namely phenolic compounds, flavonoids, alkaloids, triterpenoids, tannins, saponins, and glycosides based on temperature and time.

## 2. RESEARCH METHODOLOGY

### 2.1. Experimental design

The experimental design used in kecombrang flower powder is a Completely Randomized Design (CRD) with 9 treatment combinations and was repeated 3 times, as far as the liquid extract of kecombrang flowers using the Responsse Surface Method (RSM) method with Expert Design 10 software. Central Composit Design (CCD) with two factors and two analytical tests tested produced 13 optimization formulas. The factors that were tried in this study are temperature and time. The flower drying temperature consists of three levels, 50°C, 60°C, 70°C, and time consists of three levels, 4, 5 and 6 hours. The liquid extract temperature consists of three levels, namely 40°C, 50°C, 60°C and time consists of three levels namely 3, 4 and 5 hours.

### 2.2. Sample preparation of kecombrang flower powder

The stage before conducting research is to prepare a sample of kecombrang flower powder. The kecombrang flower is first selected and a flower crown is taken and then washed with water. Furthermore, kecombrang flower crowns are reduced in size so that the drying process is more optimal. After that, the flower slices are dried using a cabinet dryer with the appropriate treatment temperature that is 50°C, 60°C and 70°C with a time of 4, 5 and 6 hours. After the flowers are dried, they are ground using a disk mill. Flower powder that is formed into 60 mesh is used as material in the study of temperature optimization and liquid extraction time.

### 2.3. Liquid Extraction of Kecombrang Flower Powder

The extraction method used in this study is maceration using a water solvent. At first, a 60 mesh flower powder was put into the extractor. Then, it is mixed with a water solvent whose temperature is adjusted to the treatment factor which is 40°C, 50°C and 60°C through the extractor cover pipe. Comparison of powder: water used is 1:14. After that, the extractor is closed then turned on, set the temperature according to the treatment and agitator speed of 60 rpm. The extraction time used in accordance with the treatment factor is 3, 4 and 5 hours. After extraction is complete, the extract is then allowed to stand in a closed container and dark conditions for 19 to 24 hours. Next, the sample is filtered using a press to separate the yield of the powder and the filtrate. Separate powder, then dried using a cabinet dryer with a temperature of 50°C and a time of 4 hours. After drying, the extraction residual powder is first extracted again in a ratio of 1:14 using the same temperature and time. The filtrate sample used is a combination of liquid extract in the first extraction and replicates.

## 2.4. Quantitative Test

### 2.4.1. Total Phenol

400 μL supernatant was added with 1.5 mL of Folin Ciocalteu and allowed to stand for 5 minutes at room temperature. Next, added 1.5 mL sodium bicarbonate (NaHCO3) 0.556 M, shaken and left in a dark room for 90 minutes, then absorbance was measured using a spectrophotometer at a wavelength of 725 nm [10].

### 2.4.2. Total Flavonoid

100 mL or 0.1 mL sample supernatant was added with 1 mL of 2% AlCl 3 (2 g AlCl 3 in 100 mL of 5% glacial acetic acid solution) and 1 mL of 120 mM potassium acetic acid solution (1,176 g of potassium acetate in 100 mL of distilled water). After that, it was incubated for 1 hour at room temperature and the absorbance was measured using a spectrophotometer at a wavelength of 435 nm [11].

### 2.4.3. Water Content

Samples were weighed as much as 0.5 g in a moisture analyzer cup. Moisture analyzer is set at 105°C. Moisture analyzer is closed and awaited the results of its water content. The results of the water content are recorded. Moisture content of 10% [12].

### 2.4.4. pH Value

The pH measurement is carried out using a pH meter. Before use, the pH meter is calibrated first using a buffer solution of pH 4 and pH 7. The way to measure pH is to insert the pH meter electrode in the sample, wait a while until the pH stabilizes, so that the pH value is measured. After completion, the electrodes are removed and rinsed with distilled water [13].

## 2.5. Qualitative Test

### 2.5.1. Phenolic Compounds

As much as 2 mL of the extract sample was added 2 drops of 5% FeCl3 solution. The formation of green or blue green indicates phenol compounds in the ingredients [14].

#### 2.5.2. Alkaloid

0.1 gram extract was added with 10 mL chloroform and a few drops of ammonia were added. The chloroform fraction is separated and acidified with a few drops of concentrated H2SO4. Acid fraction is taken and divided into 3 tubes, then Dragendorff, Meyer and Wagner reagents are added. The presence of alkaloids is characterized by the formation of white deposits in Meyer reagents, red deposits in Dragendorff reagents, and brown deposits in Wagner reagents [14].

#### 2.5.3. Steroid/triterpenoid

A total of 1 g of sample was dissolved with 25 mL of 50 ° C hot ethanol, then filtered into a Porcelain cup and evaporated to dryness. The residue is dissolved with ether and transferred into a test tube, then added 3 drops of anhydrous acetic acid and 1 drop of concentrated H2SO4 (Lieberman Burchard Test). Red or purple indicates triterpenoids and green or blue indicates steroids [14].

#### 2.5.4. Saponin

Saponins can be detected by foam testing in hot water. The foam is stable for 10 minutes and does not disappear with the addition of 1 drop of HCl 2 N indicating saponin [14].

#### 2.5.5. Tannin

Extract as much as 10 mL boiled. After cooling, the filtrate was added with 5 mL FeCl3 1% (w / v). If the color changes to dark blue, it means that the sample contains tannin [14].

#### 2.5.6. Flavonoid

A sample of 2 mL was put into a test tube. Then 100 mg of magnesium powder and 3 mL of amyl alcohol (a mixture of 37% hydrochloric acid and 95% ethanol of equal volume) and 2 mL of hydrochloric acid are added, then the mixture is shaken. The formation of red, yellow or orange in the amyl alcohol layer shows the results of flavonoids [15].

#### 2.5.7. Glycoside

It is done with the Lieberman-Buchard reaction. Kecombrang flower bubk extract was dissolved in ethanol solvent, evaporated on a water bath and then dissolved in 5 mL acetic acid anhydride and then added 10 drops of concentrated sulfuric acid. The formation of blue or green indicates glycosides [16].

### 2.6. Data Analysis

Kecombrang flower powder data were analyzed using the F test (analysis of variance) at a 95% confidence level with ANOVA. If the analysis results show a real effect, then proceed with the DMRT test (Duncan Multiple Range Test) at the level of 5%. Kecombrang flower liquid extract data were analyzed using the Response Surface Method (RSM) to obtain the optimal response of liquid extraction methods. The results of the analysis of quantitative optimization variables are visualized in the form of two-dimensional images to show the effect of factors on responses. The recommended optimum formula is then verified and validated using an independent t test to find out whether there is a difference between the predicted value and the product after treatment.

## 3. RESULT AND DISCUSSION

### 3.1. Quantitative

#### 3.1.1. Total Phenol

The average value of total phenol kecombrang flower in powder and liquid extract can be seen in Figure 1.

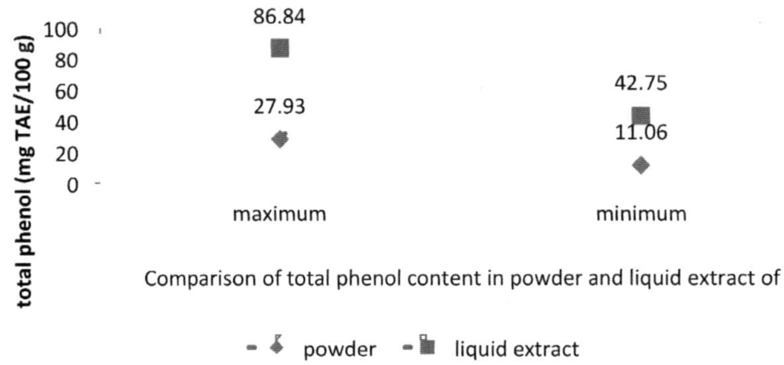


Figure 1. Comparison of total phenol content in powder and liquid extract of kecombrang flowers.

Based on Figure 1, the highest value of total phenol liquid extract of kecombrang flower is 86.84 mg TAE / 100 g and the lowest value is 42.75 mg TAE / 100 g. While the highest total value of phenol kecombrang flower powder was obtained in the amount of 27.93 mg TAE / 100 g and the lowest value was 11.06 mg TAE / 100 g. The total value of phenol liquid extract of kecombrang flower is better than kecombrang flower powder. This is seen based on the total phenol value of liquid extract which is higher than the total value of phenol in kecombrang flower powder, the higher total value of kecombrang flower phenol in the form of liquid extract because of the bioactive component especially phenol has separated from the insoluble material with liquid solvent while still in powder form. Therefore, the total phenol content of kecombrang flowers in liquid extracts is higher than when it was still in powder form.

According to research conducted by [17, 18] differences in total phenol levels due to phenol damage can be caused by environmental factors such as light, temperature and oxygen. This statement is reinforced by research [19] that the total levels of phenol compounds decreased due to oven drying. The higher drying temperature results in an increase in the enzyme inactivation process of the polyphenol oxidase enzyme, so that the enzyme activity will be lower and the damage to the polyphenol compound less, but if the drying temperature exceeds the optimum temperature, the stability of the polyphenol compound will be disturbed so as to cause a decrease in the content of polyphenol compounds in the material [20].

According to [21], extraction is the activity of withdrawing soluble chemicals so that they are separated from insoluble materials with liquid solvents. During extraction, optimal use of temperature and time is also one of the reasons for the higher total phenol content produced. According to [22] the longer the extraction time, the greater the chance of the solvent interaction to dissolve the material so that the results also increase to the saturation point of the solution. Contact between the sample and the solvent can be improved if it is assisted with shaking. This is useful so that contact between the sample and the solvent is more frequent, so that the extraction process is more perfect.

### 3.1.2. Total Flavonoid

The average value of total flavonoids of kecombrang flowers in powder and liquid extract can be seen in Figure 2.

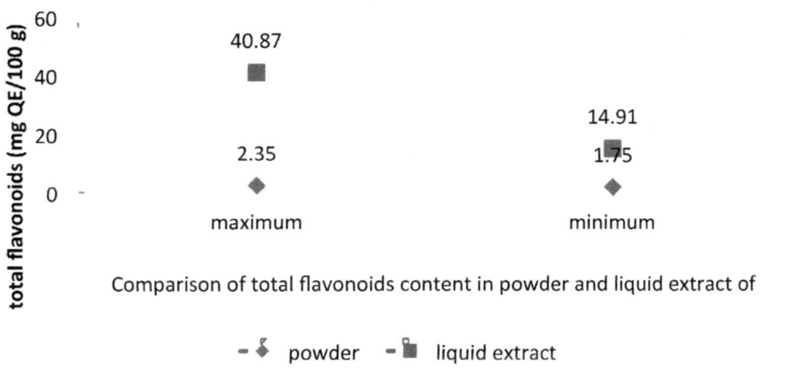


Figure 2. Comparison graph of flavonoid content in powder and liquid extract of kecombrang flower

Based on Figure 2, the highest value of total flavonoid kecombrang flower powder was obtained at 2.35 mg QE / 100 g and the lowest flavonoid value was 1.75 QE / 100 g. The highest value of total flavonoids of kecombrang flower extract is 40.87 mg QE / 100 g and the lowest total value of flavonoids is 14.91 mg QE / 100 g. The value of flavonoids in liquid extract of kecombrang flower powder is higher than in kecombrang flower powder, the higher total value of kecombrang flower flavonoids in liquid extract form is presumably because the flavonoid compound has separated from the insoluble material with liquid solvent while still in powder form. Therefore, the total flavonoid content of kecombrang flowers in liquid extracts is higher than when it was still in powder form.

Flavonoids are polyphenols with the basic structure of phenols which have already oxidized and sensitive to heat treatment so that the drying process will affect the levels of flavonoids in the sample [23]. Flavonoid analysis was carried out using ethanol as a polar solvent. According to [21], extraction is the activity of withdrawing soluble chemicals so that they are separated from insoluble materials with liquid solvents. Polar flavonoids can be extracted well using solvents so that the kecombrang flowers produce flavonoids.

An increase in temperature and time extraction need to be considered. Extraction temperatures that are too high and extraction times that are too long and exceed the optimum time limit can cause the loss of compounds in the solution due to evaporation. Therefore, the extraction treatment temperature of 60°C with a length of 3.735 hours can provide maximum results because between the temperature and extraction time are not maximized both [24].

#### 3.1.3. Water Content

The average water content of kecombrang flower powder can be seen in Figure 3.

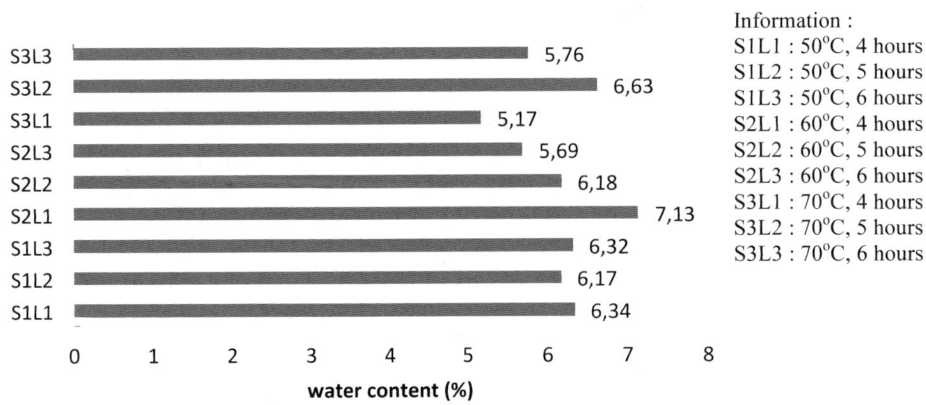


Figure 3. Water content of dried simplicia kecombrang flowers with various treatments

Based on Figure 3, it shows that the value of the water content of simplicia kecombrang flowers ranges between 5.17 - 7.13%. According to the Indonesian Herbal Pharmacopoeia [25] and the Minister of Health Decree No. 661 / Menkes / SK / VII / 1994 regarding Requirements for Traditional Medicines, the standard water content of calcium at simplicia is 10%. Fresh kecombrang flowers have high water content which is around 90%, so they need to be dried to extend their shelf life. The purpose of drying is to reduce the water content so that the material is not easily overgrown with molds and other microorganisms and stop enzymatic reactions that can further decompose the active ingredient [26].

Decreased water content in kecombrang flowers is influenced by water evaporation from temperature and time drying [27]. The longer the drying process causes the evaporation of water which is found in the kecombrang flowers, the lower. According to the literature, the higher the temperature and duration of drying, have a great influence on the speed of water transfer in the material. The ability of the material to release water from the surface will be even greater with the increasing temperature of the drying air used and the length of the drying process [28].

#### 3.1.4. pH Value

Temperature and time did not affect the pH value according to [29] that if the concentration (%) of ethanol increases, the pH obtained also increases, but if the extraction temperature increases the resulting pH remains. The pH value in the optimization formula of liquid extract of kecombrang flower powder (60°C for 3.735 hours) was 3.9. According to [2] kecombrang flowers have a pH value of 3.89.

The decrease in pH is characterized by an increase in color density. Increasing the pH value can cause a decrease in color intensity and concentration of *flavilium cations*. There are several factors that can affect the color of the extract [30]. The process of extracting spices, the composition, color, aroma and rendeman produced will be influenced by the type, size, level of maturity of the raw material, type of solvent, temperature and time of extraction and the extraction method [31].

### 3.2. Qualitative

Qualitative variables in this study consisted of phenols, flavonoids, alkaloids, triterpenoids, tannins, saponins, and glycosides. The results of testing the qualitative variable of powder and liquid extract of kecombrang flower are presented in Table 1.

Table 1. Qualitative test results of kecombrang flower powder and liquid extract

| Treatment | Phenolic | Flavonoids | Alkaloids | Triterpenoid | Tannins | Saponins | Glycosides |
|---|---|---|---|---|---|---|---|
| Powder | ++++ | +++ | +++ | +++ | - | ++ | - |
| Liquid extract | ++++ | +++ | +++ | +++ | ++ | + | - |

Information : - = negative, + = weak positive, ++ = positive, +++ = strong positive, ++++ = very strong positive

#### 3.2.1. Phenol

Phenol compounds can function as antioxidants because of their ability to eliminate free radicals and peroxide radicals so that they are effective in inhibiting lipid oxidation [2]. The presence of phenolic is marked in blue to green which is formed when the sample is subjected to FeCl3. The more phenols contained, the more dense the resulting color. This is in accordance with [32] which explains that phenol testing is carried out by reacting 1 mL samples with 5% FeCl. The presence of phenol compounds is characterized by the formation of a bluish-green color to black.

Based on the phenolic test data in Table 1, it shows that the liquid powder and extract of kecombrang flower contains phenol which is very strong due to the change in bluish-green to blackish color. The use of high temperatures and a long time in extraction can reduce levels of phenolic compounds. [33] explained that phenolic compounds are thermosensitive substances, thus enabling hydrolysis and percentage reduction at high temperatures. The profile of phenolic compounds in sample extracts can change with differences in the degree of polarity of the solvents used [34]. This is in accordance with [35] which states that the phenol component contained in the fraction of kecombrang plant parts is thought to have a polarity close to ethanol polarity, so that the use of ethanol solvents is more effective to dissolve phenol compounds.

#### 3.2.2. Flavonoid

Flavonoids are compounds consisting of C6-C3-C6. Flavonoids are commonly found in plants as glycosides. The sugar group has 11 compounds in one or more phenolic hydroxyl groups. The purpose of flavonoids for plants is to attract insects that help the pollination process and to attract the attention of animals that help spread seeds. For humans, flavonoids in small doses work as stimulants in the heart and capillaries [36].

Based on data from the flavonoid test results in Table 1, it shows that the liquid powder and extract of kecombrang flowers contain strong flavonoids because they are marked by the red color formed. The more concentrated the color, the more flavonoid content is. This is also in accordance with [37] that a positive result is indicated by the appearance of dark red (magenta) within 3 minutes. From the analysis, according to [38], the addition of magnesium powder and hydrochloric acid in the testing of flavonoids will cause the reduction of existing flavonoid compounds, causing a red reaction which is a characteristic of the presence of flavonoids.

### 3.2.3. Alkaloid

Alkaloids are chemical compounds produced by secondary metabolites that are formed based on the principle of formation of a mixture. Alkaloids are divided into three parts, namely N-containing elements involved in the formation of alkaloids, elements without N found in alkaloid molecules and reactions that occur for the typical binding of elements to the alkaloids [36].

Based on the alkaloid test data results in Table 1, it shows that the powder and liquid extract of kecombrang flowers contain strong alkaloids because it produces a red dregs when dropped with Dragendorff reagents. Alkaloid testing is done by adding HCl before adding reagents. This is because alkaloids are alkaline so they are extracted with solvents that contain acids [16]. Alkaloid test results are characterized by the formation of orange to red dregs at the bottom of the tube. The positive alkaloid results in the Dragendorff test were marked by the formation of orange dregs. These dregs are metal complexes with alkaloids. In the manufacture of Dragendorff bismuth nitrate reagents react with potassium iodide to form bismuth (III) iodide dregs which then dissolve in excess iodide to form potassium tetra iodobismunate. In the alkaloid test with Dragendorff's reaction, the lone pair of electrons in nitrogen is used to form covalent bonds with bismuth producing an orange to red dregs [39].

### 3.2.4. Triterpenoid

Triterpenoids are compounds whose carbon skeletons originate from six isoprene units and are biosynthetically derived from acyclic C30 hydrocarbons, namely squalene. Triterpenoids are colorless, crystalline compounds, often with high melting points and optically active, which are generally difficult to distinguish because there is no chemical reactivity. Triterpenoids are classified into four groups, namely actual triterpenes, steroids, saponins, and cardiac glycosides [16].

Based on the alkaloid test results in Table 1, it shows that the liquid powder and extract of kecombrang flowers contain strong triterpenoids. Test results of liquid powder and extract of kecombrang flower show the formation of red color. This indicates that the liquid extract of kecombrang flowers contained compounds namely terpenoids. The test of terpenoids and steroids using Lieberman Burchard reagents [40].

#### 3.2.5. Tannin

Tannins are polyphenol compounds that can form polyphenols that form complex compounds that do not dissolve with protein. The formation of green or blue color in the sample after the addition of FeCl3, possibly because tannin compounds will form ion $Fe^{3+}$ complexes [16].

Based on the phenolic test results in Table 1, it shows that the kecombrang flower powder does not contain tannin while the liquid extract of kecombrang flower contains tannin. Kecombrang flowers basically do not contain tannins. However, the results of tannin green to blackish blue test were formed in all samples of liquid extract of kecombrang flower extract. This is because FeCl3 can react with other compounds with the same color. The main properties of tannins depend on the phenolic-OH group contained in tannins. Color reactions occur when combined with iron salt. Iron salt ($FeCl_3$) will give a green or blackish blue color when reacted with tannin. But this test is not good because in addition to tannins that can provide a color reaction, other substances can also provide the same color reaction [20].

Green to blackish blue color that is formed is thought to be due to $FeCl_3$ which reacts with phenolic compounds. On kecombrang flowers, the results of phenolic compounds test produce positive results with the formation of green to blackish blue. Phenol testing is carried out by reacting 1 mL samples with 5% FeCl. The presence of phenol compounds is characterized by the formation of a bluish-green color to black [32].

#### 3.2.6. Saponin

Saponins are glycosides that are found in plants, consisting of sugar groups that bind to sapogenin aglycones. Saponin has the characteristics of a foam because when reacted with water and shaken will make a foam. Saponins are classified into two types, namely steroid saponins with carbohydrate molecules, whereas triterpenoid saponins are composed of a triterpenoid core with carbohydrate molecules [41].

Saponin testing is characterized by the formation of foam when the sample is shaken. formation of a stable foam as high as ± 1.5 cm, which indicates a positive result. In general, if the results are positive, the addition of 2N HCl aims to increase polarity so that the hydrophilic group will bind more stable and the froth formed becomes stable [42].

Based on the alkaloid test results data in Table 1, it shows that the positive kecombrang flower powder and liquid extract contain saponins, but the saponin content in flower powder is more than liquid extracts. Saponins are compounds that have hydrophilic and hydrophobic groups. Saponins are formed when foam is formed due to the presence of hydrophilic groups that bind to water, while hydrophobes will bind to air. In micelle structures, the polar group faces outward while the non-polar group faces inward. This situation affects the formation of foam, but the liquid saponin extract produced is very little because of the foam form [43].

#### 3.2.7. Glycoside

Glycosides are natural compounds consisting of carbohydrate parts and non-carbohydrate parts. The most non-carbohydrate parts found are triterpenes, steroids, and flavonoids, while the most commonly found carbohydrate molecules are glucose, galactose, xylose, and arabinose. These monosaccharides can be bound to one or more C atoms in the non-carbohydrate part. The word glycoside means carbohydrate or sugar which is generally an oxidizing agent called glycone, while non-sugar is called aglycone. Chemical bonds formed by glycosides resemble ether so that chemically in the process of its formation always releases water or $H_2O$ [44].

The presence of glycoside compounds is characterized by the formation of blue or green rings. The glycoside test is carried out with the Lieberman-Buchard reaction. Itchy leaf extract is dissolved in ethanol solvent, evaporated on a water bath and then dissolved in 5 mL acetic acid anhydride then 10 drops of concentrated sulfuric acid are added. The formation of blue or green color indicates the presence of glycosides [45]. However, the testing of powder and liquid extract of kecombrang flower powder did not form the color ring. This is presumably due to the glycoside compounds found in kecombrang flowers, namely the type of sugar glycoside compounds. [46] explains, testing glycosides using anhydrous acetic acid and concentrated sulfuric acid can only test non-sugar glycoside compounds and cannot detect the presence of sugar glycoside compounds.

## 4. CONCLUSION

Optimization, verification and validation results of the optimum formula showed that the liquid extract of kecombrang flower powder had a phenol response value of 64.458 mg TAE / 100 g, flavonoids 35.120 mg QE / 100 g, and a pH value (3.9). While the total phenol value of kecombrang flower powder is 27.93 mg TAE / 100 g, and the flavonoid value is 2.35 mg QE / 100 g. The compounds contained in the optimum product formula are phenolic, flavonoids, alkaloids, terpenoids and tannins.

**Acknowledgements**

Authors wish to thanks to the DRPM Kemenristek Republic of Indonesia which has provided research funding through Applied Research Grant Scheme 2019 also to Institute for Research and Community Service of Jenderal Soedirman University for the support.

## REFERENCES

[1] Naufalin, R., Wicaksono, R., Arsil, P., Salman, M.F. 2018. Antimicrobial coating on quality attributes of sausage during refrigerated storage. E3S Web of Conferences 47, 01002 (2018) https://doi.org/10.1051/e3sconf/20184701002
[2] Naufalin, R., 2019. Natural preservation opportunities and challenges in improving food safety. In *AIP Conference Proceedings*. Vol. 2094, No. 1, p. 020032. AIP Publishing.
[3] Naufalin, R. and Rukmini, H.S. 2018. Antibacterial activity of kecombrang flower extract (Nicolaia speciosa) microencapsulation with food additive materials formulation. IOP Conference Series: Earth and Environmental Science 102 (1): 012035

[4] Putri, F.A., R. Naufalin, R. Wicaksono. 2019. Antimicrobial edible coating application of Kecombrang flower concentrate to reduce microbial growth on gourami fish sausage. IOP Conf. Series: Earth and Environmental Science 250 (2019) 012056: 1-8.
[5] Latifasari, N., R. Naufalin, R. Wicaksono. 2019. Edible coating application of kecombrang leaves to reduce gourami sausage damage. IOP Conf. Series: Earth and Environmental Science 250. 012055: 1-7.
[6] Adawyah, R. 2014. Pengolahan dan Pengawetan Ikan. Sinar Grafika Offset, Jakarta.
[7] Naufalin, R., Herastuti, S. R. 2017. Antibacterial activity of *Nicolaia speciosa* fruit extract. *International Food Research Journal* 24.1:379-385.
[8] Hanifah, M.R., R. Naufalin, R. Wicaksono. 2019. The effect of edible coating contained Kecombrang leaves concentrate on gourami fish fillet quality. IOP Conf. Series: Earth and Environmental Science 250 (2019) 012057: 18.
[9] Budiyanto, A., Yulianingsih. 2008. Pengaruh suhu dan waktu ekstraksi terhadap karakter pektin dari ampas jeruk siam (*Citus nobilis L*). *Jurnal Pascapanen,* 5(2): 37-44.
[10] Singleton, V. L., Rossi, J. A. 1965. Colorimetry of total phenolics with phosphomolybdic-phosphotungstic acid reagents. *American Journal of Enology and Viticulture, 16:144-158*
[11] Aminah, D. N., Tomayahu, N., Abidin, Z. 2017. Penetapan kadar flavonoid total ekstrak etanol kulit buah alpukat *(Persea Americana Mill.)* dengan metode spektrofotometri Uv-Vis. *Jurnal Fitofarmaka Indonesia,* 4(2).
[12] Kumalasari, H. 2012. Validasi Metoda Pengukuran Kadar Air Bubuk Perisa Menggunakan Moisture Analyzer Halogen HB43-S, sebagai Alternatif Metoda Oven dan Karl Fischer. *Skripsi*. Institut Pertanian Bogor, Bogor.
[13] AOAC. 1984. *Official Methods of Analysis of The Association of Official Analytical Chemist.* AOAC Int, Washington.
[14] Harborne. J. B. 1987. *Metode Fitokimia.* Institut Teknologi Bandung, Bandung.
[15] Noer, S., Pratiwi, R. D. 2016. Uji kualitatif fitokimia daun *Ruta angustifolia. Faktor Exacta*, 9(3): 200-206.
[16] Departemen Kesehatan RI. 1995. *Farmakope Indonesia.* Edisi IV. Departemen Kesehatan RI, Jakarta.
[17] Ho, C.T. and Chen, C.W. 1995. Antioxidant properties of polyphenols extracted from green tea and black tea. J. Of Food Lipids. 2: 35-46.
[18] Farida. 2002. Proses pengolahan air sungai untuk keperluan air minum. *Tesis.* Universitas Sumatra Utara, Medan.
[19] Hikmah, A.F., Budhiyanti, S.A., Ekantari, N. 2009. Pengaruh pengeringan terhadap aktivitas antioksidan *Spirulina platensis*. Prosiding Seminar Nasional Tahun VI Hasil Penelitian Perikanan dan Kelautan. PA-04: 11.
[20] Susanti, D.F. 2008. Efek suhu pengeringan terhadap kandungan fenolik dan kandungan katekin ekstrak daun kering gambir: 1-13. Makalah disampaikan dalam *Prosiding Seminar Nasional Teknik Pertanian Yogyakarta, 18-19 November 2008.*
[21] Dirjen POM. 2000. *Parameter Standar Umum Ekstrak Tumbuhan Obat: Cetakan Pertama.* Departemen Kesehatan RI, Jakarta.
[22] Koirewoa, Y. A., Fatimawali, Wiyono, W. I. 2012. Isolasi dan identifikasi senyawa flavonoid dalam daun beluntas (pluchea indica l.). *Laporan Penelitian.* FMIPA UNSRAT, Manado.
[23] Syafrida, M., Darmanti, S., Izzati, M. 2018. Pengaruh suhu pengeringan terhadap kadar air, kadar flavonoid, dan aktivitas antioksidan daun dan umbi rumput teki (*Chyperus rotundus* L.). *Bioma.* Vol 20 (1): 44-50.

[24] Ibrahim, A. M., Yunianta, Sriherfyna, F. H. 2015. Pengaruh suhu dan lama waktu ekstraksi terhadap sifat kimia dan fisik pada pembuatan minuman sari jahe merah (*Zingiber officinale var. Rubrum*) dengan kombinasi penambahan madu sebagai pemanis. *Jurnal Pangan dan Agroindustri,* 3(2): 530-541.
[25] Departemen Kesehatan, 2008. *Farmakope Herbal Indonesia.* Departemen Kesehatan Republik Indonesia, Jakarta
[26] Rivai, H., Nurdin, H., Suryani, H., Bakhtiar, A. 2010. Pengaruh cara pengeringan terhadap perolehan ekstraktif, kadar senyawa fenolat dan aktivitas antioksidan dari daun dewa (*Gynura pseudochina* L.). *Majalah Obat Tradisional*. Vol 15 (1): 26-33.
[27] Yamin, M., Dewi F. A., Faizah F.H. 2017. Lama pengeringan terhadap aktivitas antioksidan dan mutu herbal dan ketepeng cina (*Cassia alata L.*). *JOM Faperta* 4(2): 1-15.
[28] Riansyah, A., Supriadi, A., Nopianti, R. 2013. Pengaruh perbedaan suhu dan waktu pengeringan terhadap karakteristik ikan asin sepat siam (*Trichogaster pectoralis*) dengan menggunakan oven. *Fishtech.* Vol 2 (1): 46-51.
[29] Aji, A., Meriatna, Ferani, A. S. 2013. Pembuatan pewarna makanan dari kulit buah manggis dengan proses ekstraksi. *Jurnal Teknologi Kimia Unimal,* 2(2): 1-15.
[30] Novitriani, K., Hasanah, H. N., Zulfa, A. 2017. Ekstrak bunga kecombrang (*Etlingera elatior*) sebagai indikator alternatif pada media gula-gula. *Jurnal Kesehatan Bakti Tunas Husada,* 17(1): 81-86.
[31] Farrel, K. T. 1990. *Species, Condiments and Seasonings*. Edisi kedua. Editor Van Vostrand. Reinhold, New york.
[32] Syafitri, N. E., Bintang, M., Falah, S. 2014. Kandungan fitokimia, total fenol ekstrak buah harendong (*Melastoma affine D.Don*). *Current Biochemistry,* 1(3): 105-115.
[33] Wenjuan, Q., Pan, Z., Ma, H. 2010. Extraction modeling and activities of antioxidants frompomegranate marc. *Elsevier Journal of Food Engineering,* 99: 16-23.
[34] Kahkonen M.P., A.L. Hopia., H.J. Vourela., J.P. Rauha., Pihlajak., T.S. Kujala & M. Heinonen. 2001. Antioxydant activity of extract containing phenolic coumpounds. *J. Agric Food Chem.* Vol. 47(10): 3954-62.
[35] Naufalin, R., Rukmini, H. S. 2010. Potensi antioksidan hasil ekstraksi tanaman kecombrang (*Nicolaia speciosa horan*) selama penyimpanan. Makalah dipresentasikan pada *Seminar Nasional Membangun Daya Saing Produk Pangan Berbasis Bahan Baku Lokal*, Surakarta 8 Juni 2010
[36] Sirait, M. 2007. *Penuntun Fitokimia dalam Farmasi.* ITB, Bandung.
[37] Sangi, M., Runtuwene, M. R. J., Simbala, H. E. I. & Makang, V. M. A. 2008. Analisis fitokimia tumbuhan obat di Kabupaten Minahasa Utara. *Journal of Chem. Prog,* 1(1): 1-3.
[38] Robinson, T. 1995. *Kandungan Organik Tumbuhan Tinggi*, (terjemahan oleh Padmawinata, K.). ITB, Bandung.
[39] Marliana, S. D., Suryanti, V., Suyono. 2005. Skrining fitokimia dan analisis kromatografi lapis tipis komponen kimia buah labu siam (*Sechiumedule* Jacq. Swartz.) dalam Ekstrak Etanol. *Biofarmasi,* 3(1): 26-31.
[40] Ningsih, D. R., Zusfahair, Kartika, D. 2016. Identifikasi senyawa metabolit sekunder serta uji aktivitas ekstrak daun sirsak sebagai antibakteri. *Jurnal Molekul*, 11(1): 101–111.
[41] Samarang, Isnawati, R., Murni. 2015. Potensi kandungan karondo (*Etlingera elatior*) sebagai obat cacing tradisional masyarakat kulawi di Sulawesi Tengah. *Jurnal Penyakit Bersumber Binatang*, 2(2): 1-8.

[42] Simaremare, E. S. 2014. Skrining fitokimia ekstrak etanol daun gatal (*Laportea decumana (roxb.) wedd). Pharmacy*, 11(1): 98-107. ISSN 1693-3591.

[43] Rijai, L. 2016. Senyawa glikosida sebagai bahan farmasi potensial secara kinetik. *J. Trop. Pharm. Chem*, 3(3): 213-218. p-ISSN: 2087-7099, e-ISSN: 2407-6090.

[44] Departemen Kesehatan RI. 1995. *Farmakope Indonesia.* Edisi IV. Departemen Kesehatan RI, Jakarta.

[45] Jannah, R., Husni, M. A., Nursanty, R. 2017. Uji daya hambat ekstrak methanol daun sirsak (*Annona muricata Linn.)* terhadap bakteri *streptococcus mutan. Jurnal Natural,* 17(1): 23-30.

# Effect of fermentation temperature and culture concentration on microbial and physicochemical properties of cow and goat milk yogurt

**Ibrahim Aldaw Ibrahim[1]* Rifda Naufalin[2], Erminawati[2], Hidayah Dwiyanti [2]**

[1]Department of Food Science & Technology, Faculty of Agriculture, Omdurman Islamic University, Sudan
[2]Faculty of Agriculture, Jenderal Soedirman University, Purwokerto, Indonesia
*Corresponding authors: ibrahim.sudann@gmail.com

**Abstract**. This present study aimed to assess the effect of fermentation temperature and culture concentration on LAB growth, physical and chemical properties of fermenting cow and goat milk. The International Official Analytical Chemistry Association (AOAC) and the Indonesian National Standard (SNI) methods were used to analyze the physical and chemical properties of yogurt. Yogurt was prepared using fermentation temperature (40, 43, 46°C), a combination of *Streptococcus thermophilus and Lactobacillus acidophilus* strains. (1: 1), culture inoculation (5, 10, 15persent) and fermentation times (8 hours). The results showed that the cultures concentration and temperature have a significant effect on physicochemical properties, and viable amount of fermented milk samples. There were a significant difference ($P \leq 0.05$) between all yogurt samples in LAB growth, total solids, pH, viscosity, and there were no significant difference ($P \leq 0.05$ between , titeriatble acidity, color and density of all samples. The general results of this present study proved that the ideal state of fermentation of cow and goat milks fermentation were: fermentation 46°C, culture concentration 10, 15%, for 8h. Thus, the benefit from these results are developing physical and chemical properties and sensory characteristics of cow and goat milk yogurt.

**Keywords**: *Cow and goat milk, temperature, culture concentration, fermentation*

## 1. Introduction

Yogurt is one of the oldest known fermentation products for humans, originating in the Middle East and Asia. Yogurts are one of most popular fermented dairy products widely consumed throughout the world [12, 10]. Yogurt is the product produced during reaction of lactose fermentation (milk sugar) in milk and bacterial enzymes under certain conditions. During the manufacture of milk, lactose is broken down by the enzyme lactase. During action of bacterial enzyme on lactose, lactic acids and acetaldehyde are produced, which in turn reduces pH of milk causing it to have an acidic or pungent taste [7]. Yogurt is a fermented food product which is increasing its consumption by year, due to its benefit for human health [2].

The pH of milk about 6.7. However, during the fermentation process the pH decreasing between (3.6 - 4.5), depending on fermentation conditions. Low pH affects casein (milk protein), making it coagulate and precipitate, forming a solid or dense curd that forms milk. The most common bacteria in the milk industry are *Lactobacillus bulgaricus, Streptococcus thermophilus* [9]. Cow and goat yogurt has antioxidant activity and which is increasing during fermentation, and can developing produced yogurt and increasing nutritional values of its [15,

IOP Conf. Series: Earth and Environmental Science **406** (2019) 012009 doi:10.1088/1755-1315/406/1/012009

The total solids content, heat treatment, and incubation temperatures influence the textural properties of yogurts [11]. The quality of yogurt depend on three factors that are; the quality of raw milk, process conditions and starter culture [13]. The yogurt quality is effected by different factors such as heat processing, incubation temperature, amount of culture inoculated and time of incubation. more study mentioned that the incubation temperature is increased the viscosity of yogurt is increase, the ph reduce and titeritable acidity increases [11]. This present research investigated the optimum fermentation temperature and culture concentration of cow and goat milk yogurt, and effect of incubation temperature on quality of produced yogurt.

## 2. Material and method

The milk samples were bought from baduraden animal farmer, Central Java – Indonesia. Amount of cow and goat milk samples were weight separetly into empty beakers (5litters). Each milk samples was heated process 85°C for 5sec. Then samples were cooled to 45°C. Cow and goat milk were mixed (25:75v/v). There are make 9 samples depending on different temperature and culture concentration (inoculation), then coded them that are A = (40°C/0.05persent), B = (40°C/0.10persent), C = (40°C/0.15persent), D = (43°C/0.05persent), E = (43°C/0.10persent), F = (43°C/0.15persent), G = (46°C/0.05persent), H = (46°C/0.10persent), and I = (46°C/0.15persent). The culture freezed (*Lactobacillus aciduphilus and Streptococcus thermophilus* (mixted 1:1)) was activited by inouclated in skemmed milk and incubated at 37°C for 24h. The milk samples were inoculating by different culture concentrations (5, 10, and 15persent). Then were incubated all samples at temperature ranging (40, 43, 46°C) for 8h.

**The samples analysis:**

**The determination of pH** of yogurt samples were conducted using standard procedures. was recorded all reading of samples by pH meter.

**The titritable acidity** was determinated by titration. 1M Sodium hydroxide was prepared and titrated against 10ml from all yogurt samples using 2 or 3 drops of phenolphthalein indicator untill it was neutralized and turned light pink. each samples was titrated three times.

**The viscosity** of yogurt was determinated by digital viscometer using standard procedures. the sample was put into the cup of viscometer (250ml). checked viscosities of samples at revolution per minute (60 rpm).

**The density** was determined according to (SNI: 06- 2385-2006), Official method 920.212.

**The total solids** was determinated according method of (A.O.A.C, 2005).

**The color** was determined according to method of (Chugh et al., 2014) using spectrophotometer according to color measured are color (L*) Lightness, (a*) Greenness (-), and (b*) yellowness (+). Each samples of yogurt checked thrice.

**Microbiology Analysis:** A total analysis of lactic acid bacteria for the samples was performed by the method described in [6]. Each sample was diluted sequentially in sterile and distilled water to obtain the vaccine. Aliquot each dilution will be cultured on (MRSA) and incubated at 37°C for 48 hr.

**Panel test:** The sensory assessment test was determined according to the 7-point hedonic scale, and was used to illustrate team members (15 participants from food science and technology students), acceptance of color, taste, smell, texture, and overall acceptability.
**Statistical analysis:** All data on statistical analysis were performed using ANOVA by Microsoft Excel version 13. Important differences between the means were determined by Duncan's Multiple Range test at (P≤ 0.05).

## 3. Results and Discussion

Table1 and Fig 1 showed effect of incubation temperature and culture concentration on cow and goat milk yogurt quality. The viscosities of fermented milk samples were 97.9, 99.3, 67.26, 99.3, 99.3, 77.5, 99.3, 99.3, and 99.3 respectively. From these results showed the highest values of viscosity is 99.3, and lowest is 67.26. There were significant difference (P ≤0.05) between C, E samples and others group, and no significant difference between A, B, D, F, G, H and I samples. These results showed that the incubation temperature has significant effect on viscosity values, the values viscosity increased during ingreasing incupation temperature from 40 to 46°C. These results agree with Aguirre-Ezkauriatza et al., [1], and Wardani et al., [18], they reported that fermentation temperature up to 40°C and culures concentration ranged 1 to 5persent affected on yogurt properties, decreased ph of yogurt and increased amount of bacteria cells. The total solids values of yogurt samples were 13.21, 13.26, 14.28, 13.04, 12.63, 13.55, 12.69, 12.59, 12.86 respectively. From these results the total solids of sample C is highest and sample H is lowest. There were significant difference (P ≤0.05) between each yogurt samples. From these results the incubation temperature has high significant effect on decreased total solids of yogurt samples and culture concentration has significant effect on increased total solids of fermented cow and goat milk. The pH results were 5.8, 5.7, 5.2, 5.6, 4.6, 4.5, 4.4, 4.4, and 4.43 respectively. The C sample is has upper pH value is 5.8persent, and G, H, I samples have lower pH value is 4.4 percent. this result showed there were significant difference (P ≤0.05) between fermented cow and goat milk. From these results the incubation temperature and culture concentration have significant effect on decreased pH, This result agree with [11] they reported that the incubation temperature increased titriatble acidity, viscosity, density and decreased pH, of yogurt). The titriatble acidity result were 0.23, 0.30, 0.33, 0.26, 0.28, 0.34, 0.30, 0.33, and 0.38 respectively. The upper titriatble acidity value is 0.38 and lowest 0.23. there were no significant difference (P ≤0.05) between each yogurt samples. these results showed that the incubation temperature significant effect on increased titriatble acidity of fermented milk products. This result agree with Lee and Lucey [11] they reported that the incubation temperature increased titriatble acidity, viscosity, density and decreased pH, of yogurt, and compartable acidity ranged between 0.61 to 0.66 percent in yogurt. And compatibly with Oladimeji et al., [9], they reported the incubation temperature ranged (35 – 45°C) significantly affected in both physicochemical (pH, and Titritable Acidity). The color of fermented milk values showed in Table2. The result of (a*) average were -8.30, 7.83, 6.73, 2.93, 4.56, -0.166, -7.0, -1.66, and -3.66 respectively. Where are no significant difference (P ≤0.05) between each samples. The average of (b*) were -18.5, -25.5, -19.36, -4.4, -1.26, -1.83, 4.10, -023, -17.96 respectively. This result showed significant difference (P ≤0.05) between G sample and others group, and no significant difference (P ≤0.05) between all others fermented milk samples. The average of (L*) were 28.2, 28.93, 31.0, 24.3, 23.56, 23.63, 26.5, 42.33, and 42.70 respectively. These results showed there were significance difference (P ≤0.05) between I, C, H Samples and no significant difference (P ≤0.05) between A, B, D, E, F, H samples. From these results stablished that the fermentation temperature at 46°C and culture concentration 15 percent significantly affected on color of fermented cow and goat milk. This result agree with Oladimeji et al. [9], they reported the incubation temperature ranged (35 – 45°C) significantly affected in both physicochemical properties of yogurt.

Table1. Effect of incubation temperature and culture concentration on cow and goat milk yogurt quality

| Temperature | LAB Concentration (%) | | | | | | | | | | | |
|---|---|---|---|---|---|---|---|---|---|---|---|---|
| | Viscosity | | | Total solids | | | Ph | | | Titriatble Acidity | | |
| | 5% | 10% | 15% | 5% | 10% | 15% | 5% | 10% | 15% | 5% | 10% | 15% |
| 40°C | 97.9$^{a}$ | 99.3$^{a}$ | 67.26$^{c}$ | 13.21$^{d}$ | 13.26$^{c}$ | 14.28$^{a}$ | 5.8$^{a}$ | 5.7$^{b}$ | 5.2$^{d}$ | 0.23$^{a}$ | 0.30$^{a}$ | 0.33$^{a}$ |
| 43°C | 99.3$^{a}$ | 99.3$^{a}$ | 77.5$^{b}$ | 13.04$^{e}$ | 12.63$^{h}$ | 13.55$^{b}$ | 5.6$^{c}$ | 4.6$^{e}$ | 4.5$^{ef}$ | 0.26$^{a}$ | 0.28$^{a}$ | 0.34$^{a}$ |
| 46°C | 99.3$^{a}$ | 99.3$^{a}$ | 99.3$^{a}$ | 12.69$^{g}$ | 12.59$^{i}$ | 12.86$^{f}$ | 4.4$^{f}$ | 4.4$^{f}$ | 4.4$^{f}$ | 0.30$^{a}$ | 0.33$^{a}$ | 0.38$^{a}$ |

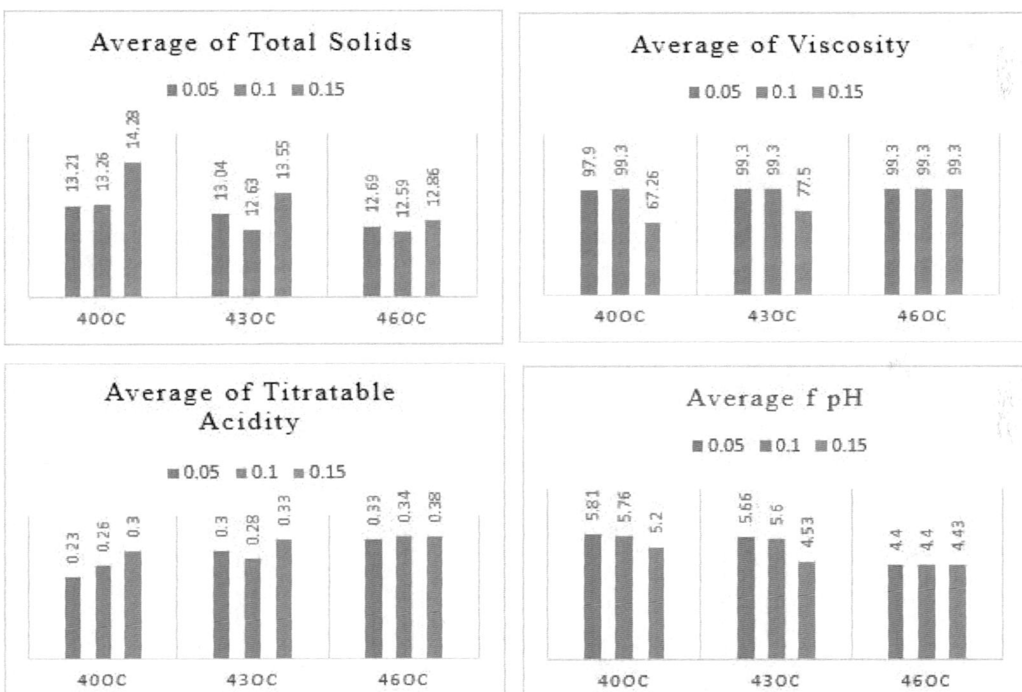


Fig1: Effect of incubation temperature and culture concentration on cow and goat milk yogurt quality

Table 2. Effect of incubation temperature and culture concentration on cow and goat milk yogurt color

| Temperature | LAB Concentration (%) | | | | | | | | |
|---|---|---|---|---|---|---|---|---|---|
| | a | | | B | | | L | | |
| | 5% | 10% | 15% | 5% | 10% | 15% | 5% | 10% | 15% |
| 40°C | -8.30[a] | 7.83[a] | 6.73[a] | -18.5[b] | -25.5[b] | -19.36[b] | 28.2[c] | 28.93[c] | 31.0[ac] |
| 43°C | 2.93[a] | 4.56[a] | -0.166[a] | -4.4[b] | -1.26[b] | -1.83[b] | 24.3[c] | 23.56[c] | 23.63[c] |
| 46°C | -7.0[a] | -1.66[a] | -3.66[a] | 40.1[a] | -0.23[b] | -17.96[b] | 26.5[c] | 42.33[ab] | 42.7[a] |

The total of Lactic acid bacteria showed in Fig2. The total lactic acid bacteria cells were 5.0, 7.0, 7.3, 6.3, 7.4, 8.3, 8.3, 10.0, and 12 $10^6$cfu /ml respectively. These results showed that there were significant difference (P $\leq$0.05) between all yogurt samples in amount of lactic acid bacteria. The amount of total lactic acid bacteria in I sample is highest than other group, and in A sample is lowest than others samples. From these results showed that the incubation temperature and culture concentration has significant effect on increased growth of lactic acid bacteria when increasing incubation temperature degree and concentration of start culture. These result compatible with [4] them reported the incubation temperature and time storage significantly inflonced overall properties of yogurt samples. During incubation at 42°C titreitable acidity and total lactic acid bacteria increased or higher than amount of its in incubation at 37-42°C. Agree with Østlie et al. [14], they mention that when the incubated the strains probiotic bacteria at the different temperatures, the amounts of ce produced after incubation at 45°C is highest than incubation at 30°C. These results agree with Wardani et al., [18], they reported that fermentation temperature up to 40°C and cultures concentration ranged 1 to 5 percent affected on yogurt properties, decreased ph of yogurt and increased amount of bacteria cells.

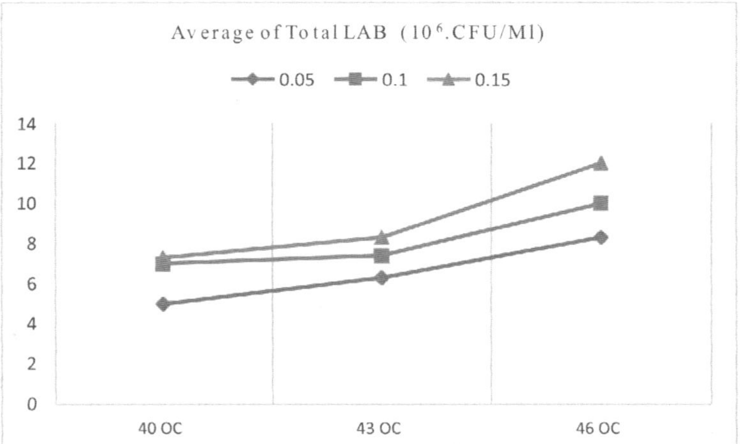

Fig2: effect of fermentation temperature and cultures concentrations on amount of lactic acid bacteria.

**Conclussion**

Inoculum concentration of *Streptococcus thermophilus and Lactobacillus acidophilus* and fermentation temperature had significant effect on viable cell, and physicochemical properties of fermented milk. the results showed that increase cultures concentration from 5 to 15% and incubation temperatures from 40 to 46% increased the viable amounts, viscosity, titriatble acidity and decreased pH. the use of high fermentation temperature resulted in the decease the gelation time and pH values in yogurt produced. The optimum temperature for cow and goat milk fermentation by was 46°C, and optimal culture concentration was 10 and 15% for 8h.

**Acknowledgment :**

The authors are grateful to the Ministry of Research, Technology and Higher Education, and are interested in the Research and Community Service Institute at Jenderal Soedirman University for their funding and support for the completion of this study. our thanks to staff of the Agricultural Technology Laboratory for supporting and guiding.

**References:**

[1] Aguirre-Ezkauriatza, E. J., Galarza-González, M. G., Uribe-Bujanda, A. I., Ríos-Licea, M., López-Pacheco, F., Hernández-Brenes, C. M., & Alvarez, M. M. (2008). Effect of Mixing During Fermentation in Yogurt Manufacturing. Journal of Dairy Science, 91 (12),4454-- 4465. doi: 10.3168 / jds.2008-1140.

[2] Agustini T W, D Soetrisnanto, and W F Maruf. 2017. Study on chemical, physical. Microbiological and sensory of yogurt enriched by spirulina platensis. International food research journal 24(1): 367-71.

[3] A. O. A. C. 2005. Official methods of analysis. Association of Official Analytical Chemists. 16th edition, North Frederick Gaithersburg Street, Maryland, USA.

[4] Akin M S, and Guler-Akin M B. 2005. Effect of different incubation temperatures on the microflora, chemical composition and sensory characteristics of bio-yogurt. Ital. J. Food Sci. n. 1, vol. 17 - 67.

[5] Arioui, F., Ait Saada, D., & Cheriguene, A. 2016. Physicochemical and sensory quality of yogurt incorporated with pectin from peel of Citrus sinensis. *Food Science & Nutrition,* 5(2), 358–364. doi:10.1002/fsn3.400.

[6] Ehirim, F.N., Onyeneke,E.N. 2013. Physico – Chemical and Organoleptic Properties of Yoghurt Manufactured With Cow Milk and Goat Milk. Natural and Applied Sciences. ISSN-L: 2223-9553, ISSN: 2223-9944.

[7] Eneger E F. 1997. The chemistry and ecology of yogurt production. In E.F. Eneger Laborotory manual: concepts in biology, eightedition (pp.95-97). West minster: Brown publishers.

[8] Haneen, H. Mouminah. 2019. Physicochemical, Microbiological and Sensory Evaluation of Yogurt Prepared with Date paste. *Asian Journal of Applied science and Technology*. Volume 3, Issue 1, Pages 234-248.

[9] Oladimeji T E, E Lyi-Eweka, R Oyinlola. 2016. Effects of Incubation Temperature on the Physical and Chemical Properties of Yoghurt 3rd international conference on African development issuse (CU-ICADI 2016). SSN: 2449 – 075X.

[10] Igbabul B, Shember J, and Amove J. 2014. Physicochemical, Microbiological and sensory evaluation of yogurt sold in makurdi metropolis. African Journal of Food Science and Technology 5(6):129–35. Retrieved (http:/dx.doi.org/10.14303/ajfst.2014.052).

[11] Lee w, and Lucey J. 2010. Formation and physical properties of yogurt. Asian-Australian journal of animal science. 23: 1127 – 1136.

[12] Loveday, Simon M., Anwesha Sarkar, and Harjinder Singh. 2013. Innovative Yoghurts: Novel Processing Technologies for Improving Acid Milk Gel Texture. *Trends in Food Science and Technology* 33(1):5–20.

[13] Paskov, V., M. Karsheva, and I. Pentchev. 2010. Effect of Starter Culture and Homogenization on the Rheological Properties of Yoghurts. Journal of the University of Chemical Technology and Metallurgy 1(January 2010):59–66.
[14] Østlie H M, Treimo J, Narvhus J A. 2005. Effect of temperature on growth of probiotic bacteria in milk. International Dairy Journal. VOL. 15: 989–997. doi:10.1016/j.idairyj.2004.08.015.
[15] Rahmawati, I. S. and Suntornsuk, W. (2016). Effects of Fermentation and Storage on Bioactive Activities in Milks and Yoghurts. Procedia Chemistry. Elsevier Ltd., 18(Mcls 2015), pp. 53–62. doi: 10.1016/j.proche.2016.01.010.
[16] Standard National Indonesia. 2006. Patchouli Oil (SNI 06-2385: 2006), Jakarta.
[17] Supavititpatana, P., Wirjantoro, T. I. and Raviyan, P. (2010). Characteristics and shelf-life of Corn Milk Yogurt', Chiang Mai University Journal of Natural Sciences, 9(1), pp. 133–150.
[18] K S Wardani, Cahyanto M N, S E Rahayu, Utami T. 2017. The effect of inoculum size and incubation temperature on cell growth, acid production and curd formation during milk fermentation by Lactobacillus plantarum Dad 13. International Food Research Journal 24(3): 921-926.

# AUTHOR INDEX